Praise for A Historical Scholarly Collection of Writings on the Earth Liberation Front

"[This book is] [a]n important and timely collection of essays that help expand our understanding of the ELF, specifically, and nonhuman liberation, more generally. The essays avoid some of the more common knee-jerk reactions and engage the issues in thoughtful, critical, and intellectually stimulating ways. This collection will no doubt spark important discussion and debate!"

Dr. Jason Del Gandio, author of *Rhetoric for Radicals: A Handbook for 21st Century Activists*

"For those interested in the history of radical and revolutionary organizations whose strategies and tactics involve both violent and non-violent action, this book is indispensable. While readers may not agree with the arguments put forward in this book, there is too much at stake for them to remain ignored. It's time to share this planet together, not blow it up. Read this book."

Dr. Peter McLaren, Distinguished Professor in Critical Studies, College of Educational Studies, Chapman University

"It is good to finally have in one edited volume the contributions of leading scholars and activists who not only clearly articulate the perspectives and actions of the Earth Liberation Front, but also validate them as legitimate approaches to environmental degradation, exploitation, and oppression."

Dr. Scott Hurley, Associate Professor of Religion, Luther College

"The collection is theoretically intriguing and practically applicable to contemporary social movements and political dynamics. The range of contributions provide a broad yet highly informative set of writings that, in their diversity, give the reader an in-depth sense of the many important and timely facets of the Earth Liberation Front."

Dr. Lauren Eastwood, Associate Professor and Chair of Sociology, SUNY College at Plattsburgh

"*A Historical Scholarly Collection of Writings on the Earth Liberation Front* is one of the most important and powerful texts to be published about the Earth liberation movement. If you are concerned about social justice, environmental justice, animal liberation, and anarchism then this scholarly text of classic writings on the ELF is a must-read. You will not be disappointed.

Madelynne Kinoshita, Social Media Coordinator, Save the Kids

"*A Historical Scholarly Collection of Writings on the Earth Liberation Front* is an outstanding book speaking about a group that goes beyond nonprofits and fancy banners over the White House to do what is needed and must be done to defend the planet against those that want to profit off of it. Read this book and ignite a revolution."

Chris Mendoza, Durango Food Not Bombs

"The Earth Liberation Front remains one of the most important social movements in all of human (and more-than-just-human) history. It is also one of the most misunderstood and maligned social change agents of our time. This collection of writings provides a crucial scholarly foundation for understanding this movement's history, significance, implications, and future possibilities. The Elves are alive and kicking, and making much mischief on every page of this invaluable book!"

Dr. David Pellow, Endowed Chair, Dehlsen Professor of Environmental Studies, University of California, Santa Barbara

"The sheer breadth of this work might be enough to recommend consideration. The extraordinary depth of its treatment demands attention. Of particular relevance are the multiple entry points into this monumental movement. Irrespective of one's political or ecological perspective, these sometimes profound explorations of and engagements with ELF allow readers to form their own opinions, to experience some of the thinking that helped shape a movement."

William Shanahan, Curry College

"This outstanding interdisciplinary scholarly text takes us past the media and political propaganda of eco-terrorism and examines what the Earth Liberation Front is and how and why they exist."

Dr. Erik Juergensmeyer, Editor, *Green Theory and Praxis Journal*

"A powerful text in defense of a revolutionary environmental group. Chapters are interwoven and interdisciplinary build off of each other. This book will save you time in searching for the most pivotal articles on the subject of eco-terrorism and the Earth Liberation Front."

Arash Daneshzadeh, Editor, *Transformative Justice Journal*

"A wonderful academic collection examining eco-terrorism and the Earth Liberation Front from a social justice, sociological, criminological, and critical perspective. This is one of the most fundamental texts within the field of critical animal studies, green criminology, and environmental studies."

Arissa Media Group

A Historical Scholarly Collection of Writings on the Earth Liberation Front

RADICAL ANIMAL STUDIES AND TOTAL LIBERATION

Anthony J. Nocella II
Series Editor

Vol. 4

The Radical Animal Studies and Total Liberation series
is part of the Peter Lang Education list.
Every volume is peer reviewed and meets
the highest quality standards for content and production.

PETER LANG
New York • Bern • Berlin
Brussels • Vienna • Oxford • Warsaw

A Historical Scholarly Collection of Writings on the Earth Liberation Front

EDITORS
Anthony J. Nocella II, Sean Parson,
Amber E. George, and Stephanie Eccles

PETER LANG
New York • Bern • Berlin
Brussels • Vienna • Oxford • Warsaw

Library of Congress Cataloging-in-Publication Data

Names: Nocella II, Anthony J., editor.
Title: A historical scholarly collection of writings on the
Earth Liberation Front / edited by Anthony J. Nocella II, Sean Parson,
Amber E. George, and Stephanie Eccles.
Description: New York: Peter Lang, 2019.
Series: Radical animal studies and total liberation; vol. 4
ISSN 2469-3065 (print) | ISSN 2469-3081 (online)
Includes bibliographical references and index.
Identifiers: LCCN 2018033816 | ISBN 978-1-4331-5992-3 (hardback: alk. paper)
ISBN 978-1-4331-5993-0 (paperback: alk. paper)
ISBN 978-1-4331-5994-7 (ebook pdf) | ISBN 978-1-4331-5995-4 (epub)
ISBN 978-1-4331-5996-1 (mobi)
Subjects: LCSH: Earth Liberation Front. | Ecoterrorism. | Environmentalism.
Classification: LCC GE197.H57 2019 | DDC 363.325—dc23
LC record available at https://lccn.loc.gov/2018033816
DOI 10.3726/b14465

Bibliographic information published by **Die Deutsche Nationalbibliothek**.
Die Deutsche Nationalbibliothek lists this publication in the "Deutsche
Nationalbibliografie"; detailed bibliographic data are available
on the Internet at http://dnb.d-nb.de/.

29 Broadway, 18th floor, New York, NY 10006
www.peterlang.com

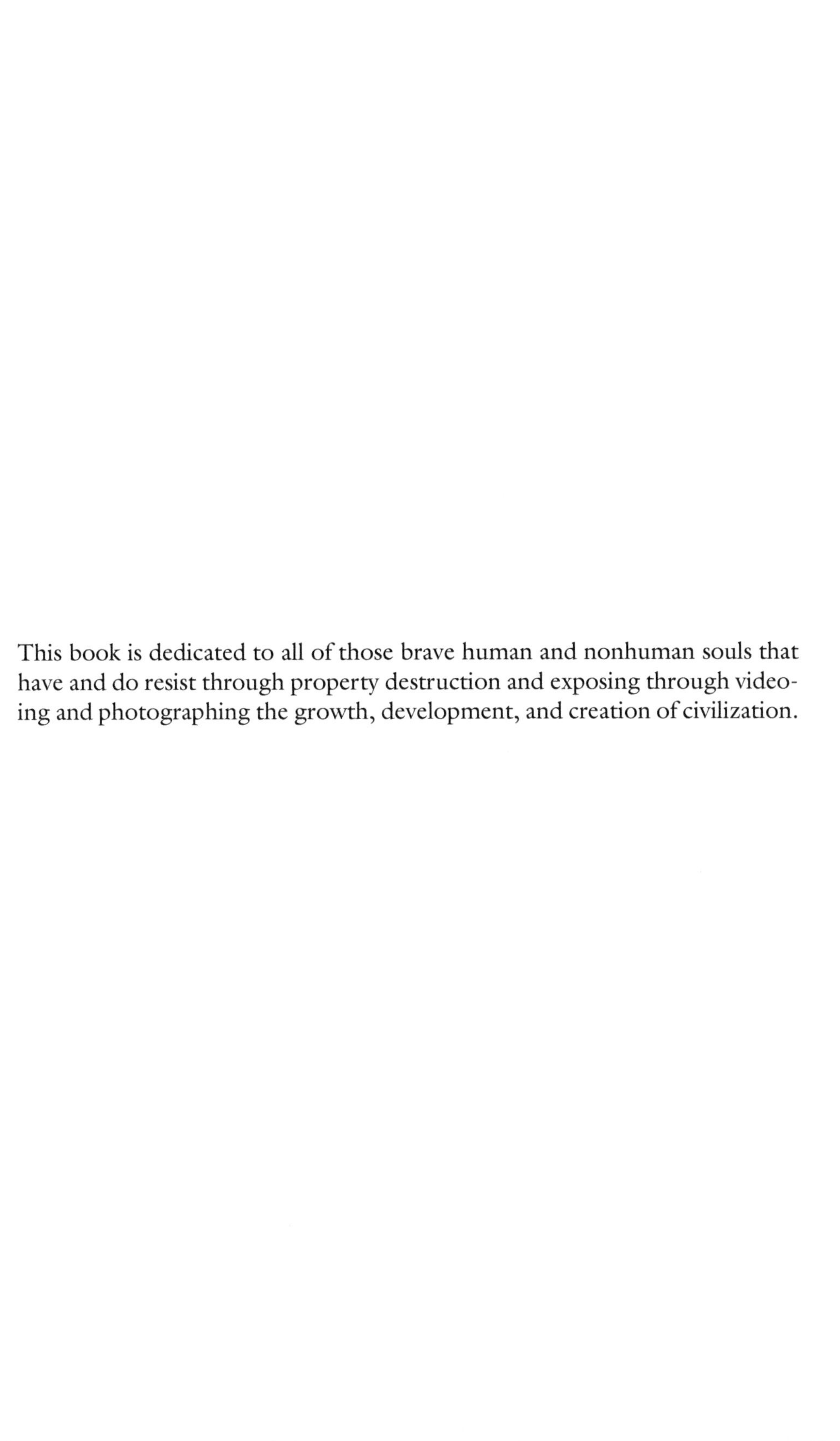

This book is dedicated to all of those brave human and nonhuman souls that have and do resist through property destruction and exposing through videoing and photographing the growth, development, and creation of civilization.

Contents

Tables ix

Acknowledgments xi

Foreword xiii
CAROLYN DREW

Introduction: A Call to All Scholars to Defend Revolutionaries 1
SEAN PARSON, ANTHONY J. NOCELLA II, AMBER E. GEORGE, AND STEPHANIE ECCLES

Part I: Classic Writings on Revolutionary Environmentalism

1. *Rhizomatic Resistance: The Zapatistas and the Earth Liberation Front* 13
MICHAEL BECKER

2. *Understanding the Ideology of the Earth Liberation Front* 39
SEAN PARSON

3. *Nihilism and Desperation in Place-Based Resistance* 59
MARK SEIS

Part II: Classic Writings on Ecoterrorism Rhetoric

4. *Ecoterrorism? Countering Dominant Narratives of Securitization: A Critical, Quantitative History of the Earth Liberation Front (1996–2009)* 79
MICHAEL LOADENTHAL

5. *Activism as Terrorism: The Green Scare, Radical Environmentalism and Governmentality* 103
COLIN SALTER

6. The Myth of "Animal Rights Terrorism" 127
JOHN SORENSON

7. Leaderless Resistance and Ideological Inclusion: The Case of the Earth Liberation Front 157
PAUL JOOSSE

Part III: Classic Writings on Political Repression

8. Standing Up to Corporate Greed: The Earth Liberation Front as Domestic Terrorist Target Number One 179
ANTHONY J. NOCELLA II AND MATTHEW J. WALTON

9. Mapping Discursive and Punitive Shifts: Punishment as Proxy for Distinguishing State Priorities Against Radical Environmental Activists 191
LAWRENCE J. CUSHNIE

10. Speaking About "Ecoterrorists": Terrorism Discourse and the Prosecution of Eric McDavid 217
JOSHUA M. VARNELL

Part IV: Current Perspectives

11. Radical Environmentalism as Teacher: A Pedagogy of Activism 253
MENEKA REPKA

12. Those Mischievous Elves of Lore: The Legend and Legacy of Earth Liberation 269
ALEXANDER REID ROSS

13. Magic Kills Industry: Reclaiming ELF and Witch Deviance as Ecoqueer and Anticapital 295
MARA PFEFFER AND BETHANY RICHTER

14. Problematising Non-violent "Terrorism" in an Age of True Terror: A Focus on the Anarchic Dimensions of the Earth Liberation Front 315
RICHARD J. WHITE

Contributors 335

Index 341

Tables

Table 4.1. Most Commonly Attacked Targets. 88

Table 4.2. Comparison of 1st, 2nd, 3rd Most Commonly Utilized Tactics. 91

Table 4.3. Comparison of 4th, 5th, 6th Most Commonly Utilized Tactics. 91

Table 4.4. Comparison of the Presence of ELF Communiqué or "calling card". 93

Table 4.5. Comparison of Group Claims. 94

Table 4.6. Comparison of Location. 94

Table 4.7. Comparison of US Attack Location Regionally. 95

Table 9.1. Sentence Lengths, in Months, of Environmental Activists—Property Crimes. 200

Acknowledgments

The editors of this book would like to thank all our fellow editors for the fun and challenging important experience in putting this book together. We would also like to thank all the contributors of the book—Paul Joosse, Michael Becker, Sean Parson, Bron Taylor, Mark Seis, Michael Loadenthal, Colin Salter, John Sorenson, Anthony J. Nocella II and Matthew J. Walton, Lawrence J. Cushnie, Josh Varnel, Meneka Repka, Alexander Reid Ross, Mara Pfeffer, Bethany Richter, and Richard J. White. Everyone that wrote pre-published reviews of the book, for the book—Jason Del Gandio, Joe Leeson-Schatz, Peter McLaren, Scott Hurley, Lauren Eastwood, Madelynne Kinoshita, Chris Mendoza, David Pellow, William Shanahan. We would also like to thank Peter Lang Publishing especially—Sarah Bode, Sara McBride, Timothy Swenarton, Heather Boyle, Megan Madden, Jackie Pavlovic, Janell Harris, and Sophie Appel. We would also like to thank Save the Kids, *Transformative Justice Journal*, Eco-ability Collective, Durango Food Not Bombs, Durango Peace and Justice, Arissa Media Group, Institute for Critical Animal Studies, Total Liberation Working Group, *Peace Studies Journal*, *Green Theory and Praxis Journal*, Hampton Institute, Houston Animal Rights Team, Dirty Hands Collective, Durango Prisoner Letter Writing, Poetry Behind the Walls, *Journal for Critical Animal Studies*, Wisdom Behind the Walls, Outdoor Empowerment, Academy for Peace Education, Institute for Hip Hop Activism, Terrorism and Political Violence, University of Hong Kong Department of Sociology, Anarchist Developments, and Anarchist Studies.

Foreword

Carolyn Drew

The domination of civilization over nature has been the underlying story of humankind, from our first steps to dominate the world around us with our hand-made tools to a post-industrial future where our understanding of the earth is mediated by the idea to such an extent that the idea is all and the actual a mere shadow. Each step has taken us further away from our animality, our wild. Each step has seen us, in turn, try to de-animate the wilderness. Each step has seen us trash the environment as we pursue a Platonian world where idea-over-object attempts to dominate the very breath of life itself, while around us shattering ice shelves, the destruction of once great forests, the desertification of once fertile soils, the dying rivers, threats to marine life and the rapid extinction of land-based wildlife remind us of the precarious state of the earth.

Considering this, the collection contains important writings on the Earth Liberation Front, a group dedicated to the end of this destruction. It tells the story of its roots, through its birth and the road it walks. The collection gives the reader an opportunity to understand and engage in the problems that we all must face as the earth is battered with often irrevocable consequences.

Many will argue, when faced with problems such as these, for turning to the very same system of alienation driving the present destruction. Thus, we scrabble around in our glass and steel caves searching for ever more sophisticated tools repeating the behavior that has brought us to this tipping point.

However, the Earth Liberation Front takes a different approach. Since its inception there has been much discussion about who, and what, it is. There has been speculation about its purpose, motives and endgame. There has been much debate over the value of its approach and tactics, when fighting to end the destruction that marks our time, the Anthropocene. Indeed, this naming is as Narcissus falling in love with his own reflection and is indicative of the peril in which the earth and its children are placed and the challenges that confront those of the Earth Liberation Front. Often condemned as terrorists by those in power and sometimes disparaged by those they would consider their own, the Earth Liberation Front, like those who have come before, has been involved in raising awareness of, and attempts to stop, the destruction of the wilderness, the destruction of the earth.

Images of clandestine figures moving through the night setting fires, gluing locks, spiking trees, burning SUVs are what most people think about when they hear of the Earth Liberation Front. Chaos and random acts of senseless crime are images that often flood the media and its readers. Of course, this is exactly what the corporations, the various industrial complexes, want people to imagine. But, as the collection will show, the group is much more than this. Though its actions may indeed incorporate property destruction and other seemingly irrational ways to stop the corporate industrial obliteration of the natural landscape, what drives this, what underpins these and other actions is what this collection seeks to clarify and record.

And to this point, the history of groups like the Earth Liberation Front is rarely recorded on their own terms. What is often presented as history is typically from the point of view of those who feel threatened by groups such as these. This collection, instead, seeks to draw a fuller, richer account of the group, its grounding, in ways that allow the reader to gain a window into a world they seek to understand. And thus the importance of this book. It is a much-needed collection which analyses the roots of the Earth Liberation Front. It analyses its inception and its philosophy. The collection looks at perceptions and misunderstandings. The reader is given insight into the rhetoric and the push back from the system as it struggles to contain ideas it fears may spark a revolution. Further, the collection analyses the ties with the Animal Liberation Movement and anarchism. Then, as a way of drawing breath after this historical ride it finishes with the current perspectives. Importantly, it gives the reader a clear understanding of the milieu out of which the Earth Liberation Front was born and hence a deeper appreciation of its clarion call for the earth first.

Introduction: A Call to All Scholars to Defend Revolutionaries

Sean Parson, Anthony J. Nocella II, Amber E. George, and Stephanie Eccles

The Time Is Now

The planet is facing a myriad of ecological crises from climate change and ocean acidification to species die off and the depletion of topsoil. Even though activists, scientists, and artists have been warning the public about these issues for decades the political and economic elite throughout the entire world have been slow to act. When it comes to climate change, the only major international agreement, the Paris Accords, developed a volunteer climate mitigation plan that is designed not to harm large corporations or alter the global economic and political order. Even so, President Donald Trump, in one of his first major moves as presidents, signed an executive order calling for the United States to leave this American and corporate-friendly agreement. Trump is not the only major leader who has expressed disdain for international climate agreements. Canadian Prime Minister Justin Trudeau, a figure loved by liberals throughout the Western world, has sided with the Canadian tar sands industry over environmentalist concerns for the climate and indigenous first nations activists anger toward the neocolonial practices at the root of resource extraction in Canada.

While political leaders dither, delay, and deny, the scientific community wonders if there is anything that can be done to stop catastrophic climate change from occurring. This February, the United Nations Climate Chief stated, "The transformation has started. I think it's unstoppable" and they are right. The impacts of climate change are here and we, as a planet, experience them, albeit in uneven and unequal ways. Droughts, flooding, heat waves, chaotic weather, and the like are becoming commonplace. For the first time since the last ice age, the past is no longer a valuable metric to understand current and future weather patterns, and this is only going to get more pronounced and troubling as time goes on.

As the reputed anarchist poet Utah Phillips famously said: "The earth is not dying, it is being killed, and those who are killing it have names and addresses." The killing that Phillips referenced has only accelerated in recent decades, putting all life on the planet in peril and foreclosing any possibility for liberal or reformist politics. At a moment like this, when the planet and human civilizations are on the precipice, we must think, theorize, and act in defense of life. We need to reimagine fanatical environmentalism and agitate for revolutionary and radical action. We cannot save the planet while acting reasonably in the legal parameters of the law. If the world is going to be defended, it is going to take mass civil disobedience, including the possibility of armed struggle against governments and corporations.

The first major U.S. Earth Liberation Front (ELF) communiqué in 1996 started off with the phrase: "We are the Burning Rage of a dying planet." Since 1996 the ELF has served as the most well-known revolutionary environmental group, using ecotage and property damage as a tool to undermine and resist the destructive actions of advanced industrial capitalism. Since the first ELF actions in North America, the right wing think tanks and press, as well as fearful and milquetoast liberal environmental and media organizations, have dominated the discussion of the group, framing them as eco-terrorists and denouncing their actions as a threat to mainstream environmental efforts. This book is an attempt to change that and focus on academic articles that explore many facets of the group—from their ideology to their strategic importance—and the governments' response to them. This collection of articles serves as a critical, academic, response to the partisan and polemical right wing and liberal denouncements of the group. One should not assume this means the authors in this volume fail to critically analyze the ELF; they all do. Critical engagement is essential at a moment like this, when the fate of the planet is on the precipice. What we need right now is a new, empowered, and strategically effective environmental movement to resist the power of capital. In order to do this, we, as activists and scholars, need to step back and learn from and reevaluate the past.

The ELF (Best & Nocella, 2006) influenced by the Animal Liberation Front (Best & Nocella, 2004; Colling & Nocella, 2011) are both decentralized nonhierarchical anarchist-influenced clandestine underground groups who have very similar guidelines (Amster, DeLeon, Fernandez, Nocella, & Shannon, 2009; Nocella, White, & Cudworth, 2015). There is no membership and leadership, and no one publicly claims to be a member of this underground organization. If you obey the guidelines, your actions can be claimed as associated with the ELF and the Animal Liberation Front. The only representation of the ELF is from communiqués, which risk their privacy and safety by communicating since their correspondence can produce an internet fingerprint if sent via e-mail. The guidelines of the ELF are:

1. To cause as much economic damage as possible to a given entity that is profiting off the destruction of the natural environment and life for selfish greed and profit,
2. To educate the public on the atrocities committed against the environment and life,
3. To take all necessary precautions against harming life.

The Animal Liberation Front guidelines are:

1. To liberate animals from places of abuse, that is, laboratories, factory farms, fur farms, etc., and place them in good homes where they may live out their natural lives, free from suffering;
2. To inflict economic damage on those who profit from the misery and exploitation of animals;
3. To reveal the horror and atrocities committed against animals behind locked doors, by performing nonviolent direct actions and liberations;
4. To take all necessary precautions against harming any animal, human and nonhuman.

These guidelines represent the only leadership guidelines of the ELF and the Animal Liberation Front. These two groups have proven themselves highly effective and successful in their goals, so much so that law enforcement have identified them as top domestic terrorist groups due to the threat they pose to domination, capitalism, and fascism.

A Call to Scholars

This book, which discusses the ELF and is edited by critical animal studies scholar-activists (Best, Nocella, Kahn, Gigliotti, & Kemmerer, 2007; Nocella, Sorenson, Socha, & Matsuoka, 2013), is part of a broader intellectual project that seeks to develop revolutionary and radical environmentalism for the 21st century. As editors, have put together a collection of the most important scholarly articles about the ELF from the last decade. We are calling all scholars willing to risk their plush academic jobs as professors to write, organize, and speak out for the ELF. No more should academics and public scholars write liberal critics about the government and President Trump. We cannot keep asking students to learn how to write papers, read their textbooks, and prepare for their finals. Professors and teachers must, as Paulo Freire argued, educate students on what they need, rather than what the system wants (hooks, 1994). Teachers need to teach students how to liberate and achieve justice, not in abstraction, but in real, tangible actions such as how to blockade roads, fight against Nazis and the KKK (Nocella II, Bentley and Duncan, 2012). Teachers need to organize in the face of academic repression (Nocella, Best, & McLaren, 2010). The time for total liberation and revolution for all, human and nonhuman, is now. This world is being destroyed by capitalist-driven fascists such as Donald Trump. His supporters for racism, anti-Semitism, hate, and oppression need to retreat into their holes and never come out.

Unfortunately, the 1st Amendment in the U.S. Constitution only defends those in power and domination, it never and will never include the marginalized and oppressed such as People of Color, women, people with disabilities, economically disadvantaged, LGBTTQQIA people, immigrants, noncitizens, dissenters, and the economically disadvantaged. There is law after law repressing, suppressing, and oppressing these individuals to assure they will have the freedom of assembly or freedom of speech. Any one that is for social justice that argues for freedom of speech for all is supporting fascism and is morally bankrupt.

If you are reading this text because you want to become an activist for social change and liberation, there are a few suggestions the editors of this book recommend you follow. These guidelines include: (1) be organized in life and tactically like a chess game, (2) be sober, drug free, and healthy so you are prepared to take on physical challenges from sabotage to defense, (3) be networked and build community to gather the support you need to promote a cause or defend those in prison, and (4) finally always expand your diversity of methods you outreach locally and globally from the media to society. Furthermore, activists need to (1) take risks, (2) go beyond the nonprofit industrial complex, (3) do constant self-reflection, (4) listen more than speak, (5) not take credit for their work, (6) challenge not just one form of oppression, but all forms of oppression, (7) support total liberation, (8) strive to be decentralized, and (9) oppose hierarchical organizational structures.

Outline of the Book

This book is broken up into four sections. The first section "Classic Writings on Revolutionary Environmentalism" explores the structural and social movement dynamics of the ELF, while the second section "Classic Writings on Ecoterrorism Rhetoric" primarily explores the ideological and intellectual underpinnings of the movement. The third section, "Classic Writings on Political Repression" shifts the focus away from the ELF to how the media and state have criminalized and repressed the group. In the last section "Current Perspectives" contemporary scholars reflect back on the ELF and explore aspects and strands of the group's thoughts and actions and explore their contemporary relevance.

The first chapter, "Rhizomatic Resistance: The Zapatistas and the Earth Liberation Front," written by Michael Becker, links the ELF and the Zapatistas' revolutionary by exploring the way that both use rhizomatic resistance networks. He argues that the EZLN and the ELF, in their ideological bricolage, their anarchical and underground organization, and their now you see them now you don't tactics mark a form of organized resistance unique to the conditions of the new corporatized, globalized, surveilled, (para)militarized, and neoliberal/neo-fascist world order.

The second chapter, "Understanding the Ideology of the Earth Liberation Front," written by Sean Parson, attempts to patch a hole in the current research by analyzing the ideology of the ELF as stated in the group's key communiqués from 1996 and 2003. He argues that unlike what most critics have stated, the ELF has a complex and multivariant group ideology, one that shifts away from the deep ecology perspective of the ELF in favor of its own unique perspective of "revolutionary environmentalism." This revolutionary environmentalism incorporates components of deep ecology, social ecology, and, increasingly over the last decade, green anarchist thought.

The third chapter, "Nihilism and Desperation in Place-Based Resistance," written by Mark Seis, examines how cultural nihilism threatens to influence environmental activists engaged in a defense of place (specific political, legal, and other actions taken to protect a place that is threatened). In doing so, he develops a conception of cultural nihilism and the nihilist bind in relation to two popular environmental texts. His analysis explores cultural nihilism and individual place-based resistance present in the communiqués from the ELF.

The fourth chapter, "Ecoterrorism? Countering Dominant Narratives of Securitization: A Critical, Quantitative History of the Earth Liberation Front (1996–2009)," written by Michael Loadenthal, explores the movement's attack history through an in-depth analysis of statistics and its above-ground support network. To counter claims asserted by many academic and government sources, he uses quantitative data to critically contrast this rhetoric. Ultimately, his work presents an incident-based historical analysis of the ELF that is not situated within a logic of securitization.

The fifth chapter, "Activism as Terrorism: The Green Scare, Radical Environmentalism and Governmentality," written by Colin Salter, argues that following 9/11, the U.S. government turned to the discourse of "terrorism" as a tool to undermine resistant groups. He argues that a significant implication of the ideological rhetoric of terrorism, patriotism, and national (in)security is the self-regulation it has fostered: a form of "regulated freedom." This chapter explores the implications of governmentality, focusing on radical and revolutionary dissent which seeks to delegitimize capitalism, the property status of nonhuman animals and the environment more broadly.

The sixth chapter, "The Myth of 'Animal Rights Terrorism,'" written by John Sorenson, argues that the assumed connection between animal rights advocacy frequently linked with terrorism needs to be critically examined. In his chapter, he challenges this linkage, suggesting that accusations of violence are greatly exaggerated and argues that the terrorist image is the product of corporate propaganda.

The seventh chapter, "Leaderless Resistance and Ideological Inclusion: The Case of the Earth Liberation Front," written by Paul Joosse, examines the development of the leaderless resistance strategy by the radical right and more recently by the radical environmentalist movement. He argues that while both movements use leaderless resistance to avoid detection, infiltration, and prosecution by the state, environmental groups like the ELF benefit additionally because of the ideological inclusiveness that leaderless resistance fosters.

The eighth chapter, "Standing up to Corporate Greed: The Earth Liberation Front as Domestic Terrorist Target Number One," written by Anthony J. Nocella II and Matthew J. Walton, focuses on the police repression of radical environmental and animal rights activists. In their chapter, they examine the actions and philosophy of the ELF, particularly in relation to global capitalism. Their goal is to provide insight into why the ELF does what it does, and why its actions have situated it atop the FBI Domestic Terrorist list, despite ELF guidelines specifically prohibit inflicting any harm to human or nonhuman animals. They argue that the ELF actions contain a compelling critique of capitalism, which is much more of a threat to "American values" and to the consumer-driven U.S. way of life, than other potential threats that seek to harm humans such as Christian pro-life or right-wing groups.

The ninth chapter, "Mapping Discursive and Punitive Shifts: Punishment as Proxy for Distinguishing State Priorities Against Radical Environmental Activists," written by Lawrence J. Cushnie, explores why over the past decade sentencing rates have climbed steadily for environmental activists who choose property destruction as their form of protest. Cushnie contends that the courts, in sentencing radical environmental activists, adopt clear signals from the federal government. Literature on judicial behavior is helpful toward addressing some of these questions. However, the most important questions revolve around the theoretical implications concerning a state, which, in certain cases, punishes the destruction of property at levels comparable to the destruction of sentient life.

The tenth chapter, "Speaking About 'Ecoterrorists': Terrorism Discourse and the Prosecution of Eric McDavid," written by Joshua M. Varnell, explores how terrorism discourse was employed to investigate and prosecute Eric McDavid as a domestic terrorist. By investigating the use of the terrorism discourse in McDavid's trial, Varnell illustrates how hegemonic terrorism discourse was used to prosecute McDavid. First, how the terrorism discourse has been used to justify law enforcement investigative tactics, specifically the use of informants in terrorism investigations. Secondly, this chapter demonstrates how the terrorism discourse was reproduced in McDavid's trial to prosecute him as a dangerous domestic terrorist.

The eleventh chapter, "Radical Environmentalism as Teacher: A Pedagogy of Activism," written by Meneka Repka, examines how the second ELF guideline, which advocates "educating the public on atrocities committed against the environment and life" can be used (perhaps counterintuitively) within the system of capitalist public education to destabilize and ultimately dismantle the system itself. This chapter proposes a Trojan horse of sorts: that ELF tactics can be introduced to youth activists by hiding radical ideas and forms of education in plain sight—within the existing system of public schooling. A classroom can act as an individual ELF cell by realizing three significant positions of ELF. First, students will formally be exposed to the atrocities that have historically and are currently being committed against the environment and life. As well, corporate ties between the school and their sources of funding will be discussed openly with students. Finally, students will be encouraged to oppose dominant practices that directly or indirectly harm the earth, such as dissection.

The twelfth chapter, "Those Mischievous Elves of Lore: The Legend and Legacy of Earth Liberation," written by Alexander Reid Ross, is a unique chapter that relates the ELF to elves, spirits, and fairies of the medieval to Earth First!, Black Panther Party, Zapatistas, and Rising Tide. This detailed theoretically passionate article has a needed critique of Deep Green Resistance of their anti-transgender perspective, which divided the radical environmental movement and pushed out Derrick Jensen and Lierre Keith. Moreover, the author is extremely knowledgeable about the philosophy, purpose, and the history of the ELF. Ross goes on to not only relate the ELF to historical groups, but examines them from an anarchist, green anarchist, and critical theory perspective.

The thirteenth chapter, "Magic Kills Industry: Reclaiming ELF and Witch Deviance as Ecoqueer and Anticapital," written by Mara Pfeffer and Bethany Richter, explores the connection between the ELF, magic, and folklore, in an attempt to unravel the insurrectionary potential of the radical environmentalism activism. Looking to adrienne maree brown's new book *Emergent Strategies*, the authors develop a radical queer environmentalism that undermines not just the governing logic of capitalism, but the heteronormative and colonial logic implicit within it. Looking to the myth of the elves and their connection with witches, the authors support a politics of magical solidarity that elicits a radical politics that focuses on creativity and play.

The fourteenth chapter, "Problematising Non-violent 'Terrorism' in an Age of True Terror: A Focus on the Anarchic Dimensions of the Earth Liberation Front," written by Richard J. White, is an excellent summary and introduction of the ELF. White examines how the ELF is stigmatized as a violent terrorist group, even though they have never strived or have harmed anyone since the organization's development. He also examines how other movements, such as anarchist groups, have been stigmatized in similar ways. White argues these groups have been targeted for bias because of their critique of property, which challenges the whole capitalist system and argues that everything in our society has a specific value, which these groups are against.

In conclusion, we hope this book ignites a global passionate strategic total liberation revolution and burns bright against fascism, hate, and oppression (Del Gandio & Nocella, 2014). The goals of the ELF include burning bridges, not building them. They do so for the end of colonialism and civilization and in defense of those oppressed such as transgender people, People of Color, LGBTTTQQIA people, nonhuman animals, elements, the air, the water, the land, the mountains, people with disabilities, people that are economically disadvantaged, women and girls, youth, elderly, and those that are not Christian.

References

Amster, R., DeLeon, A., Fernandez, L., Nocella, A. J., II, & Shannon, D. (2009). *Contemporary anarchist studies: An introductory anthology to anarchy in the academy*. New York, NY: Routledge.

Best, S., & Nocella, A. J., II. (2004). *Terrorists or freedom fighters? Reflections on the liberation of animals*. New York, NY: Lantern Books.

Best, S., & Nocella, A. J., II. (2006). *Igniting a revolution: Voices in defense of the earth*. Oakland, CA: AK Press.

Best, S., Nocella, A. J., II, Kahn, R., Gigliotti, C., & Kemmerer, L. (2007). Introducing critical animal studies. *Journal of Critical Animal Studies*, *5*(1), 4–5.

Colling, S., & Nocella, A. J., II. (2011). *Love and liberation: An animal liberation front story*. Temple: Arissa Media Group.

Del Gandio, J., & Nocella, A. J., II. (2014). *Educating for action: Strategies to ignite social justice*. Gabriola Island, BC: New Society Press.

hooks, b. (1994). *Teaching to transgress: Education as the practice of freedom*. New York, NY: Routledge.

Nocella, A. J., II, Bentley, J. K. C., & Duncan, J. (2012). *Earth, animal, and disability liberation: The rise of the eco-ability movement*. New York, NY: Peter Lang.

Nocella, A. J., II, Best, S., & McLaren, P. (2010). *Academic repression: Reflections from the academic industrial complex*. Oakland, CA: AK Press.

Nocella, A. J., II, Sorenson, J., Socha, K., & Matsuoka, A. (2013). *Defining critical animal studies: An intersectional social justice approach for liberation*. New York, NY: Peter Lang.

Nocella, A. J., II, White, R., & Cudworth, E. (2015). *Anarchism and animal liberation: Essays on complementary elements of total liberation*. Jefferson, NC: McFarland.

Part I: Classic Writings on Revolutionary Environmentalism

1. *Rhizomatic Resistance: The Zapatistas and the Earth Liberation Front*

MICHAEL BECKER

Ana Carrigan's proclamation that the Zapatista rebellion in southeastern Mexico is the world's "first post-modern revolution" can be taken in as many ways as there are definitions of the hackneyed term "postmodern" (Palaez, 2001). Certainly, the rebellion marks a liberation of many "others." Subcomandante Insurgente Marcos describes the uprising as "all the minorities... untolerated, oppressed, resisting, exploding, saying 'Enough'" (Vahabzadeh, 2004). The postmodern rebellion brings to the attention of the privileged ignorant the face of the other, not to make it known but to present it as it is, largely unknowable unless the familiar becomes unfamiliar. The now astonishingly familiar, but still unknown faces of "all the minorities" in Chiapas have even come to be seen only because they are masked and defiant. The mountain areas of southeast Mexico are in Zapatista rebel hands. "And these Zapatistas are very otherly.... These Zapatistas neither vanquish nor die, but nor do they surrender, and they despise martyrdom as much as capitulation. Very otherly, it's true.... They are rebel indigenous. Breaking...the traditional conception, first from Europe and afterwards from all those who are clothed in the color of money that was imposed on them for looking and being looked at" (Subcomandante Marcos, 2003).

Still, it is not merely bringing attention to the other but the "resisting, exploding, saying" way that Chiapas illumines the face of the other that is new. This decentralized and proliferating discourse and set of tactics links up with many "others" under the Zapatista umbrella of "civil society." When Marcos allegedly was outed by the Mexican government in 1995 as Rafael Guillen, an unemployed Communication Philosophy professor, Marcos responded with his own version of his identity. "I'm gay in San Francisco,

Black in South Africa...an Asian in Europe,...a Chicano in San Ysidro... an anarchist in Spain...a pacifist in Bosnia...a Palestinian in Israel...a *chava banda* in Nezahuacoyotl...an Indian in Chiapas" (Subcomandante Marcos, 1994). If Marcos fragments into "all the minorities" the minorities conversely unite into Marcos. In response to the supposed identification of Marcos and an attempted police/military roundup of Zapatista leaders hundreds of thousands of students, activists, laborers, and others filled the Zocalo in Mexico City and "*Todos somos Marcos!*" ("We are all Marcos") became a rallying cry in support of the Zapatistas. The phrase then echoed around the globe among antiglobalization resistance fighters of many sorts.

It is this expansive manner of resistance in random order that distinguishes the EZLN. Like its revolutionary counterparts in North America and Europe—Earth First!, Sea Shepherd, the Earth Liberation Front (ELF), the Animal Liberation Front (ALF), SHAC, The Animal Rights Militia, the Revolutionary Cells, the Justice Department, and others—the Zapatista rebellion takes root, expands, and erupts as rhizomatic resistance. The EZLN and the ELF, in their ideological bricolage, their anarchical and underground organization, and their now you see them now you don't tactics mark a form of organized resistance unique to the conditions of the new corporatized, globalized, surveilled, (para)militarized, and neoliberal/neo-fascist world order. Certainly, Deleuze and Guattari's rhizome is among the apt literary-biological metaphors for describing the postmodern character of revolutionary environmentalism.

These movements liberate by rupturing the conceptual foundations, the received organizational forms, and the [il]legitimacy of everyday action in the corporate–state world order. We will trace some of the roots of the EZLN and the ELF. Specifically we seek to link up two interrelated figures that mark main points of divergence of revolutionary environmentalism from mainstream groups and convergence with the EZLN: a Heideggerian critique of Western technology and indigenous biocentrism. At first glance these themes may appear to be completely at odds. The first moves from near the end point of the Western philosophical tradition, while the second is rooted in an ancient oral and spiritual tradition that survives despite an onslaught of Westernizing forces. Yet, Heidegger's work too is retrospective. And the Zapatistas incorporate indigenous traditions in the context of a critique of a globalized, technological–capitalist "Fourth World War" that resonates with Heidegger's interpretation of nihilism. More importantly, both Heidegger's critique of technology and the indigenous themes we trace here illuminate an ontology of openness wherein Being, far from being determined and defined, remains an open space within which the spiritual basis for a surmounting of globalization might unfold. Human freedom understood as care taken that each event of creation might unfold according to its own limits and hope that such a world might emerge within the present are the words of both Heidegger and the indigenous. In these and other discursive figures and their related actions the EZLN and the ELF are clarifying a set of revolutionary principles rooted in indigenous decentering of subjective identity. These particular traces are part of an incalculably larger rhizome, its branches and leaves rooted in an ultimately untraceable root-mass. It is a grassfire; it is a wind gathering the force of a hurricane.

Rhizome

Just as the identity of Marcos fragments into "all the minorities," in their introduction to Thousand Plateaus (1987), Deleuze and Guattari move quickly to take apart various conceptions of unity—of themselves as subjects and authors, of the text they write, of literatures, of the disciplines and subdisciplines of linguistics, and psychoanalysis. Of course, there are themes that hold these fields together as coherent wholes. But beneath the surface, unity dissolves into irreducible multiplicity. "In a book as in everything else, there are lines of segmentarity, strata, territorialities; but also lines of flight, movements of de-territorialisation, and destratification" (Deleuze & Guattari, 1987, pp. 3-4). The rhizome is an expression of the underside and the tension within disparate elements that always already exists in any sort of organized whole. Deleuze and Guattari's overall project is to constantly call attention to this play of unity and multiplicity, consolidation and rupture. They contest, for example, the accepted idea of evolution of social and political relations from nomadic hunter-gatherer, to agrarian village with its surpluses, to state form, developing from the agricultural surplus of the village. Rather, forces of organization and dis-aggregation are always in tension. "[N]omads do not precede the sedentaries; rather, nomadism is a movement, a becoming that affects sedentaries, just as sedentarization is a stoppage that settles the nomads" (Deleuze & Guattari, 1987, p. 430).

Today, nomadism erupts in green anarchy and primitivism, a countermove to "civilization" present from Diogenes and the Cynics in the classical world to the Brothers and Sisters of the Free Spirit in medieval times to the Diggers and Levelers of modernity. Rather, forces of organization and disaggregation are always in tension. "[N]omads do not precede the sedentaries; rather, nomadism is a movement, a becoming that affects sedentaries, just as sedentarization is a stoppage that settles the nomads" (Deleuze & Guattari, 1987, p. 430).

Both Deleuze and Guattari and Foucault emphasize the significance of Nietzsche in breaking apart metaphysical fictions and restoring a recognition of the accidental, the contingent, and the singular forces that underlie that which evolves into apparent unities. Lines of descent from an alleged origin fragment into lines of dispersion that mark the unique and unremembered chaos of events from which an apparent unity first emerged. Foucault's genealogy, like Deleuze' and Guattari's cartography trace "passing events in their proper dispersion; it is to identify the accidents, the minute deviations—or conversely, the complete reversals—the errors, the false appraisals, and the faulty calculations that give birth to those things that continue to exist and have value for us" (Foucault, 1984, p. 81).

The strange phrases that Deleuze and Guattari (1987) use name the dynamic of shifting substrata that constantly challenges a unified, organized whole: a system without center or borders, lines of flight and intensities, bodies without organs, units of density and convergence. Over against the metaphor of the tree with its taproot and trunk unifying the roots, branches, stems, and leaves Deleuze and Guattari posit the proliferation of the rhizome.

> The multiple *must be made* [emphasis original] by subtracting a singular characteristic from the whole expressed symbolically: "write to the power of n-1'. Such a system is called a rhizome...an absolutely distinct type of underground stem-system. Bulbs and tubers are rhizomes.... Even animals are, in their pack form: rats are rhizomes. So are warrens, in all their functions as habitat, provision, passage, evasion and disappearance. (Deleuze & Guattari, 1987, pp. 6–7)

The EZLN, suddenly emerging to occupy towns, their infiltrators just as suddenly disappearing and dissolving Mexican army units, and then fading into their jungle redoubts are rhizomes. The anonymous and autonomous cells of the ELF erupting in sudden arson attacks across the United States and as rapidly disappearing are rhizomes. Rhizomes threaten an established order; they often operate unseen; they are irrepressible and cannot be eradicated as their root stem system allows proliferation at each of its nodes. One may break off and analyze a section of bamboo in a mature bamboo stand. But the system to which any part of a rhizome may be attached is ultimately untraceable. No matter the number of segmentations, one is lost in a prodigious maze of branches and stems, not to mention a bewildering and unyielding mass of hidden, densely tangled roots. There is always n–1, with the singular part open to consideration. But the sum of the parts is ultimately incalculable and, as such, the parts cannot be summed up in an ostensible whole.

Deleuze and Guattari (1987) isolate six characteristics of the rhizome. Four of these are of particular concern here. First and second, they describe a rhizome's connection and heterogeneity. By reference to language, the authors argue that rhizomatic language features "semiotic chains of biological, political and other kinds, bringing into play not only different regimes of signs, but also different orders of states of affairs" (Deleuze & Guattari, 1987, pp. 6-7). Rhizomatic language disrupts the alleged fixed point of unity and order, the mother tongue or grammatical rules that arrest the multifaceted points of connection in a prodigious rhizome of language, referents, language games, and their corresponding states of meaning and states of affairs.

> A rhizome endlessly connects semiotic chains, power organizations, occurrences relating to the arts, the sciences or to social struggles. A semiotic chain is like a tuber agglomerating quite different types of acts—linguistic, but also perceptual, mimetic, gestural, cognitive ones: there is no language in itself, nor any universal language, but a concourse of dialectics, patois, slangs, special languages.... Language...is "an essentially heterogeneous reality." There is no mother tongue, but a seizure of power by a dominant tongue within a political multiplicity. (Deleuze & Guattari, 1987, p. 53)

The explicitly political character of the authors' analysis here should not be overlooked. Naturally, much Deleuze criticism centers on psychological questions about subjective and personal identity. But the connection between language, signs, political forms, and social struggles is made quite explicit here (Patton, 2014). Additionally, in regard to language, Derrida's famous essay "Differance" is of considerable use here. Signs have meaning based on their difference from other signs, not based on an ostensible referent of the sign. Additionally, signs defer to other signs in order for an ultimate meaning to be reached, a meaning which, of course, is never in fact reached. Instead there is simply the constant deferring and differing that is "différance." The same might be said of the rhizome: neither the sovereignty of the "One," the unified whole is ever complete any more than the dispersive effect of the multiple. Rather each largely gains its "identities" from its ongoing tension with the other (Derrida, 1982).

With revolutionary environmentalism, both in its refiguring of the language of globalization and the radical challenges to it—especially in indigenous philosophy and its anarchic, consensus-based democracy—the uniformity of conventional language/social order is exploded. The play of language that connects what, according to the existing constraints of language and social form, are dissonant elements marks the connective heterogeneity of rhizomatic resistance. Zapatista discourse recalls revolutionary heroes, provisions of the Mexican constitution, features of Mayan oral tradition and practice, and neo-Marxism. Revolutionary environmentalism connects both shallow and deep ecology, primitivism, anarchism, indigenous spirituality, and the Zapatistas!

Third, Deleuze and Guattari (1987) emphasize the notion of multiplicity. Multiplicity must be accepted as such, as substantive. In this manner, the multiple "loses all relationship to the One as subject or object, as natural or spiritual reality, as image and world" (p. 14). Multiple determinations absent any substantive, unifying signifier, are themselves to be traced, not to a common source but in their heterogeneity. Writing n–1 takes the subtracted element as it is not as having meaning only in its connection to an alleged common denominator.

Fourth, and most importantly perhaps for tracing the ELF and the EZLN, is the characteristic of a rhizome termed "asignifying rupture." Here again, Deleuze and Guattari (1987) emphasize the irreducibility of rhizomatic segments to any ultimate organizing principle. Break off a section of a rhizomatic plant or detach or kill a part of a rhizomatic animal population and the plant or pack shoots off in other directions continuing to proliferate. No qualifiers of good or bad, positive or negative can be attributed to this eruptive growth of the rhizome; new shoots take their course in a deterritorialization or destratification of any schema by which they would be contained or controlled. Describing the "symbiotic" relationship between orchid and wasp, Deleuze and Guattari (1987) claim: "There is no imitation or resemblance, but an explosion of two heterogeneous series into a line of flight consisting of a common rhizome that can no longer be attributed or made subject to any signifier whatever" (p. 9).

In demonstrating certain parallels between the EZLN and ELF we must resist the temptation to reduce them to a single explanatory framework. The rhizome metaphor is precisely intended to prevent such a reduction. We are merely tracing moves in the context of similar struggles. All we leave is a trace of rhizomatic ruptures that radically confront the state corporate apparatus. Similarly, focus on the two pathways identified here is not meant to obscure, take precedence over, or downplay other lines of segmentarity in these movements. Heterogonous connectivity in the EZLN links up "civil society"—indigenous rights groups, environmentalists, labor groups, women's rights groups, anarchists, human rights and democracy activists, and other left political activists—with appeals to nationalism, the Mexican Constitution (especially Articles 27 and 39), significant figures in Mexican history, both Catholic and Mayan communal and spiritual traditions, and the dancing of cumbias. ELF communiqués are rife with appeals to deep ecology, social ecology, animal rights, anarchism, concern for natural ecosystems, and a sheer, liberating sense of mischief in monkeywrenching the corporate machine. In the realm of tactics and organization there are a variety of roots one might trace, not least of which is that both movements have been able to maintain anonymity and yet broadly extend their connective heterogeneity by an extraordinary use of the internet. In short, while we trace Heideggerian and indigenous, biocentric roots of revolutionary environmentalism we must not lose sight of the "resisting exploding, saying 'Enough'" that is these movements. Heidegger's "ontological anarchism" (his attempt to avoid the reduction of the question of Being to some particular kind of being) complements indigenous philosophy of an ineffable mystery of Being. We might go so far as saying that the ontological space described by Heidegger and indigenous philosophy is parallel to the political space opened up by rhizomatic resistance to the war machine.

Critical Discourses of the ELF and EZLN

Constantly a wider set of cultures and persons are asked to behold the spectacle of the enormous productive forces constituted by the corporate–state apparatus. Both the ELF and EZLN turn this Roman Triumph on its head, demonstrating the equally colossal destructiveness of globalization. It is not a dialectic that operates here; rather, it is a deconstructive proliferation of counterclaims to neoliberal propaganda. One element that runs through both ELF and EZLN discourse is a profound critique of the technological character of globalization. The critical discourse in ELF and EZLN communiqués reflects the same basic critical interpretation of technology—namely, that technology is a historical and ontological formation—rooted in Western metaphysics and centering on synthesizing entities including, ultimately, people into cybernetic systems. Technology is not simply a neutral set of tools and methods but a cultural imperative that everything yields to efficient systematization.

Both ELF and EZLN communiqués reveal a critical interpretation of technological praxis similar to Heidegger's conception of "challenging forth" wherein the Earth is assaulted and provoked to yield up "natural resources" to interlocked, increasingly cybernetic systems. The command character of challenging forth is revealed in the ELF activists' sense of the provocative nature of those business ventures targeted for direct actions. In a series of communiqués from Long Island in 2000 and 2001 concerning the torching of luxury homes that threatened sensitive pine barren habitat and an important aquifer, activists spoke of the virtual assault mentality of the developers. The communiqué held that the "Earth is being murdered." The writers speak of the "rape of the Earth" by the Earth's "oppressors." The activists vowed to continue to stop such destruction as long as the "Earth is butchered." ELF activists specifically identify with the EZLN rebels and those they defend in Chiapas; both groups' actions are rooted in a defense against a corporate–

state apparatus that wages "a war against the environment" as well as "a war against the people who live sustainably within it." One of the leading figures in the direct action movement, Rod Coronado (2003), whose Pascua Yaqui ancestors had engaged in a rebellion against Spanish conquistadores, speaks of the insight he had early on in his activism when trying to stop the slaughter of harbor seals in Canada. It suddenly dawned on him that the genocide against animal nations, including the seals, was part of the same process of the genocide of the Spanish against his people.

In our view these pronouncements are not mere ideological hyperbole. Emanating from the communiqués is a profound sense of misgiving about and defense against the inherently destructive, "command and control" technological orientation toward nature. In public forums activists cite statistics on the extinction crisis but not with clinical detachment. Instead these statistics help to bear witness to the violent appropriation of nature that they are contesting. Such concern is further demonstrated in videos produced by the press offices of ELF and ALF. One can speak of a terrorism of the Earth as it is bulldozed, a forest of trees splintered by the chainsaw and crashing to the forest floor, and the subsequent gaping wound to the Earth of a clear-cut. It is the same terror expressed in the eyes of an elephant that has rampaged in fear and anger against its keepers. The testimony to the terrorizing of nature was expressed in the communiqués and was first moved toward an absolutist animal rights position by his seeing chained and caged animals "backstage" at the zoo. He sensed the deep fear and rage in the animals' eyes and witnessed the neurotic behavior induced by their confinement. That activists recognize the assault as a systematic form of destruction is clear in their repeated reference to "genocide" against many animal nations. What is necessary is a thoroughgoing and constant recognition of the inherently violent nature of technological assault on the Earth.

Bringing to conscious awareness the violent character of modern technics comprises, in large part, the truly revolutionary character of the ELF and ALF. These are the only "environmental" organizations that have full grasped that the current integration of modern technics and corporate capital results in systematic violence against all of nature, including human nature. Consistently ELF and ALF spokespersons critique single-issue environmental policies for failing to understand the universally homicidal/suicidal assault of corporate technics on the Earth. A statement from the ELF press office reads:

> The Earth Liberation Front does not commit merely symbolic acts to simply gain attention to any particular issues. It is not concerned merely with logging, genetic engineering, or even the environment for that matter. Its purpose is to liberate the earth. The earth, and therefore all of us born to it, are under attack.

> We are under attack by a system which values profit over life, which has, and will, kill anything to satisfy its never ending greed. We have seen a recent history rich in the destruction of peoples, cultures, and environments. We have seen the results of millions of years of evolution destroyed in the relative blink of an eye. (Lesliejames, 2000)

The radicalism decried by critics of direct action environmentalism is a counterpoint to the terrorism against the Earth which continues unabated. It is only in the context of a technological assault on the Earth that one can grasp ELF and ALF arguments that revolutionary direct actions are a form of self-defense. Self, as we allude to below, must be considered in the context of the Indian word "mahatama"—the wider self which includes all that self relates to itself as itself. Ultimately, this self-relation must move beyond the relation of particular things to a relation to the whole, to nature or Being. The true, wider self is a relation involving acknowledgment of the whole to which one is constantly related (Naess, 1985). It is largely in defense of indigenous people that the EZLN continues to oppose the destructive element of the essence of technology, particularly as it is manifested in such globalizing events as NAFTA.

Ten years before the Zapatistas burst onto the world stage the EZLN was established in the Lacandon jungle. Two years before the uprising, Marcos penned "The Southeast in Two Winds: A Storm and A Prophecy." The destruction described by Marcos is couched in more explicitly neo-Marxian terms than the ELF communiqués (specifically, Wallerstein's conception of a world capitalist economy in which core countries prosper upon the extraction of cheap labor value and cheap resources from peripheral countries). Also, Marcos is concerned to a far greater extent with the impact of capitalist exploitation on the indigenous people of Chiapas, the ELF centering more on the destruction of the Earth. While this should not be overlooked, the simple recognition and unmistakable foreboding and loathing regarding the corporate–technological juggernaut draws the ELF communiqués and Marcos's "prophecy" into a network of rhizomatic flows. Marcos describes foreign and comprador class exploitation as a "beast [that'] feeds on the blood of the people." Foreign and domestic businesses as well as the Mexican state (Pemex) "take all the wealth out Chiapas and in exchange leave behind their mortal and pestilential mark." Recognizing the ecological dilemma, Marcos notes that legalities allow the destruction of the jungle for oil extraction and large-scale logging but disallow cutting in the Lacandon by indigenous people. "The poor cannot cut down trees, but the petroleum beast can, a beast that every day falls more and more into foreign hands. The campesinos cut them down to survive, the beast to plunder." Thousands of barrels of petroleum and billions of cubic feet of natural gas are sucked out of Chiapas; "ecological destruction, agricultural plunder, hyperinflation, alcoholism, prostitution, and poverty" are left behind.

Chiapas, according to Marcos, "bleeds" coffee, beef, 55% of Mexico's hydroelectricity, 20% of Mexico's total electricity, hardwoods, and a wide variety of agricultural products from corn to honey to avocados, tamarind, and mameys. It leaves behind a third of municipal seats without paved road access, the people in 12,000 communities on foot, following mountain trails. The railroads and the single port in Chiapas move products not people. Seventy-two percent of children do not finish first grade—the richest state in natural resources has the worst schools. There are .2 clinics for every 1000 Chiapanecos, .3 hospital beds, one operating room per 100,000, .5 doctors and .4 nurses per 1000. Fifty-four percent in Chiapas are malnourished, 80% in the highlands and forests.

The Zapatista uprising began on January 1, 1994, the day NAFTA took effect. As predicted, US government-subsidized corn imports are undercutting Chiapaneco farmers' corn, deepening the oppression in the state. "The fee that capitalism imposes [on Chiapas] oozes, as it has since the beginning, blood and mud." From the beginning of colonialism to the present, Chiapas exports its natural resources, "it continues to import capitalism's principal products: death and misery" (Subcomandante Marcos, 2004).

Marcos's "prophecy" (the indigenous elements of prophecy are discussed below) is a powerful, neo-Marxist indictment of capitalism. It is also something more. While Marx considered capitalism to be beneficial in creating the infrastructure from which lower and higher stages of communism would spring, Marcos depicts the malevolence of a form of expropriation from the Earth unhinged from any sense of indigenous reality. To give some sense of the indigenous reality that is other to globalization, consider corn. As for many meso-Americans, the Mayans hold that corn—red, black, yellow, and white—is the original ancestor of all humans. That bio-confinement of the bio-engineered corn flooding Mexico after NAFTA does not work is evident in the very place where corn first emerged as one of the most vital of human food sources. Researchers have discovered genetically altered material in native corn varieties. UC Berkeley plant scientists discovered that 4 of 6 varieties of native criollo corn grown in fields in the mountains of Oaxaca contained a genetic "switch" commonly used in genetically modified crops. The Zapatista rebellion was an act of defense, in this case (in part) against genetically modified corn. Even before the uprising, the Tzotzil Mayan people of Chiapas took steps to protect their centuries-old heirloom corn from Monsanto's Frankencorn by creating a seed "safe house" where heirloom variety seeds could be preserved. The hybrid (pun-intended) "modern/postmodern" nature of the seed saving juxtaposed high-tech conditions in which the seed is protected (carefully regulated temperature in which the seed is frozen) with a ceremony accompanying the seed-saving project. Zapatista autonomous school board members joined students in praying in their native Tzotzil for the survival of the mother seeds of corn. Illuminating the violence of globalization, one of the Zapatista teachers explained "We have to protect these little seeds because they are under attack just like our communities. My grandfather was killed because he defended the traditions of our community and he believed in justice and democracy. Now even if I am an indigenous woman I have to defend our corn so that our traditions can continue." Drawings by students represented the safe houses for the seeds and for the indigenous knowledge that surrounds and gives the seed and the Mayan people their eternal cycle of life. As a Mayan elder put it "you see the seed that cannot survive without its people, and we cannot survive without our corn" (Organic Consumers, n.d.).

It is this depiction of the extent and nature of the systematic destruction wrought against the Earth and its masses that is reflective of Heidegger's depiction of the essence of technology as an assault on the Earth. As we discuss below, it is the necessity of being guided by an indigenous heart in order to liberate the Earth and the Earth's creatures from this force that marks the profoundly heterogonous character of the ELF and the EZLN nations are destroyed (The Indigenous Revolutionary Clandestine Committee, 1998).

A second characteristic of the essence of technology, according to Heidegger, is the "standing" or visible aspect of natural entities set up through challenging forth. It is the way in which things commonly appear "when they are wrought upon" by challenging forth (Heidegger, 1971a, p. 17). *Bestand* or "standing reserve" expresses the way in which entities within a technological framework appear as constantly ready or on standby. Entities appear to constantly avail themselves to courses of action oriented toward maximum efficiency. Everything is ready to be used; everything is available for instantaneous manipulation. It is "the whole objective inventory in terms of which the world *appears* [emphasis added]" (Heidegger, 1971a, p. 111). With the development of modern technology nature and works literally appear differently to us. "The world now appears as an object open to the attacks of calculative thought.... Nature becomes a gigantic gasoline station, an energy source for modern technology and industry [emphasis added]" (Heidegger, 1966, p. 50). "The Earth itself can show itself only as the object of assault.... Nature appears everywhere...as the object of technology" (Heidegger, 1954, p. 100). Ultimately the Earth appears as "a giant gasoline station," that is, set up for the pumping out of resources.

The awareness that technology has reduced entities to the level of standing reserve, on call for instantaneous use, is widely revealed in revolutionary environmental communiqués, especially from ALF activists. A communiqué from 1999 explains the grounds for the liberation of beagle puppies from Marshall Farms in upstate New York. Marshall is a breeder for Huntingdon Life Sciences. The thirty liberated puppies were among "hundreds of beagle puppies waiting to be shipped to vivisection labs." Within the essence of technologically ordered cybernetic systems, these animals literally do not appear as animals at all or even as distinct objects. They are factors within a giant corporate–scientific research system ordered for corporate profits. That puppies are often slammed against walls or otherwise abused is obviously shocking and disgusting. That they are subject to live vivisection is simply horrific. But these actions and the entrapment of the beagles in the first place occur within a technological context in which nature, in whatever particular form, disappears and is able to show itself only as standing reserve. These are not puppies but factors of production in the corporate research/commodification system.

Language in the communiqués that express the conversion of minerals, plants, and animals into materiel and commodities reflects Heidegger's notion of the "ordering of the orderable." A tree-spiking action in Brown County and Monroe County, Indiana, state parks was "a warning to all those who want to turn the beings of the earth into cash" (Resistance, p. 3). Similarly, a Wisconsin communiqué concerning genetic modification of white pine trees notes that forest "management" treats "wildlife as some numbers on a graph." The Forest Service coordinates with timber companies in "an insane desire to make money and control Life" (Resistance, p. 4). Direct action tactics are, on the one hand, self-defense against the assault on the Earth (challenging forth). They are, additionally, motivated by a reaction to a form of state–corporate technics that characterizes humankind's conversion of nature into standing reserve as the only "natural" relationship of humankind to the Earth. That nature is set up as standing reserve, on call for integration into extensive technological networks, is also prevalent on EZLN discourse.

In one of the most powerful and remarkable of all EZLN documents, the "First Declaration of La Realidad for Humanity and Against Neoliberalism," Marcos describes the distribution of world power as "concentrating power in power and misery in misery." "Dispensible" minorities are arrayed against a "modern army of financial capital" and corrupt governments. "The indigenous, youth, women, homosexuals, lesbians, people of color, immigrants, workers, and peasants; the majority who make up the world's basements are presented, for power, as disposable.... Men, women, and machines become equal in servitude and in being disposable [emphasis added]" (Subcomandante Marcos, 1998, p. 12). The description of the leveling effect that Marcos invokes here is remarkably similar to Heidegger's. The latter's account would seem to involve a clear dichotomization of subject and object, a core principle of Western philosophy at least since Descartes. But the advent of standing reserve as an "inclusive rubric" actually undermines even the objective character of individual entities. Entities within standing reserve are reduced to a manipulable homogeneity, losing even their identity as distinguishable objects. The standing reserve is "mere material...a function of objectification."

If even the objective quality of an entity disappears within the standing reserve it is obvious that its unique qualities will similarly be eclipsed. What the Zapatistas point out is that disappearance by integration of entities within globalized markets includes human beings. In speaking of Europe's negotiating a free trade agreement with the Zedillo administration at the height of the oppression against EZLN and its supporters (not long after the Acteal massacre), Marcos points out that the logic of the market is superior to the logic of human rights or even the recognition of peoples and cultures.

> In the great fraud called the "North American Free Trade Act" (product of the great Salinas lie), the future is now being projected with the signing of a free trade agreement with the European Union...the European governments are extending their hands to Zedillo without caring that his is covered with indigenous blood.... The European Union's flexibility can be understood, what is at stake is a slice of the pie that is called, still, "Mexico." Due to the marvels of globalization, a country is measured by its macro-economic indices. The people? They do not exist, there are only buyers and sellers. And, within those, there are classifications: the small, the large and the macro. These latter ones buy or sell countries. At one time they were governments of Nation States, today they are only merchants [emphasis added]. (Subcomandante Marcos, 2000)

People as individual human beings or in their collective cultural or national respects do not exist. Signifiers such as Tztotzile, Zapatista, or even Mexico—if what is meant by that term is a cultural designation—literally are obliterated except insofar as they might denote something of market value as standing reserve (the sign exchange value of an exotic vacation destination, perhaps). Any such nonentities who threaten corporate state hegemonic control must be wiped out. Regarding the Zapatistas, this crucial and absolute fact of international capital was most dramatically brought out in a dry, matter-of-fact Chase Manhattan Bank memo of January 13th leaked by a banking insider to Counter-Punch magazine: "The government will need to eliminate the Zapatistas to demonstrate their effective control of the national territory and security policy.... While Chiapas, in our opinion, does not pose a fundamental threat to Mexico's political stability, it is perceived to be so by many in the investment community" (Silverstein & Cockburn, 1995).

The third and most complex of the terms Heidegger uses to discuss the essence of technology is Enframing (*Gestell*). Heidegger refers to the essence of technology as a "way of revealing." By this phrase Heidegger has in mind an epoch as defined by a historically conditioned response of human beings to Being. In each epoch the response to Being is rooted in fundamental words (*Grundworte*) that the most important thinkers of that period have coined to orient human beings toward Being. The pre-Socratics conveyed a poetical experience of the mystery of Being: they grasped how the unity of Being concealed itself to allow the coming to presence of beings in their particularity. The elemental forces described by the Milesians are not literally meant to represent the "stuff" of the universe but rather the ultimately unnamable process of unity diversifying into plurality and reuniting into one-ness. The same is true for Heraclitus's notion of the "ever-living fire."

But, since Plato, Western metaphysics has been marked by an increasing tendency to neglect the question of Being. Instead, Western philosophers have consistently tried to represent Being in terms of a specific kind of being—the Platonic form of the Good, Aristotelian substance, Augustinian will of God, Leibnizian monad, Cartesian, res cogitans, etc. The foundational words of Western metaphysics have always served to obscure rather than to illuminate Being. For Heidegger, this "errant" characteristic of Western metaphysics, the increasing turn away from Being, marks the inherent nihilism of the West (and with the worldwide extension of Enframing, most of the globe).

The essence of technology—Enframing—is the extreme point of the development of Western nihilism. Being has become completely obscured in a metaphysics of subjectivity worked out in the technological practice of total control. The term "Enframing" is meant to characterize the historical–ontological factors conditioning a cybernetically centered, nihilistic response to the question of Being. Modernity is marked by a technological imperative, a will to integrate all beings into cybernetic systems or "enframe" them within the orderable. As it stands at present, Enframing is our destiny. Heidegger claims that Nietzsche's doctrine of will to power (the foundational words of Enframing) epitomizes the subject-centered, nihilistic extreme of Western metaphysics. Ontologically, Nietzsche's doctrine prepares the way for understanding Being as defined by force vectors oriented toward continually increasing power.

We obviously do not expect to find a comprehensive ontology in ELF and EZLN discourse. Implicit in their communiqués, however, is a clear awareness of the manner in which corporate, mass-consumer capitalism continually integrates nature into technical production systems. A communiqué regarding the firing of two USDA Animal Damage Control Buildings in Olympia, Washington, refers to facilities "which make it a technological praxis and that such practice is rooted in the unfolding of a distinctively western European oriented history. Moreover, ELF and AZLN discourse clearly is suffused with the sense that, as it unfolds through globalization, this historical process has lost all fundamental sense of human meaning, moral clarity, or ultimate purpose. In a single page introduction to one of its volumes, the ELF press office refers to the system or the systematic destruction of nature eleven times daily routine to kill and destroy wildlife" [emphasis added] (Resistance, p. 1). It is precisely the tendency toward the routinization of a technical orientation toward life that is named by Enframing. The positions outlined in the communiqués reflect an implicit awareness of the manner in which Enframing and the metaphysics of cybernetic will increasingly define technical, corporate practice. Activists write that animals are being turned into machines for human consumption. In fact, the description of natural entities as machines is becoming increasingly frequent, not just in core countries but in peripheral countries as well. Even a cursory scan of the internet generates numerous references to animals and plants as machines. According to James Robl, president of Hematech LLC, "Cows are ideal factories." Hematech works in partnership with Kirin Brewing Co. to produce human immunoglobulins in cows. Paul Elias, AP Biotechnology writer, notes that this has involved 672 attempts at cloning cows with six live births, two of which died within forty-eight hours. For us the significance lies in the manner in which cows have been reduced to research units in a systematic attempt to turn them from their essence as bovine creatures into manufacturing facilities. A recent *New York Times* article describes how Malaysia is conducting research to "engineer palm oil trees genetically to serve as factories of specialized plastics for medical devices" (Barboza, 2003). But it is not merely the commodification of nature to which revolutionary environmentalists object; it is the setting upon nature, the setting of it within systems that reduce nature to useable bits of material. More importantly, they recognize this as an impersonal force that is only gathering strength. "This world is dying. All that is beautiful about the world is being destroyed." Anger and rage is specifically directed "at this system."

The nihilistic character of the extension of Enframing across the globe is more specifically conveyed in Zapatista discourse. Marcos characterizes "globalization" as a "Fourth World War...against all humanity." Against that which provides a sense of human meaning and dignity, globalization offers a reduction of life to calculable, cash value. "Instead of humanity, [globalization] offers us stock market value indexes, instead of dignity it offers us the globalization of misery, instead of hope, it offers us an emptiness, instead of life it offers us the international of terror" (Subcomandante Marcos, January 1996, pp. 12–13). Marcos links up the specter of war with the increasing militarization taking place within nations and societies:

> From the stupid course of nuclear armament, destined to annihilate humanity in one blow, it has turned to the absurd militarization of every aspect in the life of national societies, a militarization destined to annihilate humanity in many blows, in many places, and in many ways... What were formerly known as "national armies" are turning into mere units of a greater army, one that neo-liberalism arms to lead against humanity...armies, supposedly created to protect their own borders from foreign enemies, are turning their cannons and rifles around and aiming them inward. (Subcomandante Marcos, August 1996, p. 38)

This characterization of militarization resonates with anyone whose form of dissent goes in any way or form beyond carrying a sign in a legally designated "protest zone." An overwhelming presence of paramilitary jackboots with their armored personnel carriers, assault weapons, and swat tactics is a given at any IGO gathering anywhere in the world as is the beating, arrest, and incarceration of dissidents.

But what Marcos is pointing to here is the command and control character of everyday life under globalization: its standardization, routinization, constant surveillance, performativity, and military-style discipline. The Fourth World War is the "most brutal, the most complete, the most universal, the most effective" for this is a modality of power that "administer[s] life and decide[s] death" (Subcomandante Marcos, August 1996, p. 43). Part of the lie by which globalization extends its power is by insisting that "everything is under control, including everything that isn't under control" (Subcomandante Marcos, August 1996, p. 57). In the logic of total control that which cannot be controlled must be eliminated. "Accompanying the government's war strategy is State terrorism. The utilization of the army and the war against Zapatismo represents the possibility of reestablishing political and economic control. The logic is that of a modernization which dictates the elimination of those social groups who have neither the capacity, nor the desire, to consume the products offered by the neoliberal market" (Subcomandante Marcos, 1998, p. 11). It is the impersonality of this logic that reflects Heidegger's notion of Enframing. Moreover, note that Marcos describes this process as "destined," recognizing that it flows inevitably from the dual logic of economic and political "control" inherent in Western capitalism. On the one hand, this can be a source of strength; resistance is able to thrive beyond the area of control that neoliberalism attempts to extend. On the other hand, the imperative to drive forward the will to control is precisely what is most dangerous in the project of globalization.

For Heidegger, the fundamental danger presented by Enframing is that human beings will become incapable of grasping their essence as a being that can attain a thoughtful awareness of the relationship to Being. Inasmuch as the only kind of worthwhile activity appears to be securing, locking, interconnecting, and enhancing technical power (i.e., manifesting the will to will), Enframing threatens the utter loss of meditation on and solidarity with Being. Human beings now stand at "the brink of the possibility of pursuing and pushing forward nothing but what is revealed in ordering" (Heidegger, 1954, p. 26). Such an exclusively technological life threatens to block the experience of human essence—"the needed belonging to revealing."

Heidegger writes of humans as the beings who, early on, hearkened to Being, but who emerge, in the end, as "the laboring animal…left to the giddy whirl of its products so that it may tear itself to pieces and annihilate itself in empty nothingness" (Heidegger, 1973, p. 85). Confident talk of values is always already part of "the armament mechanism of the plan," and that which is esteemed as progress is really an "anarchy of catastrophes" confirming "the extreme blindness to the oblivion of Being" (Heidegger, 1969, p. 71). Direct action events, similarly, reflect "the rage of a dying planet." Activists are motivated by a commitment to divert us from a "path towards annihilation;" recognizing that the ultimate effect of destroying biotic diversity is "suicide." It is morally impossible for ELF activists to "allow the rich to parade around in their armored existence, leaving a wasteland behind in their tire tracks" (Rosebraugh, 2004, p. 189).

Yet, for both Heidegger and revolutionary environmental activists there exist possibilities for transformation. In the midst of technological peril—indeed, because of that peril—there emerges a sense of solidarity with nature, understood as the living spiritual whole of the natural world including human beings. For Heidegger, it is from within the destiny of Enframing that the world must collapse, that the earth must become desolate, that human work must be reduced to sheer labor power. It is in the context of nihilism that "Being can occur in its primal truth." Heidegger describes the possibility of a return of Being as a refigured humanism. It is the possibility of suspending the will and attaining a lucid sense of the free play of Being. A human being, like any entity, is—s/he stands forth as present. But "his distinctive feature lies in [the fact] that he, as the being who thinks, is open to Being.… Man *is* essentially this relationship of responding to Being" (Heidegger, 1969, p. 31).

Heidegger uses the word *Gelassenheit*—a free comportment toward nature and technics alike—to describe this transformation. Releasement concerns the process of Being, the openness within which beings emerge from absence into presence through their genesis, maturation, and finite perfection and back again into the draft of the concealed—Being. Meditating on the essence of a thing involves acknowledgment of the unique limits that govern its appearing and disappearing. In this sense, an ethos of care allows a human or nonhuman being to become what it is.

Releasement toward things thus expresses the opportunity of human beings to correspond with Being through saving. It is in this sense that Heidegger writes "Mortals dwell in that they save the earth.... Saving does not only snatch something from a danger. To save really means to set something free into its own presencing" (1971b, p. 150). The audacious phrase "saving the earth" might come to mean simply allowing the creatures of the Earth to live as nature and millions of years of DNA development intended them to live. That is, to save means to allow a plant, river, or animal to be freed "into its own presencing" rather than being channeled into a human technical system. Through this ethos of meditation and care humility is attained. Control gives way to the awareness of Being or "life" as primary. From there, a simple relationship to technology can ensue. Instead of deluding ourselves as supposed masters of the Earth, we can easily move from using technics—itself never allowed to undermine the essence of a thing—into a more exalted and higher participation in the realm of our belonging with Being. In fact, dwelling authentically will substitute for much facile technological willing.

The possibilities of an emerging new humanism rooted in a meditative reflection upon and awareness of Being may arise from different contexts. Heidegger can rightly be criticized for a tendency to emphasize an alleged inner connection between Greek and German language as the sole path to a recovery of a sense of Being. On the other hand, in some instances Heidegger points to non-Western traditions and language as actually better exemplifying the human belonging together with Being (Heidegger, 1971b, pp. 1–56).

Revolutionary environmentalism also centers on a spiritual reawakening revolving around the mystery of Being. But a predominant theme in ELF and EZLN communiqués is Native American and indigenous spiritual philosophy and practice. A reverence for the sacred power of nature pervades ELF and ALF communiqués (as well as other radical environmental organizations including Earth First). Bron Taylor's (2001) work has been instrumental in documenting the diversity and pervasive influence of Native American religious themes in revolutionary environmentalism. Totem animals and other Native American religious symbols are encountered frequently, especially among Earth First activists. Spiritual identity with animal "nations" is a recurrent theme in ELF communiqués. A November 1997 communiqué concerns an arson event against the BLM horse corral at Burns, Oregon, and an earlier ALF arson event at a Redmond, Oregon, slaughterhouse and horse meat processing plant. The focus of liberation was wild horses on BLM lands—classified as invasive and non-native—that are rounded up and auctioned off for slaughter. In defending the arson activists speak of the "genocide against the horse nation." The Vail arson event occurred, in part, in defense of the "mink and fox nations." More generally activists speak of "wildlife nations" and abhor the destructive forces that hate and kill off the spirit of that which is wild.

Spiritual identity with the Earth's creatures understood as "relations" of different "nations" is central to traditional Native American practice. In a sweat lodge ceremony even the rocks are acknowledged as the old ones who know everything because they have been here from the beginning. The closing prayer of the sweat lodge invokes "all my relations," meaning a prayer to all one's relatives with whom one is constantly connected. The prayer is an acknowledgment and reminder of that connection. Linda Hogan (1995), a Chickasaw poet, powerfully evokes the living-remembering connection forged in the sweat lodge:

> The entire world is brought inside the enclosure…smoking cedar accompanies this arrival of everything.… Young lithe willow branches remember their lives rooted in the ground, the sun their leaves took in…that minerals rose up in their trunks…and that planets turned above their brief, slender lives.… Wind arrives from the four directions. It has moved through caves and breathed through our bodies. It is the same air elk have inhaled.… Remembering is the purpose of the ceremony.… It is the mending of a broken connection between us and the rest.… The words "All my relations"…create a relationship with other people, with animals, with the land. To have health it is necessary to keep all these relations in mind. (pp. 227–228)

Obviously it is difficult for a person from a Western, rational-scientific-technological context—i.e., who is destined from within Enframing—to grasp the notion of a willow pole "remembering." The point is that the willow has an essence as a willow. But as a natural being it is also connected to other beings (the sun processed through photosynthesis, the river and rain nourishing it, the minerals flowing in its sap). Precisely the same is true for human beings, and ceremonies like a sweat lodge or a bear dance enable a spiritual identity with specific relations or with Being. In such ceremonies the reflexive association with oneself as ego is often surmounted by a more authentic prayerful voice. Such a voice in song or prayer can attain a simultaneity of self and "relation." Ego and other is surmounted by a spiritual connection of beings. The identity with horse nations in the communiqué stems from this kind of remembering/acknowledging spiritual relationship.

Bron Taylor (2001) notes that many wilderness defenders have experienced a variety of spiritual epiphanies while in the places they seek to protect, all of them involving some sense of profound spiritual solidarity to the place and creatures they hope to defend. Paul Watson's lifelong activism to defend whales and other sea life was "cemented by a vision in an Oglala Sious sweat lodge…. A bison appeared to Watson [and] told him that he should 'concentrate on mammals of the sea, especially whales'" (Scarce, 1990, p. 97). Similarly, Rod Coronado (2000) relates an experience on the Great Plains while on the run from federal authorities. His fear of being captured and constantly keeping a gun at his side had brought him to the breaking point. "That's when she spoke. I cannot describe it as anything other than love. A flow of energy that reduced me to tears as I awakened to the spirit around me. 'We are here. We have always been here. We will always be here, but there is nothing we can do for you until you believe in us more than you believe in them'" (Coronado, 2000, p. 88). Coronado was strengthened by the solidarity he experience with the despised and the hunted and by the knowledge that everything he had been taught in his traditions was true.

In fact, this is the very meaning of the "Earth *Liberation*" and "Animal *Liberation*" Fronts. They seek to literally free plant and animal species as well as natural environments from a cultural-political-economic construct that would convert them from what they essentially are into commodities for exploitation and profit. "Welcome to the struggle of all species to be free."

At the same time these efforts are oriented around a spiritual practice of identity with the species and environments being liberated. The close of the aforementioned communiqué is a petition for others to "stop the slaughter and save our Mother Earth." Mother Earth, in a traditional Native American context, is the first mother, the life-generating and life-sustaining force from which all creatures live. The act of saving as restoring lies both in deed and in spiritual recognition. This is a restorative surmounting which unites actor and the fullness of the life-giving ground from which all our relations thrive.

Though it is not born out in the communiqués it can be argued that traditional Native American spirituality draws even closer to Heideggerian ontology in its evocation of an unseen and unnameable but all-encompassing spiritual power. There is an extraordinary dialogue between J. R. Walker, a physician who lived among and was accepted by Oglala *wicasa wakan* (or "medicine men") during the early part of the 20th century, and a number of such Oglala figures including Finger. Finger describes how there are eight separate elements—the sky, the Sun, the Earth, the rocks, the moon, the winged, the wind, and the beautiful buffalo calf woman who brought the Lakota the pipe and the first ceremonies to the Lakota people. Yet each of these elements is one—Wakan Tanka, the Great Spirit or *Taku Skanskan* which is the living spirit in each thing giving it its essence and causing it to behave in its own unique fashion. Walker asks whether the sun and *Taku Skanskan* are the same. Finger responds that this is not so, that the sun is in the sky only half of the time. But Finger adds that it is the sky which symbolizes *skan* because *skan* "is a Spirit and all that mankind can see of him is the blue of the sky" (Tedlock & Tedlock, 1975, pp. 210–211). What is fascinating is the idea that that which, unlike the sun or even the Earth, cannot be delineated as a thing—namely the sky—symbolizes the ever-present, pervasive, and ineffable spirit. The vault of the sky, a continuum within which everything unfolds, is taken to represent the unity of spirit which is itself unseen but through which every being takes its course. The timeliness of revolutionary environmentalism stems from its elucidation of an ethic rooted not in subjectively centered values but in spiritual unity, grounded in an ontology which itself cannot be ascribed. A similar bridge to indigenous biocentrism exists for the EZLN.

Two mythic figures in Marcos's discourse exemplify the indigenous spiritual/political elements of the EZLN: Old Antonio and Votan–Zapata. In fact, according to Marcos, it was *viejo Antonio* who first explained to the rebels the real meaning of Zapata. As Marcos recounts the story, the first village that the Zapatistas entered in the mid-1980s was that of old Antonio. Antonio asked Marcos about the rebels, and Marcos told the elder Antonio about the Mexican Revolution, Pancho Villa, and Emiliano Zapata. Antonio, whose gaze had never left Marcos's eyes, replied simply "that's not how it was" and proceeded to tell the real story of Zapata. The story begins at the beginning when the first two gods were making the world. These two were Ik'al and Votan—opposites, night and day, dark and light, cold and heat. They were two as one, but their movements were uncoordinated. However, they found that if they sought together how to move and what to do they could move together as one. Soon their laughing and dancing exhausted them, and they agreed that "who moved first and how they moved was irrelevant—they moved together, separated and in agreement." That is how the true men and women learned that the questions help us to walk, not to just stay stuck in one place. Zapata is Ik'al and Votan appearing as one person; they had come to Chiapas at the end of their journey to find out where the road led. The sacred Votan–Zapata said that "sometimes there would be light and sometimes there would be darkness, but that they were all the same, Votan–Zapata,

Ik'al Zapata, white Zapata and black Zapata, and that the two were the same road for all real men and women" (Stephen, 2002, pp. 158–161).

Votan–Zapata links a great hero of the Mexican revolution with the spiritual traditions of the Mayan people of Canada. The rhizomatic nature of this hybridization is suggested by Lynn Stephen in her description of the potential impact of the EZLN on the conceptualization of Mexican national identity. She describes "the possibility of multiple levels of sovereignty" involving communities, regions, and ultimately a genuinely pluralistic, multiethnic nation. Similarly, in reference to the struggles of the Miskito people of Nicaragua, Charles Hale describes a "strategic multiplicity" not "a unified discourse" but instead a "hybrid politics" (Stephen, 2002, pp. 335–337). But we must not overlook the radical rupture that occurs when introducing the indigenous aspect. The figure of Votan–Zapata decisively transforms the nationalist issue by grounding Mexican tradition in the spiritual traditions of the people who have lived in the land for millennia. This tradition underscores the rhizomatic element of difference. Night and day, heat and cold, one and the other in their multiplicity must be accepted as valid in that and in what they are within the balance of life. That is, the other must appear authentically, without obscuring or oppressing the actuality of the other as other. Each moves together in their separateness because they seek and discover together. The indigenous rebellion echoes and re-echoes in a way that "recognizes the existence of the other and does not overpower or attempt to silence it" (Subcomandante Marcos, August 1996, p. 47). Such recognition, as it recognizes the reality of the other, is what constitutes "the real men and women."

Votan and Zapata help bridge the gap between the indigenous and nationalist elements of the Zapatista rebellion. In a sense it is a figure for the reemergence of the indigenous in a way that links it to the position of the peasant in the context of the revolutionary aspect of Mexican history (Jung, 2003, p. 433).

But Old Antonio's stories also provide a bridge from Mayan tradition to Zapatista action. It is possible to trace here the same ontological openness that links the Native American aspects of Sea Shepherd, ELF, and the ALF to Heidegger. Consider one of Antonio's stories which also involves the symbolism of the sky: the "History of the Upholder of the Sky." Old Antonio related to Marcos the story of the first gods who made the world. Their efforts left them exhausted, and each at one of the four corners of the world, they took hold of the sky to try to hold it up over the world. The sky is where "the sun and moon and stars and dreams could walk without difficulty." It is the open space to which prayers go, where the heavenly bodies take their course, and through which dreams awaken us to spiritual reality. But the danger is the sky falling—"then absolute disasters happen, because evil comes to the milpa [the Mayan, communal plot of corn] and the rain breaks everything and the sun punishes the land and it is war which rules and it is the lie which conquers and it is death which walks and it is sorrow which thinks." To prevent this, the gods left one of the upholders of the sky to remain alert and watchful and stop the sky from falling in. This upholder carries a caracol (a piral, conch shell) at his chest to warn the other gods and awaken them to do their part in upholding the sky. The spiral lines of the conch, endlessly circling toward itself while gathering the outward and the inward, is linked with the good heart that seeks the same—neither forgetting nor abandoning the other, including, certainly, the gods, or the self.

> The word of the one who does not sleep, of he who is alert to evil and its wicked deeds, does not travel directly from one side to the other, instead he walks towards himself, following the lines of reason, and the knowledgeable ones from before say that the hearts of men and women have the shape of the caracol [and they] awaken the gods and men so that they will be alert to whether the world is just and right…[they] use the caracol for many things, but most especially in order not to forget. (Subcomandante Marcos, 2003)

There is a triple symbolism at work here with the sky, caracol, and remembering with one's heart representing a humility in the face of that which is greater and ultimately unknowable. As with the above account of the Lakota "medicine man" Finger, the sky is the open, the vault within which everything takes its course; when the sky falls in, that is, when the natural order of elements—each following its own course—is upset, evil results. But the sky itself exceeds determination; it is the space within which each entity follows its limits. Similarly, the caracol, containing "the sounds and silences of the world within it," marks the spiral path by which the self turns in upon itself but simultaneously draws the external into itself in the gathering spiral. As a spiral, there is no end yet there is constant connection. Finally, the heart, spiral-shaped like the conch, connotes a felt, remembered connection with the first makers of the world and others. On this felt connection is based the diverse lines of reason that allow a human world to be present in just fashion. In this just fashion Zapatista politics emerge through "walking and asking" and good governments are seated, embodied by the paradox of ruling by obeying the interests of the Others.

Unless and until an ethos rooted in biocentrism becomes a matter of course, environmentalism will always be consigned to a series of half measures concerning humans and their need to "manage resources." In this context the inherently destructive practices of technological Enframing will never be decisively surmounted. Native American spiritual practice is fundamental to a revolutionary shift in thought and everyday behavior because, for the first time in the West, the most fundamentally destructive hierarchy, that of human dominion over nature and all nonhuman beings, is fundamentally challenged. Heidegger's refigured humanism, like deep ecology and Native American ceremonial practice, comprises an ontological anarchism. It is marked by a radical egalitarianism wherein the intrinsic worth and interdependence of all beings is acknowledged, honored, and celebrated. Moreover, in regard to revolutionary action it opens a way for the healing of an antagonistic relationship between human beings and the Earth. In the nexus enabled by a radical openness to the Other, solidarity is attained by all those struggling to bring this transformation about. In what we have described here as rhizomatic resistance "the reproduction of resistances, the 'I am not resigned' the 'I am a rebel,' continues." In becoming other oneself, one is linked in a rhizome of resistance. There is "no ultimate organizing structure, no central head or decision-maker, no central command or hierarchies. We are the network, all of us who resist" (Subcomandante Marcos, August 1996, p. 53).

References

Barboza, D. (2003, February 21). Development of biotech crops is booming in Asia. *New York Times*, p. A3.

Coronado, R. (2000, November–December). Dzil nchaa si an: In the altar of our warrior ancestors. *Earth First!*, p. 88.

Coronado, R. (2003, February 14). *Classroom discussion*. CSU Fresno.

Bron, T. (2001). Earth and nature-based spirituality: From deep ecology to radical environmentalism. *Religion*, *31*, 175–203.

Deleuze, G., & Guattari, F. (1987). *A thousand plateaus: Capitalism and schizophrenia*. Minneapolis, MN: University of Minnesota Press.

Derrida, J. (1982). Différance. In A. Bass (Ed.), *Margins of philosophy*. Chicago, IL: University of Chicago Press.

Foucault, M. (1984). Nietzsche, genealogy, history. In P. Rabinow (Ed.), *The Foucault reader*. New York, NY: Pantheon Books.

Heidegger, M. (1954). *The question concerning technology and other essays*. New York, NY: Garland Science.

Heidegger, M. (1966). *Discourse on thinking*. New York, NY: Harper and Row.

Heidegger, M. (1969). *Identity and difference*. New York, NY: Harper and Row.

Heidegger, M. (1971a). *Poetry, language, thought*. New York, NY: Harper and Row.

Heidegger, M. (1971b). *On the way to language*. San Francisco, CA: Harper Collins.

Heidegger, M. (1973). *The end of philosophy: Translated by Joan* Stambaugh.
New York, NY: Harper & Row.

Hogan, L. (1995). Sweat lodge ceremony. In A. Hirschfelder (Ed.), *Native heritage: Personal accounts by American Indians 1790 to the present* (pp. 227–228). New York, NY: Macmillan.

Indigenous Revolutionary Clandestine Committee. (1998). *The second declaration of la realidad for humanity and against neoliberalism*. New York, NY: Seven Stories Press.

Jung, C. (2003). The politics of indigenous identity: Neo-liberalism, cultural rights, and the Mexican Zapatistas. *Social Research*, *70*(2), 433.

Lesliejames. (2000). *Editorial from an ELF spokesperson*. Retrieved from http://www.farnish.plus.com/amatterofscale/mirrors/omni/directaction.htm

Naess, A. (1985). Identification as a source of deep ecological attitudes. In M. Tobias (Ed.), *Deep ecology*, pp. 256-261. Santa Monica, CA: IMT Productions.

Organic Consumers Association. (n.d.). Retrieved from http://www.organicconsumers.org/chiapas/index.cfm

Palaez, M. P. (2001). *The EZLN: 21st century radicals*. Retrieved from http://www.flag.blackened.net/revolt/mexico/reports/pomo_ezln.html

Patton, P. (2014, October 21). Metamorphologic: Bodies and powers in a *Thousand Plateaus*. *Journal of the British Society for Phenomenology*, *25*(2), 157–169.

Resistance Magazine. Volume 1.

Resistance Magazine. Volume 2.

Rosebraugh, C. (2004). *Burning rage of a dying planet: Speaking for the Earth Liberation Front*. New York, NY: Lantern Books.

Scarce, R. (1990). *Eco-warriors: Understanding the radical environmental movement*. Chicago, IL: Noble Press.

Silverstein, K., & Cockburn, A. (1995, February 1). Major U.S. bank urges Zapatista wipe-out: "A litmus test for Mexico's stability." *Counterpunch*, *2*(3). Retrieved from http://www.glovesoff.org/web_archives/counterpunch_chasememo.html

Stephen, L. (2002). *Zapata lives: Histories and cultural politics in southern Mexico*. Los Angeles, CA: University of California Press.

Subcomandante Marcos. (1994, May 31). *The majority disguised as the resented minority*. Retrieved from http://www.csuchico.edu/zapatist/HTML/Archive/EZLN%20Communiqués.htm

Subcomandante Marcos. (1996, January). *First declaration of la realidad*. Retrieved from http://www.csuchico.edu/zapatist/HTML/Archive/EZLN%20Communiqués.htm

Subcomandante Marcos. (1996, August). *The second declaration of la realidad*. Retrieved from http://www.csuchico.edu/zapatist/HTML/Archive/EZLN%20Communiqués.htm

Subcomandante Marcos. (1998). *Zapatista encuentro: Documents from the 1996 encounter for humanity and against neo-liberalism*. New York, NY: Seven Stories Press.

Subcomandante Marcos. (2000, June). *Proposal for FZLN presentation to the Encuentro by civil society against the war and repression and for democracy*. Retrieved from http://www.csuchico.edu/zapatist/HTML/Archive/EZLN%20Communiqués.htm

Subcomandante Marcos. (2003, August 6). *Old Antonio's history of the upholder of the sky*. Chiapas: The thirteenth stele—Part three: A name. Retrieved from http://www.narconews.com/print.php3?ArticleID=832&lang=en

Subcomandante Marcos. (2004, February 27). *Chiapas: The southeast in two winds, a storm and a prophecy*. Retrieved from http://flag.blackened.net/revolt/mexico/ezln/marcos_se_2_wind.html

Taylor, B. (2001). Earth and nature-based spirituality: From deep ecology to radical environmentalism. *Religion*, *31*, 175–203.

Tedlock, D., & Tedlock, B. (Eds.). (1975). *Teachings from the American earth: Indian religion and philosophy*. New York, NY: Liveright.

Vahabzadeh, P. (2004). *Zapatistas: An introduction*. Retrieved from http://www.poeticguerillas.com/articles/marcos.hml

2. *Understanding the Ideology of the Earth Liberation Front*

Sean Parson

Many in the environmental movement view their struggle as a war—a just war that holds all life on this earth in the balance. On a seemingly daily basis now, news stories and scientific research papers emerge that detail the anthropogenic role in burgeoning environmental crisis. For the movement, this gives credence to their perspective. To deal with the ongoing destruction of the natural world, the movement's most militant wing, the Earth Liberation Front (ELF), proudly proclaims that they "work to speed up the collapse of industry, to scare the rich, and to undermine the foundations of the State" (Pickering, 2006, p. 20). Though many environmental activists have engaged in ecotage—from the Fox in Chicago to Dave Foreman and early Earth First!ers (EF!'ers)—the ELF is arguably the first to move toward an eco-revolutionary program. In carrying this out the ELF rejects not only State Marxism, but also liberal, identity, or other forms of single-issue politics.

Most attempts at researching the ELF have failed to address the complexity and diversity of its members' ideology. Part of this failure rests in the fact that social scientists have spent little effort studying the radical environmental movement on the whole; and the majority of this research has dealt with either EF!, Greenpeace, or the Sea Shepard—three organizations that embrace variants of biocentrism and/or deep ecology (Ingalsbee, 1995; Manes, 1990; Scarce, 1990; Wall, 1999). By only focusing on the deep ecological influence, social scientists have neglected the historic role of social ecology and the contemporary effect of anticivilizational thought on the radical environmental movement. This academic mischaracterization has produced an image of the radical environmental movement as under the hegemonic sway of deep ecology—a view of the movement that is not shared among activists. What is required, then, is academic research that better accounts for the ideological position of anticivilizational thought within the current radical environmental movement and, more importantly for this chapter, with those who promote the ELF.

While the heightened influence of the philosophy of social ecology and of green anarchism, in particular, on ELF communiqués seems clear upon their analysis, of the few studies that seek to specifically analyze the ELF, all have more specifically dealt with the historical, ethical, and organizational components of the organization and in doing so all contend that the ELF is deeply ecological in its outlook (Leader & Probst, 2003; Liddick, 2006; Long, 2004; Somma, 2005; Taylor, 1998; Vanderheiden, 2005). This chapter attempts to patch a hole in the current research by analyzing the ideology of the ELF as stated in key communiqués as a move toward an explanation of how the ELF differs from previous environmental movements. By analyzing ELF communiqués between 1996 and 2003, a complex and multivariant group ideology emerges, one that I argue shifts away from the deep ecology perspective of EF! in favor of its own unique perspective of "revolutionary environmentalism." This revolutionary environmentalism, I maintain, incorporates components of deep ecology, social ecology, and, increasingly over the last decade, green anarchist thought.

I begin this chapter with a brief history of ecotage, or environmental sabotage, and the rise of the ELF, followed by a summary account of the dominant theories within radical environmentalism. These sections are meant to provide an historical and theoretical ground for analysis which will then be used to examine the five most detailed ELF communiqués in an attempt to map a plausible overarching political ideology for the group despite its rhizomatic, nonhierarchical structure.

The History of Ecotage and the Rise of the ELF

The founding of EF! is clouded in mystery and myth. The common story is that five environmental activists—Dave Foreman, Mike Roselle, Bart Koehler, Howie Wolke, and Ron Kezar—who had become outraged with the political compromises made by mainstream environmentalism, went on a camping trip to the Pinacate Desert in northern Mexico in 1979 and formed EF!. According to Dave Foreman, EF! was meant to be a no-compromise environmental group that put the needs of the earth and the natural world above the needs of humans (Foreman, 1991). The group openly supported the philosophy of deep ecology and radicalized the environmental movement by promoting nonviolent direct action, civil disobedience, and ecotage as legitimate political tactics in defending the earth.

The tactic of ecotage, or environmental sabotage, was the most controversial tactic that the early EF! used. Ecotage ranged from the monkeywrenching, or sabotaging, of logging equipment to spiking trees in order to destroy saw blades and its intended goal was not to radically alter society, but rather to allow individuals to actively protect the forests and wilderness they visited from the encroachment of corporate and other poachers. The early conception of ecotage is thus defense minded. Dave Foreman in *Ecodefense* (1991) writes, "MONKEYWRENCHING IS NONREVOLUTIONARY, Monkeywrenchers do not aim to overthrow any social, political, or economic system" (p. 10). By contrast, Foreman viewed monkeywrenching as a means to delay development. In his view, the tactic would cause economic damage and slow down the processes of industry in outlying areas, but it was not meant to confront and alter the economic, social, or political world in its totality. Still, as a result of the use of ecotage by EF! and like-minded groups, all Western states passed laws increasing the prison sentences for those deploying such action, and stopping ecotage became a central concern for federal employees in both the Bureau of Land Management and the Forest Service (Manes, 1990).

Around this time a new generation of activists joined EF!. These new activists embraced social justice and labor politics as well as ecological concerns. These new activists, combined with the new ecotage laws and increased media pressure that accompanied them, made EF! change its organizational stance on ecotage, thereby shifting the politics and tactics of the group. While the US EF! debated the tactic of ecotage, activists at Hasting College in East Sussex, England, formed the first lasting European chapter of EF! in 1991. One year after its formation the group engaged in its most popular campaign, the anti-roads campaign at Twyford Down. In the Twyford Down campaign, EF! (UK) occupied a controversial road being built through scenic and ecologically rich grasslands and so halted its development, thus allowing more mainstream groups time to lobby politicians and initiate litigation. By 1992, the camp had become a meeting ground for a wide variety of environmental activists, New Agers, hippies, and punks. This campaign lasted into 1994 and became a model for other anti-road campaigns that followed throughout England, even though the Twyford Down campaign itself ultimately failed in its immediate objective. During the multiyear campaign, activists utilized a wide array of tactics, ranging from nonviolent civil disobedience to covert and unreported acts of ecotage (Wall, 1999).

In 1992, at the EF! (UK) national gathering, EF! (UK) decided to abandon the tactic of ecotage. Instead EF! (UK) decided to "neither condemn nor condone" ecotage but instead allow the formation of an "Earth Liberation Front, which would promote a radical political agenda and repertoires of sabotage" (Plows, Plows, Wall, & Doherty, 2004, p. 202). The hope of the ELF founders was that "illegal action would aid the earth liberation movement in exactly the same way similar actions had helped the animal liberation movement" (Molland, 2006, p. 50).

The organizational structures of the ELF (e.g., leaderless, decentralized cells), and its guiding principles, were borrowed from the Animal Liberation Front (ALF), an organization known for successfully liberating animals from vivisection laboratories and factory farms. At the 1992 EF! (UK) meeting it was decided that the ELF would attempt:

1. To cause as much economic damage as possible to a given entity that is profiting off the destruction of the natural environment and life for selfish greed and profit.
2. To educate the public on the atrocities committed against the environment and life.
3. To take all precautions against harming life.

Yet, the ELF (UK) failed to gain the popularity and influence that the ALF had achieved throughout Europe, in part because they rarely engaged in large-scale acts of resistance. Instead, they more often committed small-scale acts termed "pixieing" (Molland, 2006). Such pixieing included diverse tactics like super-gluing locks or damaging construction machinery to the intentional spoiling of food in upscale grocery stores.

The ELF (UK)'s one large action, a "night of action" waged against Fison, an English company that was draining peat bogs through the English countryside, resulted in nearly US$100,000 worth of damage. This was also the only action for which the ELF (UK) posted a communiqué. It was published in *Green Anarchist*, stating:

> All our peat bogs must be preserved in their entirety, for the sake of the plants, animal and our national heritage. Cynically donating small amounts will do no good. The water table will drop, and the bog will dry out and die, unless it's preserved fully. FISON MUST LEAVE ALL OF IT ALONE—NOW. (Molland, 2006, p. 52)

Shortly following the Fison action, members of the ELF published the journal *Terra-ist*, which detailed ecotage happening throughout the world. Through *Terra-ist*, green anarchist zines such as *Green Anarchist* and *Do or Die*, and an organized road show across Europe, the militant focus of the ELF (UK) readily spread. By 1996, actions of ecotage had been reported in most

Western European countries. That year also marked the end of the ELF (UK) as a group, and since then there have been no actions claimed by the group, and ethnographic research has shown "no evidence of a continued ELF presence" (Plows et al., 2004, p. 203).

Coincidentally, in 1996, small groups of revolutionary environmental activists started engaging in ecotage throughout North America. The first known ELF actions in North America occurred in British Colombia, Canada, in June of 1995 by "The Earth Liberation Army." They committed vandalism against trophy-hunting stores throughout the region and committed arson against a British Columbian guide outfitter. A similarly minded group of activists committed arson on October 8, 1995 against the lumber company Weyerhaeuser's pulp mill in Alberta, Canada.

The first presence of the ELF in the United States was during the spring of 1996 when activists engaged in small acts of vandalism throughout Oregon. Quickly, the so-called "Elves" in the Pacific Northwest escalated their tactics, as they started pixieing logging equipment and engaging in arson. Before the first ELF (US) communiqué was published, in March 1997, ELF actions were reported throughout Michigan, Oregon, Washington, Northern California, and Indiana. The group's ideology had spread from the Douglas Fir forests of the Pacific Northwest to the industrial cities of the Great Lakes and beyond.

There are two interesting points to notice in the provided narrative about ecotage and the history of the ELF. First is the changing nature and dynamic of ecotage from *Ecodefense* to the ELF. As was stated earlier, Foreman and early EF! activists viewed ecotage as "nonrevolutionary." This understanding of ecotage is echoed by Steven Vanderheiden (2008) in his article, "Radical Environmentalism in the age of Anti-terrorism." In this article, Vanderheiden claims that ecotage is meant to delay and stall environmentally destructive actions only and that a strong public relations campaign and litigation are correlatively needed for a truly successful monkeywrench campaign. He further states that currently ecotage promotes a negative public image for the environmental movement and has provided opportunities for the entire movement to be cast by its opponents as "ecoterrorists." Because of this, he feels, the movement needs to discuss dropping the tactic entirely.

What Vanderheiden misses is that the logic grounding ecotage shifted with the ELF. In its hands, the tactic has become offensive, at least in theory, and has become conceived of as a bona fide stand-alone strategy. In this way, the ELF believes that if enough economic damage is done to an industry or development project, the industry or project will be eradicated. This form of revolutionary ecotage then does not require coupling with additional legal action and should not be conceived of as a mere stalling tactic. In addition,

ELF ecotage is also meant to question and confront the social, economic, and political realities of the world and to undermine them through their active problematization. This is part of what marks the move from a radical to a revolutionary environmentalism.

Secondly, we should recognize the important role that political compromise and "moving to the center" play in radicalizing activists. For example, in both the United States and the United Kingdom the ELF formed only after many environmental activists rejected ecotage as a valid tactic. Likewise, the complacency and compromise found in the environmentalism of the 1970s was itself the necessary spawn for the original EF!. Vanderheiden warns radical groups about this problem in passing when he states that,

> Moving towards the centre, in the environmental movement as in other struggles in which moderate factions exist in occasionally uneasy tandem with radical ones, can push extremists to the fringe and cause them to reject what they take to be the efficacy-limiting constraints embraced by those seen as too willing to compromise with the opposition. (Vanderheiden, 2008, p. 314)

Moving to the "center," then, appears to have the effect of further increasing militancy on the "fringe." Thus, paradoxically, as today's mainstream environmentalism moves away from the offensive conception of ecotage promoted by the ELF, it runs the risk of marginalizing and frustrating radical activists and thereby further revolutionizing them.

The Radical Ecological Tradition

This section is meant to give a basic overview of the three dominant theories that have influenced the ELF: deep ecology, social ecology, and anticivilization (or green) anarchism. These philosophical positions all believe that a radical change is required in society in order to protect the natural world from further anthropogenic destruction. They also lament the loss of natural diversity in the face of civilization, promote either the radical decentralization of power or the abolition of corporate and state power altogether, and want to restore humankind's intimate connection with the natural world. But, even though these theories share some similar short-term goals, they have historically been hostile toward each other on the whole. For example, the founding social ecologist Murray Bookchin openly rebutted and refused to support EF! during the 1980s and 1990s because he believed that the deep ecological philosophy that it waved as its primary banner was inherently racist, classist, sexist, and authoritarian (Bookchin, 1995; Bookchin, Foreman, & Chase, 1991). At the same time, many green anarchists lambaste both social ecology and deep ecology as being reformist and reactionary because of their support for civilizational progress and enlightenment sensibilities.

Deep Ecology

Deep ecology is a philosophical movement based on the works of the Norwegian philosopher Arne Naess, though in North America many environmentalists have been more directly influenced by the work of Bill Devall and George Sessions (1985) that sought to interpret Naess's insights on behalf of a syncretic worldview. As it developed, then, deep ecology mixed New Age, eastern, feral, and shamanistic notions of spirituality with concerns of liberty, freedom, and democracy. From this, deep ecology formulated an 8-point program in which the central tenet is that the natural world has intrinsic value separate from its value to humans (Point 1). To a deep ecologist, the current horrors of capitalism and Western civilization are the by-products of the human disconnection from the natural world, which is typified by anthropocentric thinking (human-centered thinking). For example, in this respect Chellis Glendinning (1995) argues that Western culture is suffering from "Original Trauma" or PTSD which was caused by "the systemic removal of our lives from nature, from natural cycles, from the life force itself" and that "the ultimate goal of recovery is to refind our place in nature" (Glendinning, 1995, pp. 37–39). Consequently, for deep ecology it is only with the return to the natural that humankind and the natural world can be saved.

One means of bridging the gap is "living as though nature mattered" (Devall & Sessions, 1985). To do this, deep ecologists want society to give moral weight to the natural world in making political and social decisions. Therefore, a deep ecological society would take into consideration the effects of a political decision not only upon humans but also on the entire ecosystem. Such a community would reject modern development protocols because of their negative impact on ecosystems unless such development could be shown to provide an essential service for the community.

The other major component of deep ecology is its promotion of small, decentralized communities. In this view, decentralized development promotes freedom and diversity (both socially and ecologically) and limits centralized power. In addition, deep ecologists argue, centralized planning cannot take into account the needs of local bioregions. For example, the Cascadia bioregion (which covers Northern California through South British Columbia) has unique regional characteristics, socially and environmentally, that only those with intimate knowledge of the area can address. Therefore, deep ecological political theory tends to believe that local decisions should be made locally and regional decisions should be federalized upward. Because of the value on keeping power situated locally, some deep ecologists like Kirkpatrick Sale and Ernest Callenbach have even become vocal supporters of secessionism believing that the centralized US government should be removed and regions should form their own independent nations based on the ecological principle of bioregions (Callenbach, 1981, 1990; Sale, 2000).

Social Ecology

Social ecology contends that environmental destruction is epiphenomenal of *hierarchical* human societies, which also generate all manner of social oppression. Therefore, in order to stop environmental destruction—logging, climate change, pollution—humans need to heal the social rifts caused by hierarchy and political domination. By contrast, Murray Bookchin argues that prior to the formation of hierarchies, human communities existed as organic components of the natural world (1991). Over time hierarchies formed—first by elders then by shamans and clerics and finally by warriors—which promoted a division of labor and other hierarchical social relationships. These developments led to increased tensions between men and women, the rich and the poor, and also created a disconnect between humankind and the natural world.

Illustrating this, Bookchin contends that technology is not inherently oppressive, as many green anarchists and at least some deep ecologists argue. Against other environmentalist camp thinking, Bookchin argues that small-scale technologies can themselves be brutal and repressive, while large-scale industrial technology may be liberating under certain conditions. What matters, according to Bookchin, are the social relationships and power dynamics surrounding the generation and use of such technology. Therefore, technological advances, like the green revolution in agriculture, are not inherently oppressive but only become so through the development of power relationships, such as private ownership and specialization that are their context. Because of this Bookchin promoted green power sources—such as solar, wind, and geothermal—as well as communally owned factory production as important components of his socially ecological sustainable society.

Yet, there has been some disagreement by other social ecologists with Bookchin over the ecological value of large-scale technology (such as industrial factory production). Perhaps the most notable of Bookchin's critics in this vein is Dave Watson. Watson contends that technological systems are inherently hierarchal and require a strict and important division of labor to maintain. For this reason, factory production inherently recreates social hierarchies. In addition, Watson argues that industrial production is always environmentally destructive. Instead, he thinks that what is required is a radical decentralization of human societies, the rejection of modernist technology, and a return to a small-village or gatherer-hunter existence. In getting to this decision, Watson contends that the only way to remove human hierarchies and thus heal our separation from the natural world is by returning to a simpler existence.

Green Anarchism

Green anarchism, or anticivilizational anarchism, is a branch of anarchist thought that contends that civilization, along with domestication, is responsible for environmental destruction and human subjugation. Unlike social ecology and deep ecology, green anarchism is generally antiacademic and the vast majority of green anarchist writings are written by activists and found in zines, such as: *Green Anarchy*, *Green Anarchist*, *Do or Die*, *Species Traitor*, *Arson*, *Fifth Estate*, and *Anarchy: A Journal of Desire Armed.* Borrowing from the radical activist movement, authors commonly use pseudonyms, such as Feral Faun, Mr. Venom, or Felonious Skunk. The use of pseudonyms is common within the radical environmental movement as a safeguard against government surveillance. Even though green anarchism does not appeal to academic authority, it has had increasing importance within anarchist communities and has influenced the radical environmental movement, the antiglobalization movement, and the youth dropout movement. According to Bron Taylor (2006), green anarchism's influence on EF! had led to a "decreasing importance of Deep Ecology" in the radical environmental movement and an increased importance for primitivism (p. 2). It is also important that there are at least two distinct strands of anticivilizational anarchism, one that is promoted by the journal *Fifth Estate* and another by the Green Anarchy Collective. For this project the divide is not crucially important but readers should be aware that for this section most of my notes will be from the Green Anarchy Collective strand of anticivilizational anarchism. This strand has been more generally influential to the development of EF!, the forest defense movement, and northwest radical activism.

Emerging from the influence of writers like Lewis Mumford, Claude Levi-Straus, Stanley Diamond, and Jacques Ellul, green anarchists contend that civilization is devouring the natural world and suppressing human desires. According to Derrick Jensen (2006), a civilization is,

> a culture—that is, a complex of stories, institutions, and artifacts—that both leads to and emerges from the growth of cities, with cities being defined—so

> as to distinguish from camps, villages and so on—as people living more or less permanently in one place in densities high enough to require the routine importation of food and other necessities of life. (Jensen, 2006, p. 17)

In this definition, one of the defining characterizes of a civilization is that it requires the importation of resources (e.g., food, oil, etc.) to continue its existence. To green anarchists, the need for external resources is why "civilization originates in conquest abroad and repression at home" (Diamond, 1974, p. 1).

In order to ensure its own survival, civilization must homogenize and domesticate life on the planet in an attempt to control the wild. This control is required to ensure a continual flow of resources, to break apart older cultures, and also to create social and political stability. This is done by military/economic force or by domestication. Domestication is the process through which animals (human and nonhuman) and plants are controlled for societal benefit. Human domestication,

> takes many forms, some of which are difficult to recognize. Government, capital and religion are some of the more obvious faces of authority. But technology, work, language with its conceptual limits, the ingrained habits of etiquette and propriety—these too are domesticating authorities which transform us from wild, playful, unruly animals into tamed, bored, unhappy producers and consumers. (Faun, 2013, p. 28)

In other words, our social system—morality, work, and education—domesticates and placates humanity for the benefit of the social order. To green anarchists, this domestication removes spontaneity, passion, freedom, and liberty from life. Domestication, according to John Zerzan (1999), requires "initiation of production, vastly increased divisions of labor, and the completed foundations of social stratification" (p. 77). Due to this, Zerzan, much like Fredrick Engels, claims that domestication is the root cause of sexism, racism, war, and capitalism. To confront the totality of civilization and return us to our natural ways of life, green anarchists support undermining and destroying civilization and modern forms of living.

As a means of resisting domestication, some green anarchists look to the process of "rewilding." Rewilding occurs when an individual rejects civilization and attempts to reconnect with the natural world by embracing the lessons and lifestyle of gatherer-hunters and other acivilizational peoples. Through learning primitive, or earth, skills people can reconnect with the natural world and embrace their lost instincts. The practical goal of rewilding "involves both accessing our present situation and looking back to what has been done before by people" in an attempt to survive in modern civilization and prepare for a postcivilizational world (Anarchy & Collective, 2004, p. 31).

Green anarchists' hostility toward civilization leads to the rejection of traditional liberal and leftist organizations as reformist. In "The Ship of Fools," Theodore Kaczynski (1999) develops the following claim: If a ship is heading toward an iceberg, worker concerns for better wages, and minorities' concerns for equal rights become insignificant. Because of this, green anarchists reject unionism, antiracism, and traditional class-based political action as reifying civilization and therefore being counterrevolutionary. The disdain for leftist groups is seen through "News from the Balcony," a common feature in the zine *Green Anarchy*. In this section, the authors—using the pseudonym's Waldorf and Stalter (the old cynics from the Muppet show)—heckle and joke about the ineffectiveness of traditional anarchist organizations and the labor union movement. This hostility to unionism and class-based movements has placed green anarchists at odds with anarcho-communists, social ecologists, and other members of the political left, limiting any collaboration between the groups.

The final component of green anarchist theory is its belief in an imminent collapse of industrial civilization. This collapse will be the result of civilization's unsustainable quest for resources and its resulting environmental damage. Authors such as John Zerzan, Derrick Jensen, and Dave Watson all argue that if we do not abolish civilization soon then the collapse will only be made worse. This desire is expressed in Dave Watson's article "We all Live in Bhopal" (1996) where he argues that "industrial civilization [is] one vast, stinking extermination camp. We all live Bhopal, some closer to the gas chambers and to the mass graves, but all of us close enough to be victims" (p. 45). To Watson, the destruction of civilization must occur abruptly. If not, he wonders, what will happen when "we all live in Bhopal and Bhopal is everywhere?" This is the worst-case scenario for him: an environment too ravaged for human life to survive. Watson, Zerzan, and Jensen all believe that ending civilization now, and not waiting for the planet to do it for us, is more sympathetic and compassionate than any technological humanist venture.

The ELF Syncretic Ideology

Between 1997 and 2002 ELF cells distributed forty-six communiqués for actions ranging from petty vandalism to animal liberation to arson. These communiqués, and statements by unofficial spokesmen Craig Rosebraugh and Leslie James Pickering (who started the Earth Liberation Front Press Office), are the only overt documents which allow for an understanding or an analysis of ELF ideology. Since the ELF is a leaderless resistance movement, without central authority, each communiqué differs in its reasons and its goals depending on the activists involved in a given action. Paradoxically, though, since the ELF is decentralized and leaderless it requires a powerful and encompassing ideology in order to attract and retain supporters.

Of course, defining an ideology is a difficult task. Ideologies can cover wide ranges of thought and are often tied together by a few guiding principles (points of unity). These points of unity are similar to the celestial bodies; they provide the needed mass and gravitational force to create and maintain the orbits of that which surround them. The ELF's ideology is no different in this respect. However, what does make the ELF's ideology unique is that being a leaderless movement anyone who wants to act and speak for the ELF theoretically can do so. Because of this, the communiqués express justifications and political philosophy freely. All of this makes ELF ideology dynamic and fluid (within the scope of the organization's guidelines). In the early years of the ELF, the communiqués ebbed and flowed—expressing a deep ecological view in one communiqué while articulating more of a social ecological vision in another. Still, by the end of the seven-year period that I am examining, the ELF communiqués started to coalesce around certain ideas and concepts. I contend that these concepts are the current ELF points of unity and so constitute its ideological centerpiece.

To show how the ELF has been forging this ideology, I will look at five of the most influential and detailed communiqués.

a. Beltane communiqué, July 1997

> Welcome to the struggle of all species to be free. We are the burning rage of this dying planet. The war of greed ravages the Earth and species die out every day. The ELF works to scare the rich, and to undermine the foundations of the state. We embrace social and Deep Ecology as a practical resistance movement. We have to show the enemy that we are serious and about defending what is sacred. Together we have teeth and claws to match our dreams. Out [Our] greatest weapons are imagination and the ability to strike when least expected.... (Pickering, 2006, p. 18)

The Beltane communiqué was the ELF's first US communiqué. It was written in connection with political actions occurring throughout Oregon during the summer of 1997. The name Beltane comes from the ancient Gaelic holiday that marks the beginning of the summer and was commonly associated with massive bonfires and the kind of elf and faerie imagery in vogue with neo-Pagan communities. The use of Beltane in the communiqué is similar to ELF (UK) appropriations of mythical and pagan images in their sabotage manual *The Book of Bells*, which was a play on a Gaelic book, *The Book of Kells*.

This communiqué introduces the ELF as a unique group that bridges the gap between social and deep ecology. The statement, "the ELF works to scare the rich, and to undermine the foundations of the state" resonates with the political philosophy of Murray Bookchin. Unlike traditional deep ecology statements, this appears to place a larger burden on the evils of the state than on a collective anthropocentric consciousness. On the other hand, the statement, "we have to show the enemy that we are serious about defending what is sacred" goes against social ecology's rejection of the rhetoric and rituals of spiritualism and sacredness more typical of deep ecology. What this communiqué does is explain how the ELF plans on being an umbrella group for all those who wish to engage in revolutionary action in defense of the Earth.

b. Rhode Island, December 19, 2000

> ...Our earth is being murdered by greed corporate and personal interests. The rape of the Earth puts everyone's life at risk due to global warming, ozone depletion, toxic chemicals, etc. Unregulated population growth is also a direct result of urban sprawl. There are over 6 billion people on this planet of which almost a third are either starving or living in poverty. Building homes for the wealthy should not be a priority.... The time has come to decide what is more important: The planet and the health of its population or the profits of those who destroy it...we are but the symptoms of a corrupt society on the brink of ecological collapse.... (Pickering, 2006, pp. 35–36)

In the second of half of 2000, the ELF repeatedly struck against housing developments throughout Long Island. The above communiqué is attached to a December 19, 2000 action, and is the most detailed communiqué associated with this string of incidents. Its argument has two facets. First, it states the environmental dangers of overpopulation. This concern is historically aligned with deep ecology and green anarchism. Followers of deep ecology and early EF! in particular viewed overpopulation as one of the main ecological problems facing the world. Some early EF! activists argued that overpopulation is depleting natural resources and is the primary cause of environmental destruction. But social ecologists and ecofeminists in turn rejected the reliance of environmentalists' use of the population model as evidence for environmental harm. Particularly they claimed that this strategy was at least implicitly racist and classist, because it criticized the poor in the developing world and did not confront the high levels of consumerism and waste in the developed world, as well as sexist, because it targeted women's reproductive cycles as the main cause of environmental degradation (Bookchin et al., 1991; Seager, 1993).

The second facet of the communiqué's argument is the claim that class and capitalism are driving urban sprawl. The ELF cell here states that "building homes for the wealthy is not a priority…the time has come to decide what is more important: The planet and the health of its populations or the profits of those who destroy it." Finally, it is claimed that the ELF are themselves but the "symptoms of a corrupt society on the brink of ecological collapse." In their view, since urban sprawl and overpopulation are destroying the world and making an ecological collapse imminent, the only acceptable response is ending urban sprawl. To them, this means abolishing capitalism and civilization.

c. Gifford Pinchot National Forest, WA, July 27, 2001

> …We want to be clear that all oppression is linked, just as we are all linked, and we believe in a diversity of tactics to stop earth rape and end all domination. Together we can destroy this patriarchal nightmare, which is currently in the form of techno-industrial global capitalism. We desire an existence in harmony with the wild based on equality, love, and respect. We stand in solidarity with all resistance to this system, especially those who are in prison, disappeared, raped, tortured…we are all survivors and we will not stop!
>
> The forest service was notified of this action BEFORE this years' logging season so we could take all precautions to assure worker safety. We must ask why they never made this public. We were trying to let them cancel this sale quietly. However, as bosses jeopardize worker's lives every day we realized that we needed to make this public…. (Pickering, 2006, pp. 50–51)

Timber sales have been a popular target for ELF cells. This communiqué concerned a tree-spiking action, during July of 2001, in the Gifford Pinchot National Forest. First, the communiqué takes up a common deep ecological concern: that humans should live in "harmony with the wild based on equality, love, and respect." This sentiment is counter to green anarchist beliefs of rewilding. In the green anarchist pamphlet, "Beyond Veganism" (2003), the author argues that eating and killing is natural and that veganism, in promoting nonviolence, is oppressive and domesticates. The concept of living "in harmony with the wild based on equality, love, and respect" does match well with the green anarchist ideological focus on the ethos of primary survival. On the other hand, this communiqué combines a deep ecological concern with a green anarchist interest in ending the "patriarchal nightmare" that is currently expressed through "techno-industrial global capitalism." This expands upon the philosophy of deep ecology to demand an environmentalism that encompasses the systemic problems of industry, technology, and, by proxy, civilization. The next paragraph in the communiqué expands the argument by claiming solidarity with workers. The quote "as bosses jeopardize worker's lives every day we realized that we needed to make this public" bears more of a resemblance to the thoughts of Judi Bari or Murray Bookchin than it does to John Zerzan or Arne Naess. This concern with workers' rights, which in other ELF communiqués goes as far as to express solidarity with Third World workers, is here combined with a green anarchist critique of techno-industrial civilization.

d. Minneapolis, MN, January 26, 2002

> ...We are fed up with capitalists like Cargill and major universities like the U of M who have long sought to develop and refine technologies, which seek to exploit and control nature to the fullest extent under the guise of progress. Biotechnology is only one new expression of this drive. For the end of capitalism and the mechanization of our lives.... (Pickering, 2006, p. 52)

Genetically modified organisms, or GMOs, have been a concern for environmental activists for decades. GMOs are impossible to control, potentially destructive for the environment, and carcinogenous. Because of this, activists have called for an immediate ban. When that does not happen, oftentimes the only option that appears immediately effective is the destruction of laboratories and test sites involved in GMO research (Plows et al., 2004, p. 205). The January 26, 2002 ELF communiqué, written to claim responsibility for an arson at a University of Minnesota research lab, is typical of ELF communiqués issued in conjunction with anti-GMO actions.

In general, the anti-GMO actions of the ELF are the most openly green anarchist in approach. The ELF argues that GMOs are an assault on nature and justifications of high technological solutions to social problems. Since biotechnology is interwoven with the dominant social hegemony of industrialism and capitalism, the ELF argue that the only way to liberate nature from this menace is to abolish capitalism and the mechanization of scientific and industrial "progress." However, the communiqué does not mention civilization in total as being a culprit. Similar to the communiqué from December 19, 2000, there is no open rejection of civilization period. Instead the civilization arguments are obfuscated and only latently visible in this communiqué because the ELF here deals with issues that anticivilizational green anarchists also often confront.

e. August 11, 2002

> ...Their blatant disregard for the sanctity of life and its perfect Natural balance, indifference to strong public opposition, and the irrevocable acts of extreme violence they perpetrate against the Earth daily are all inexcusable, and will not be tolerated. IF they persist in their crimes against life, they will be met with maximum retaliation.

> In pursuance of justice, freedom, and equal consideration for all innocent life across the broad, segments of this global revolutionary movement are no longer limiting their revolutionary potential by adhering to a flawed, inconsistent "non-violent" ideology. While innocent life will never be harmed in any action we must undertake, where it is necessary, we will no longer hesitate to pick up the gun to implement justice, and provide the needed protection for our planet that decades of legal battles, pleading, protest, and economic sabotage have failed so drastically to achieve.
>
> The diverse efforts of this revolutionary force cannot be contained, and will only continue to intensify as we are brought face to face with the oppressor in inevitable, violent confrontation. We will stand up and fight for our lives against this iniquitous civilization until its reign of TERROR is forced to an end—by any means necessary. (Pickering, 2006, pp. 54–55)

This communiqué is controversial and was immediately denounced by mainstream environmental groups. Some activists believed that the FBI forged the communiqué in order to undermine the radical environmental movement. This communiqué critiques nonviolence, one of the guiding principles of radical ecological politics, and indeed, a fundamental postulate of the original ELF guidelines themselves. In doing so, it denounces nonviolent political tactics such as tree-sits and protests (and even pixie-styled ecotage) as failures. Here the ELF argue that with the failure of nonviolent tactics to combat an overwhelming enemy bent on wreaking planetary terror, the only resistance tactic left is confrontational political violence. Note that there is a condition in this fatwa that distinguishes between innocent and noninnocent forms of life, and it suggests that such violence will protect innocent life while only targeting those who directly profit from the destruction of people and the earth. This attempt to reclaim the moral high ground, even while justifying political violence, is reminiscent of 19th-century anarchist developments of what is known as "propaganda by the deed."

This ELF communiqué is also notably the only one that openly confronts "civilization." Unlike previous communiqués, which allude to anticivilizational arguments, this communiqué openly pictures civilization, rather than capitalism or the state, as the appropriate target. This form of anticivilizational argument, because of the immense scope of its conclusion, therefore requires the movement to conceive of itself as global and so there is also a strategic attempt to portray the ELF as a member of a broader revolutionary movement still. The difficulty of traditional green anarchist theory, with its ideological defenses of personal autonomy, was to find a practical way to attack the largesse of civilization while retaining an individualist approach. This communiqué does something that traditional green anarchist theories do not in that it claims solidarity with all forces fighting injustice. It also expands the domain of traditional green anarchist politics from the "insurrectionary" to the guerilla.

Their Syncretic Ideology

These communiqués are examples of a developing political ideology that cannot be defined as deep ecological, social ecologist, or green anarchist. This ideology combines tenets of all these theories as it seeks to formulate an emergent and encompassing political worldview. The ELF ideology connects the extraction of resources and destruction of the natural environment, with the role of the state and historical oppressions that gird the progress of civilization. What the communiqués reveal is that the practical way of destroying this pathological system is through attacks upon its harmful industries, as well as their peripheral economic supports, that are essential to maintaining its sense of well-being. Because of this, the ELF strikes against forestry and resource extraction as well as research labs and housing developments. By cutting off the flow of resources and attacking destructive industries, the ELF envisions itself as striking, in however a limited fashion, at what they see as the crux of what fuels the agenda of civilization as sociopolitical project. Unlike Marxism, and unlike classical anarchism, the ELF does not portray any group as being the key actor in this revolution and instead places the impetus for change on those simply willing to act.

Overall, the ELF communiqués argue that:

1. Capitalism must be abolished in order for nature to be liberated.
2. Workers are harmed by capitalism and are not the enemy of the natural world.
3. Environmentally destructive industries—logging, mining, construction, industrial agriculture, and biotechnology—are essential for the maintenance of the state and need to be abolished.
4. Humans are animals and should relish their animal instincts and natural spontaneity.
5. All living entities should be wild and free from coercion.
6. Earth liberation, animal liberation, and human liberation are all intertwined into one revolutionary struggle.

In forming such arguments, the ELF has taken an intersectional ideological turn by seeking to find solidarity with worker rights and social justice struggles while integrating these with a general hostility toward civilization's wrecklessness. The ELF, then, rejects the green anarchist critiques of unionism and workerism, as well as Bookchin's pronouncements against technological determinism and primitivism. In this, the ELF is forging a flexible and fluid ideology. This fluidity and flexibility allow proponents of social ecology to engage in actions to protect workers, while also working in concert with attempts to undermine civilization. This flexibility might be a direct result of the ELF's organizational structure and its rejection of hierarchical authority. It also differentiates them from many failed US revolutionary movements such as the Weather Underground. The flexibility of the ELF ideology should allow their ideology to shift pragmatically according to the political climate and thereby allow them to remain politically influential far longer than they might otherwise as a militant group on the margins of mainstream environmental struggle.

Conclusion

In closing, the ELF does not wish to alter public opinion or to lobby politicians nor do they embrace Gandhian understandings of violence. What the ELF does is target environmentally exploitative industries, which they claim are essential to the maintenance of capitalism and the kind of civilization which is fueled by it. Their goal is nothing less than the destruction of the state, the abolition of capitalism as an economic reality, and the end of Western civilization as currently practiced. Derrick Jensen, in his *Endgame* series, discusses the difficulty of destroying civilization. He writes:

> Bringing down civilization is millions of different actions performed by millions of different people…it is everything from comforting battered women to confronting politicians and CEOs. It is everything from filing lawsuits to blowing up dams. It is everything from growing ones['] own food to liberating animals in factory farms to destroying genetically engineered crops and physically stopping those who perpetuate genetic engineering…it is destroying the capacity of those in power to exploit those around them. In some circumstances this involves education. In some situations this involves undercutting their physical power, for example by destroying physical infrastructure…in some circumstances it involves assassination. (Jensen, 2006, p. 252)

Jensen here realizes the enormity of the task and that it requires a wide range of tactics and individuals. The ELF is obviously unable to openly confront, let alone destroy, civilization by itself. Currently, the number of ELF actions in the United States has dropped precipitously since the terrorist attacks of September 11, 2001. Before 9/11 the ELF and ALF combined for an average of an action every 2.3 days, which has since lessened to one every 4.7 days (Somma, 2005). This drop in actions has not meant a decrease in intensity, though. The most costly action in the ELF history, an arson against a housing development in San Diego, which caused more the fifty million dollars in damages, occurred in 2003. This was well after 9/11 and the government's increasing crackdown on "ecoterrorists." With the recent arrival of a highly visible press office on the Web (http://www.elfpressoffice.org) that documents actions taken in the ELF's name across the world, it is clear that the ELF remains a viable force worthy of our attention.

It even appears that Operation Backfire, the FBI campaign against the ELF and ALF, may have backfired in eradicating environmental militants. For every member of "The Family" that is arrested and charged with terror enhancements, new alter-globalization activists around the world are engaging with ELF ideology and confronting the long histories of genocide, ecocide, and colonialism. This will only result in the ongoing transformation of the movement, moving it ever forward and onward in the fight for planetary freedom.

References

Anarchy, G., & Collective, W. (2004). Rewilding: A primer for a balanced existence within the ruins of civilization. *Green Anarchy*, p. 16.

Bookchin, M. (1991). *Ecology of freedom: The emergence and dissolution of hierarchy.* New York, NY: Black Rose Press.

Bookchin, M. (1995). *Re-enchanting humanity: A defense of the human spirit against anti-humanism, misanthropy, mysticism, and primitivism.* London & New York, NY: Cassell.

Bookchin, M., Foreman, D., & Chase, S. (1991). *Defending the earth: A dialogue between Murray Bookchin and Dave Foreman.* Boston, MA: South End Press.

Callenbach, E. (1981). *Ecotopia emerging.* Berkeley, CA: Banyan Tree Books.

Callenbach, E. (1990). *Ecotopia: The notebooks and reports of William Weston.* New York, NY: Bantam Books.

Devall, B., & Sessions, G. (Eds.). 1985. *Deep ecology: Living as if nature mattered.* Salt Lake City, UT: Peregrine Smith Books.

Diamond, S. (1974). *In search of the primitive: A critique of civilization.* New Brunswick, NJ: Transaction Books.

Faun, F. (2013). *Feral revolution.* Asheville, NC: Elephant Edition.

Foreman, D. (1991). *Confessions of an eco-warrior.* New York, NY: Harmony Books.

Glendinning, C. (1995). Recovering from Western civilization. In G. Sessions (Ed.), *Deep ecology for the 21st century.* 37-40. Boston, MA: Shambhala.

Ingalsbee, T. (1995). *Earth first! Consciousness in action in the unfolding of a new social movement* (Unpublished PhD dissertation). Eugene, OR: University of Oregon.

Jensen, D. (2006). *Endgame, Vol. 1: The problem with civilization.* New York, NY: Seven Stories Press.

Kaczynski, T. 1999. “The Ship of Fools”. *The Anarchist* Library. Retrieevd from: https://theanarchistlibrary.org/library/ted-kaczynski-ship-of-fools

Leader, S., & Probst, S. (2003). Earth Liberation Front and environmental terrorism. *Terrorism and Political Violence, 15*(4), 37–58.

Liddick, R. (2006). *Eco-terrorism: Radical environmentalism and animal liberation movements.* Westport, CT: Praeger.

Long, D. (2004). *Ecoterrorism.* New York, NY: Facts On File.

Manes, C. (1990). *Green rage: Radical environmentalism and the unmaking of civilization.* Boston, MA: Little, Brown and Company.

Molland, N. (2006). A spark that ignited a flame: The evolution of the Earth Liberation Front. In S. Best & A. J. Nocella II (Eds.), *Igniting a revolution: Voices in defense of the earth.* 47-58. San Francisco, CA: AK Press.

Pickering, L. J. (Ed.). (2006). *The Earth Liberation Front: 1997-2002.* Portland, OR, Arissa Media Group.

Plows, A. J., Plows, A., Wall, D., & Doherty, B. (2004). Covert repertoires: Ecotage in the UK. *Social Movement Studies, 3*(2), 199–219.

Sale, K. (2000). *Dwellers in the land: The bioregional vision.* Athens, GA: University of Georgia Press.

Scarce, R. (1990). *Eco-warriors: Understanding the radical environmental movement.* Chicago, IL: Noble Press.

Seager, J. (1993). *Earth follies: Coming to feminist terms with the global environmental crisis.* New York, NY: Routledge.

Somma, M. (2005). *Revolutionary environmentalism: An introduction to tactics.* Albuquerque, NM: Western Political Science Association.

Taylor, B. (1998). Religion, violence and radical environmentalism: From Earth First! to the Unabomber to the earth liberation front. *Terrorism and Political Violence, 10*(4), 1–42.

Taylor, B. (2006). Experimenting with truth. In S. Best & A. J. Nocella II (Eds.), *Igniting a revolution: Voices in defense of the earth.* 1-7. San Francisco, CA: AK Press.

Vanderheiden, S. (2005). Eco-terrorism or justified resistance? Radical environmentalism and the “war on terror.” *Politics & Society, 33*(2), 425–447.

Vanderheiden, S. (2008). Radical environmentalism in the age of anti-terrorism. *Environmental Politics, 17*(2), 299–318.

Wall, D. (1999). *Earth First! and the anti-roads movement: Radical environmentalism & comparative social movements.* New York, NY: Routledge.

Watson, D. (1996). *Against the megamachine: Essays on empire and its enemies.* New York, NY: Autonomedia.

Zerzan, J. (1999). *Elements of refusal.* Colombia, MO: Colombia Alternative Library.

3. *Nihilism and Desperation in Place-Based Resistance*

MARK SEIS

One needs something to believe in, something for which one can have wholehearted enthusiasm. One needs to feel that one's life has meaning, that one is needed in this world.

(Hannah Senesh rpt. in Jensen, 2006a, p. 361)

One of the most daunting challenges of our time is to construct a collective vision for how humans should live in nature. The dominant culture continues to persist in the destruction of our planet. Global warming, population growth, peak oil, unrelenting fossil fuel consumption, species extinction, desertification, deforestation, oceanic contamination all continue relatively unabated despite some minimal mitigation efforts. Notwithstanding the ecological decline in just about every living system on the planet there remains substantial dominant cultural resistance to establishing a sustainable, collective vision for how humans should live in nature—witness Palin chanting "drill baby drill!" at the 2008 Republican convention and more recently the resistance to basic cap and trade legislation deemed too costly in times of economic recession/depression. Despite a growing number of voices to the contrary, the dominant culture is still guided by a belief that nature is, above all, a resource for human exploitation.

The struggles of activists to preserve the integrity of place against a dominant ideology of "nature as resource" can be interpreted as an attempt to generate and affirm human meaning in connection with nonhuman nature. Another way of interpreting activists' efforts to resist the destruction of place is as a struggle against nihilism: against the obliteration of the individual's ability to experience meaning and to engage physically, emotionally, and cognitively with the natural world.

I divide this chapter into three sections examining the threat of cultural nihilism as it presents itself to environmental activists engaged in defense of place (specific political, legal, and other actions taken to protect a place that is threatened). In the first section, I sketch out a conception of cultural nihilism and the nihilist bind, as it will pertain to my analysis of two different types of environmental texts. The second section explores cultural nihilism and individual place-based resistance through communiqués from Earth Liberation Front (ELF) extracted from Jay Hasbrouck's dissertation "Primitive Dissidents: Earth Liberation Front and the Making of a Radical Anthropology." In the last section, I examine cultural nihilism and place-based resistance from the perspective of Derrick Jensen's *End Game.*

I

> *Yet our anger is impotent; if all is relative, we really have no means by which to criticize and correct others, or to entrench our own "values." Perhaps even more challenging, though less commonly addressed, is the concomitant lack of purpose that we all experience. That is, the absence of external authority that makes possible this relativistic freedom also removes any given end for the project of human existence.*
>
> (Everden, 1992, p. 7)

In this section, I am concerned with establishing a theoretical explanation for types of consciousness that propel radical environmentalists toward desperation in their defense of place. The dominant cultural perspective alluded to above has led many activists to experience a state of nihilism as I demonstrate in the following sections. For many environmental activists, meaningful experiences of place are frequently nullified by economic and political imperatives of resource exploitation. The unceasing transformation of the land bases which many individuals uniquely identify as dignified natural places is the source of desperation and a sense of nihilism that permeates the radical environmental movement.

The term "nihilism" first appeared in 1787, and then again in 1796 and 1797 (Carr, 1992), and became widely used in the 19th century. In the first half of the century nihilism was linked to the intellectual study of idealism, and in the latter half of the 19th century nihilism began to be associated with the nothingness that was created in "God's death" as Nietzsche eloquently illustrated in his essay, *The Madman* (Carr, 1992, p. 15). Nihilism has been expressed in many ways; it has been described as "a historical process, a psychological state, a philosophical position, a cultural condition, a sign of weakness, a sign of strength, as the danger of dangers, and as a divine way of thinking" (Carr, 1992, p. 27). Nihilism stems from the Latin *nihil*, which means literally nothingness. According to the Internet Encyclopedia of Philosophy the Greek Skeptics were the first to argue against any foundations of certainty, truth claims were simply matters of opinion. The Skeptic position is linked to what is referred to in contemporary discourses as epistemological nihilism or what postmodernism refers to as anti-foundationalism. These positions simply hold that there is no way to claim something is knowledge or truth because there simply is no way to know for sure.

Other philosophical categories of nihilism include aetiological nihilism which "is the denial of the reality of truth" (Carr, 1992, p. 17; Pratt, 2009). Ontological or metaphysical nihilism "is the denial of an (independently existing) world, expressed in the claim, 'nothing is real'" (Carr, 1992, p. 18). Yet another philosophical category is ethical or moral nihilism. "An ethical or moral nihilist does not deny that people use moral or ethical terms; the claim is rather that these terms refer to nothing more than the bias or taste of the assertor" (Carr, 1992, p. 18). Existential nihilism denotes a belief that life has no intrinsic meaning and therefore is pointless and absurd. Political nihilism holds that the political, economic, and social institutions of society are so corrupt that they need to be destroyed. This is the type of nihilism we see expressed in many modern environmentalists such as Derrick Jensen.

From this brief survey of usages, one can say that innuendos of nihilism as a problem confronting the truth of subjective experience have been around since the Greeks. Every "thinking" human being has probably experienced some skepticism about truth claims. Is there really a God? Does our life really have universal meaning? A healthy individual skepticism is, without doubt, a good thing. But when does nihilism become debilitating and destructive to human dignity? Friedrich Nietzsche warned that the threat of nihilism "uncanniest of all guests" represented in the death of God would create a crisis in which "everything lacks meaning" and hence "awakens the suspicion that all interpretations of the world are false" (Nietzsche, 1968, p. 7). Nietzsche foreshadowed what has become the greatest challenge of postmodernity, creating meaning in the absence of meaning but not in the absence of power.

Power is central to the study of culturally generated nihilism. Capitalist cultures represent a normalized set of objectives and behaviors which are solidified in state-sanctioned legal codes and normalized in institutional behaviors. This type of cultural power creates a type of moral and ethical nihilism for the individual—witness the "just taking orders" defense. When power manifested through economic, political, and social institutions negates individual moral and ethical action, individual nihilism becomes a permanent cultural condition. Postmodern culture places the individual into precarious and moral existence, where every individual is allowed to believe what they want to, but forced to live the way power dictates. Capitalist cultural imperatives render individual moral agency impotent, reducing ethical behavior to a series of personal decisions about consumption. Cultural power is manifested in the unquestioned acceptance of corporate and government exploitation of people and nature in the pursuit of profit. Individual nihilism exists when individual moral and ethical agency are relegated to the realm of individual consumer preferences.

Karen Carr suggests that the cheerful acquiescence of nihilism leads to the perpetuation of the status quo, a condition in which power alone determines what is ethical, moral, and intellectually worthy of pursuing (1992, p. 140). In our postmodern corporate capitalist's culture, power exercised through economic, political, and social systems and institutions does appear to be the sole determinant of how moral, ethical, and intellectual pursuits for us, as individuals, are determined. I may publicly oppose nuclear weapons, genetically modified organisms, clear-cuts, mining, and oil and gas development on public lands, yet I will ultimately be silenced by the machinery of hegemonic power which will declare such positions impractical and even extreme. Despite my declared opposition, I still subsidize such activities through my tax dollars. I may choose not to pay taxes, but I will go to jail, becoming even more socially impotent. The psychological cost of this moral precariousness is what I refer to as the nihilist bind.

The nihilist bind occurs when existing social forces deny us human agency—the ability to act on values and interpret our own subjective experiences with others in an attempt to frame an alternative collective vision. This has been the experience of all indigenous and colonized people throughout history and now it is becoming the experience of all activists attempting to alter the course of political and economic power. Jack Forbes, in his book *Columbus and Other Cannibals*, refers to this consuming of another's life by powerful people and cultures as a type of cannibalism (Forbes, 2008). Forbes writes: "Cannibalism, as I define it, is the consuming of another's life for one's own private purpose or profit.... Thus, the wealthy exploiter 'eats' the flesh of oppressed workers, the wealthy matron 'eats' the lives of her servants, the imperialist 'eats' the flesh of the conquered, and so on" (Forbes, 2008, pp. 24–25).

Using this logic, the economic and political imperatives of this culture are inherently cannibalistic of nature and people, especially of people who resist these imperatives. This nihilistic situation, as Carr denotes in the title of her book, is anything but banal. What postmodern civilization is placing beyond our reach is agency—the ability to actualize our subjective values in discourse with others in creating authentic modes of existence. I can no more live in a world where the air I breathe is healthy and the water I drink free of carcinogens than I can live in a culturally conscious world that works toward that end. In fact, working for such a world places me at odds with the political, economic, and social systems and institutions that prioritize commerce over people and nature.

The ultimate expression of this nihilistic impotence lies in the fact that in a postmodern world where all truth claims may be, like it or not, construed as on equal footing with all other truth claims, only a few individuals holding the reigns of economic and political power decide how we all will live. As atomized individuals we remain powerless, unable to act as moral agents with other moral agents in the production of our lives. Corporate capitalism and the hegemonic nature of nationalism have successfully robbed individuals of their moral and ethical agency, reducing individuals to masses generating the types of adaptations discussed by scholars like Fromm as "automatons," or Mills as "cheerful robots," and Marcuse as "one-dimensional men." Attempts by individuals to formulate alternative discourses in our postmodern world are immediately marginalized as special interest politics confined to lobbying, voting, commentary, and state-approved protest. Hold your sign in the appropriate cage or offer your one-minute, timed comment expressing your utter disgust with the Forest Service's endorsement to open another road-less area to oil and gas exploitation, mining, or logging. These prescribed modes of dissent are exercises in futility at best, humiliating and infuriating at worst.

We now turn to those who find such futile and ineffective prescriptions for ending environmental destruction as unacceptable and, hence, a source of desperation and individual nihilism. I will conduct my analysis guided by the following questions: (1) How do the activists convey the experience of a culturally generated condition of moral and ethical nihilism? (2) How do the activists convey the nihilist bind? and (3) How does engaging in defense of place mitigate the crisis of the nihilist bind? Let us now turn to the texts of the ELF who are classified as domestic terrorists by the US Congress and FBI due to their repeated use of arson and sabotage as methods of resistance.

II

> *Time is running out—change must come, or eventually all will be lost. A belief in state-sanctioned legal means of social change is a sign of faith in the legal system of that same state. We have absolutely no faith in the legal system of the state when it comes to protecting life, as it has repeatedly shown itself to care far more for the protection of commerce and profits than for its people and the natural environment.*
>
> (Hasbrouck, 2005, p. 2)

In this section I will be using select communiqués of ELF as they appear in Jay Hasbrouck's Dissertation "Primitive Dissidents: Earth liberation Front and the Making of a Radical Anthropology." Hasbrouck's focus is on examining "key discourses surrounding the actions, ideology, and motivations of a self-described green anarchist network known as the Earth Liberation Front (ELF)" (2005, p. viii). I have chosen his dissertation because it is the most comprehensive body of radical environmentalist activists' communiqués that I have encountered. My project differs from his with respect to concepts and mode of analysis and scope. His is a dissertation, mine is an academic paper. In short, the communiqués he has acquired are an excellent, in-depth look into the radical environmental movement's philosophy and actions.

As the above epigraph indicates, desperation clearly underlies ELF's motives for damaging SUVs. The words "time is running out—change must come, or eventually all will be lost" express an end of the world crisis (Hasbrouck, 2005, p. 2). The "no faith in the legal system" denotes the disingenuous nature of legal recourse as means to halting further destruction to people and the environment. In this case, SUVs were targeted because they represented the culture of overconsumption. This can be seen in how the ELF chooses their targets: the number one target of radical environmentalists' actions is housing developments and urban sprawl, followed by facilities conducting genetic engineering, followed by logging operations, and finally sports utility vehicles (Hasbrouck, 2005, p. 22).

ELF's targets are specific and they are directed at the ideological heart of corporate capitalism. They are driven by a deep repulsion with the moral and ethical nature of postmodern corporate consumer capitalism, as this ELF communiqué illustrates:

> …it is the same state structure, big business and consumer society that is directly responsible for the destruction of the planet for the sake of profit. When these entities have repeatedly demonstrated their prioritizing of monetary gain ahead of life, it is absolute foolishness to continue to ask them nicely for reform or revolution. Matters must be taken into the hands of the people who need to more and more step outside of this societal law to enforce natural law. (Hasbrouck, 2005, p. 2)

The appeal to natural law suggests that the ELF believes in higher laws, in this case they refer to "natural law." ELF members are obviously not nihilists in their beliefs; they believe in natural law on our planet and in the universe, and they believe in the inherent sacredness that all plants, animals, and facets of the natural world have. Hasbrouck demonstrates that most ELF members identify with the living philosophies outlined by green anarchists. Green anarchists reject civilization and its power relations in exchange for deinstitutionalized, "primitive" modes of subsistence or what is referred to as anarcho-primitivism (Hasbrouck, 2005, pp. 3–23). It is obvious that ELF believes in the wisdom of nature (natural law) and that humans should respect the integrity that is inherent to particular land bases. But ELF's beliefs are not shared by the status quo, and, in fact, are antithetical to the status quo. Take this ELF communiqué for instance:

> Western civilization, with its throw away conveniences, its status symbols, and its unfathomable hoards of financial wealth, is unsustainable, and comes at a price. Its pathological decadence, fueled by brutality and oceans of bloodshed, is quickly devouring all life and undermining the very life support system we all need to survive. The quality of our air, water, and soil continues to decrease as more and more life forms on the planet suffer and die as a result. We are in the midst of a global environmental crisis that adversely effects and directly threatens every human, every animal, every plant, and every other life form on the face of the Earth. (Hasbrouck, 2005, p. 185)

It is clear that the ELF rejects state-organized corporate consumer capitalism, and it is not hard to see why. ELF rejects the disconnection that capitalism has from the natural world, as capitalism shows absolute preference for capital and profit, with no regard for the consequences that extracting such a profit costs. ELF questions the sanity of state-organized corporate capitalism's persistence in destroying its ecological base, and they daily witness the relentless violence committed against human and nonhuman life to perpetuate an unsustainable existence. ELF sees the legal system as disingenuous, and they perceive mainstream environmental groups as largely ineffective. In return, ELF is rejected by mainstream environmental groups for their emphasis on property destruction, among other ideological differences. An ELF communiqué response to a mainstream environmentalist group illustrates this friction:

> Grassroots and mainstream organizations who have come out publicly against the actions of the ELF do so either due to economic reasons (they rely on donations from the public, members, or grants from charities or governmental or non-governmental organizations) and/or they have a firm belief and an exceptional amount of faith in the system of government in operation in their particular area. Either way this attitude demonstrates a clear misunderstanding and/or a great reluctance to accept the seriousness of the threats to life on this planet and to make a firm commitment to work to actually stop that destruction of life. All of us must remember that the movement to protect all life must not be a means of monetary gain for individuals and organizations but rather one that produces concrete results. (Hasbrouck, 2006, p. 201)

The ELF, along with many supporters, believe that many mainstream environmentalists are careerist and do not seek the abolition of industrial civilization but rather its regulation through technical solutions. In fact, leading environmental thinkers Michael Shellenberger and Ted Nordhaus in their article "The Death of Environmentalism—Global Warming Politics in a Post-Environmental World" noted that every environmental leader they had ever interviewed understood the immense urgency of global warming, but not one had a clear articulate vision for how to confront the problem (2005). They contend that "green groups are defining the problem so narrowly—so unecologically—that they have alienated potential allies and become just another special interest" (Shellenberger & Nordhaus, 2005, p. 21). ELF criticism of the mainstream environmental movement is shared by many mainstream environmentalists experiencing the nihilist bind from behind the walls of their nonprofit 501c3s.

So what is a faceless, alienated, eco-conscious ELF to do? As Worldwatch scientist Assadourian and Starke write, the 2005 "Millennium Ecosystem Assessment made it clear that nearly two thirds of ecosystem services have been degraded or are being used unsustainably, and indicators like the Ecological Footprint have demonstrated that human society has been living beyond its means since 1987" (Assadourian & Starke, 2009, p. 67). The article goes on to note that we "are now using the equivalent of 1.25 planets' worth of resources" (Assadourian & Starke 2009, p. 66). Yet US politicians and economists aggressively refute any large-scale changes that would jeopardize business as usual. The March/April 2009 issue of *Multinational Monitor* indicates that the greenhouse gas industry lobby outnumbers health and environment by 8-to-1, with respect to trade and cap global warming legislation (Wedekind, 2009, p. 4). The Center for Public Integrity is warning that it is going to be extremely difficult to get any meaningful greenhouse gas reduction legislation passed with this lobby effort. The law remains biased toward private interests and the mainstream environmentalists do not want to give up their iPhones nor risk alienating their wealthy granters by speaking the truth and actually attempting to enact change on the system which is set up to protect the interests of corporations—not humans, animals, or the planet. Faced with this bleak reality it is understandable why one would feel rather nihilistic about change coming from within. In fact, we might conclude that the type of deep-rooted change needed to begin to address the current environmental catastrophe is beyond the imagination of either the state-organized corporate/consumer capitalism or the futile efforts of the environmental lobbying groups.

Does the full-scale conceptual awareness of the scope of our environmental problem produce a state of individual nihilism? For the ELF, the answer appears to be yes. The following communiqué followed an ELF action of "vandalizing construction equipment and an attempted arson of four houses under construction…in Placer County, PA" (Hasbrouck, 2005):

> Psychologically speaking we are all on the verge of death, with no way out in sight. Suicide, alcoholism, and drug addiction are epidemic. Nearly everyone is on drugs be it Prozac, lithium, lattes, mochas, cigarettes, beer, pot, cocaine, or chocolate. The world we have is empty and boring us to death. WE are forced to sell our souls 8, 10, 12, 14 + hours a day 5, 6, even 7 days a week for more than half our lives, not to mention school before that, they have us work jobs we hate so we can buy shit we don't need.... We are through with the lies. (p. 186)

It is also clear from this communiqué that many engaging in ELF activities do so out of a sense of reconciling the impotence created by the nihilist bind. As one anonymous ELF writes, "you can decide to be apathetic and complacent, and hope for it all to collapse, or, you can decide to take responsibility and fight to destroy this death machine.... Either way you will have blood on your hands, it's just a matter of whose" (Hasbrouck, 2005, p. 6). It would also seem from this statement that ELF has accepted the premise that one loses either way. Complacency will end life as we know it, leaving blood on our hands, and engaging in illegal property destruction could lead to blood on ELF hands and potential incarceration. Taking responsibility for what is happening is the ELF mantra. Take, for example, this communiqué:

> There is absolutely no excuse for any one of us, out of greed, to knowingly allow this to continue. There is a direct relationship between our irresponsible over-consumption and the lust for luxury products, and the poverty and destruction of other people and the Natural world. By refusing to acknowledge this simple fact, supporting this paradigm with our excessive lifestyles, and failing to offer direct resistance, we make ourselves accomplices in the greatest crime ever committed. (Hasbrouck, 2005, p. 185)

The resolution of the nihilist bind for ELF participants is to engage in illegal property destruction, which is risking being classified as domestic terrorists and subjected to lengthy prison sentences. Their risks are rationalized by the alternative, which is being complicit in the destruction of the planet—a relegation of their agency they refuse to accept. Do ELF members really think they are going to bring down state-organized corporate/consumer capitalism with random acts of property destruction? This ELF communiqué offers some insight:

> We are not so naïve as to believe that we would have stopped development in Twelve Bridges. Though we could have caused over 2 million in damages, it was still a fairly symbolic protest and the message should have still registered; that we are exceptionally serious, the necessity of new discussions and that all of the true eco-terrorists such as JTS should consider themselves forewarned. (Hasbrouck, 2005, p. 206)

There is little doubt ELF wants to encourage other like-minded individuals to engage an eco-sabotage, but they do not appear naïve about the overall impact of their work on the culture they wish to destroy. Their intentions are more than symbolic, however; they wish to instill fear in those that perpetuate the destruction of the planet out of greed; they also seek to reclaim the term "ecoterrorist" by turning the concept against those who terrorize the natural environment for self-serving ends. It is as the old saying goes "one person's freedom fighter is another's terrorist" except in this case it is one person's environmental liberator is another's ecoterrorist.

ELF creates, if nothing else, a discourse about the state of our environment. Legislatures are more apt to not see mainstream environmentalists as radical, making environmental groups' demands more palatable. Unfortunately, the deeper message that ELF seeks to convey, which is that life is sacred and not negotiable, will fall on deaf ears in backrooms where the natural environment is bartered as a commodity for consumption. The ELF identity alleviates for many individuals the sense of nihilism that plagues many people in our culture through acting on their fears and concerns about the health of our planet. One can speculate that ELF actions create a sense of power in what is otherwise a hopeless and powerless situation. There probably is a spark of excitement and empowerment in acting in defiance of the totalitarian culture that seeks to make us blind and dumb nihilists, numb enough to watch our future dissolve in front of our eyes.

On the other hand, law enforcement also finds a new sense of purpose. The US Legislature recently created new laws with increased public funding to expand police powers to seek out "ecoterrorists" with a vengeance. Law enforcement must feel righteous in knowing that private interests to exploit the natural environment have been preserved. The message is clear: to resist is futile. After all, there is a normalized process in our country to create change.

III

> *Premise One: Civilization is not and can never be sustainable. This is especially true for industrial civilization.*
>
> (Jensen, 2006a, p. ix)

> *I advocate not allowing those in power to take resources by force, by law, by convention, or any other real or imagined means. Beyond not allowing, I advocate actively stopping them from doing so.*
>
> (Jensen, 2006a, p. 85)

Anyone who has read Derrick Jensen knows of his passion for the natural world and his lack of patience for Western civilization and its apologists.

Jensen's writing is fierce and leaves no aspect of Western civilization unturned, be it state-organized corporate/consumer capitalist society, science, technology, or any other form that violence against human and nonhuman life takes. Jensen's first book agent accused him of being a nihilist and that he should tone down his work. He writes, "I felt vaguely insulted. I didn't know what a nihilist was, but I knew from her tone it must be a bad thing" (Jensen, 2006b, p. 363). After researching the topic, Jensen decided he did not meet the first definition of nihilism; that is, he believed in truth, beauty, and love. The second definition, however, dealt with describing the current social order as being "so destructive and irredeemable that it needs to be taken down to its core, and to have its core removed—fits me like a glove" (Jensen, 2006b, p. 363).

In his book, *End Game*, Jensen defends twenty constructed premises in over 890 pages of text. He exposes the violence of civilization as it has been committed against all life, human and nonhuman: from the genocide of Native Americans and Jews, to the genocide of the Buffalo and the passenger pigeon, to vivisection of animals, to factory farms, to domestic violence, to normalized rape in war, to factory fish trawlers, to genocidal statements made throughout US history by political and economic elites, Jensen unmasks the sickness that fuels a civilization bent on destroying the land base.

"Civilization is incompatible with human and nonhuman freedoms, and in fact, with human and nonhuman life" (Jensen, 2006a, p. 13). Jensen writes "the story of civilization is the story of the reduction of the world's tapestry of stories to only one story, the best story. The real story, the most advanced story, the most developed story, the story of power and the glory that is western civilization" (2006a, p. 23). Civilization, for Jensen, is based on hegemonic control aimed at making one particular way of living the only way of living regardless of how destructive it may be. Jensen defers to Stanley Diamond's definition that "civilization originates in conquest abroad and repression at home" (2006a, p. 15). In order for civilization to thrive and continue to do relentless damage to all life and land, Jensen argues that the individual must pay a heavy psychological, sociological, and spiritual toll with respect to truncated experiences and individual agency. Take, for example, this long passage offering a painful description of the process of normalizing cultural nihilism.

A high school student bags the groceries. She's been through the mill. Twelve years of it, not counting her home life, twelve years of sitting in rows wishing she were somewhere else, wishing she was free, wishing it was later in the day, later in the year, later in her life when at long last her time—her life—would be her own. Moment after moment she wishes this. She wishes it day after day, year after year, until—and this was the point all along—she ceases anymore to wish at all (except to wish her body looked like those in magazines, and to wish she had more money to buy things she hopes will for at least one sparkling moment of purchase take away the ache she never lets herself feel), until she has become subservient, docile, domestic. Until her will…has been broken…. Until the last vestiges of the wildness and freedom that are her birthright—as they are the birthright of every animal, plant, river, piece of ground, breath of wind—have been worn or torn away. Free will at this point becomes almost meaningless, because by now victims participate of their own free will—having long since lost touch with what free will might be…. There is no longer any need for force, because the people—or more precisely those who were once people—have been fully metabolized into the system, have become self-regulating, self-policing (Jensen, 2006a, p. 285).

Most people can identify with some aspects of the drudgeries outlined above in our long and tedious endeavor to learn docility and acceptance of the fact that our life belongs to those in power. The clock teaches us that large tracts of our life belong to others, starting with school and ending with work. The leftover time you have is to live your life according to prescribed consumer behaviors. Women should love to shop after their enculturation and men should love to sit on their asses drinking corporate-brewed beer watching others perform shows and games.

Jensen wrote *End Game* to appeal to those who feel a sense of rage at what passes for their lives. He writes "we are people who are tired of living hollow lives guided by abstract moralities expressly created to serve those in power, moralities divorced from physical realities, including the land we love, including the land we rely on" (Jensen, 2006b, p. 828). He continually encourages the reader that our fate is not inevitable, "We are people who refuse to continue as slaves…. We are people who are ready to take back our own lives, and to defend our lives and the lives of those we love, including the land" (Jensen, 2006b, p. 828). He wrote this tome to encourage those experiencing the nihilist bind to stop being victims and to stop relying on an abstract hope for things to get better.

Jensen exalts individual agency in defying and resisting the civilization that is killing humans, nonhumans, and the environment. Jensen exclaims that he is in love "with salmon, with trees outside my window, with baby lampreys living in sandy stream bottoms, with slender salamander crawling through the duff" (2006a, p. 332). He entreats "if you love you act to defend your beloved…. You do what it takes. If my love doesn't cause me to protect those I love, it's not love. And if I don't act to protect my land base, I'm not fully human" (Jensen, 2006a, p. 332). It is rage against the insanity that is our lives and a passion to do something, anything about it, which makes Jensen a motivational destroyer of nihilism. Jensen encourages people to bring down civilization by "liberating ourselves" and "by driving the colonizers out of our own hearts and minds: seeing civilization for what it is, seeing those in power for who and what they are, and seeing power for what it is" (Jensen, 2006a, p. 252). But what exactly does Jensen mean by bringing down civilization?

Bringing down civilization is millions of different actions performed by millions of different people in millions of different places in millions of different circumstances. It is everything from bearing witness to beauty to bearing witness to suffering to bearing witness to joy. It is everything from comforting battered women to confronting politicians and CEOs. It is everything from filing lawsuits to blowing up dams. It is everything from growing one's own food to liberating animals in factory farms to destroying genetically engineered crops and physically stopping those who perpetuate genetic engineering.... It is destroying the capacity of those in power to exploit those around them. In some circumstances this involves education. In some circumstances this involves undercutting their physical power, for example, by destroying physical infrastructure through which they maintain their power. In some circumstances it involves assassination.... (Jensen, 2006a, p. 252).

Most people are willing to go along with most of what Jensen says until he discusses the need to counter the forces of civilization with violence. Throughout the book Jensen engages and counters common pacifist arguments, using analogies such as self-defense which ends in killing a potential rapist, the many assassination attempts of Hitler, the Jews whose survival rate was greatly increased from resisting in the Warsaw Ghetto uprising, and a mother grizzly bear's defense of her young. Jensen is relentless in rebutting the essentialist pacifist position. He loses many potential sympathizers on these points even though he relentlessly reminds the reader of the endless violence that is committed against life daily by states and corporations. Jensen forces us to confront this fact frequently by asking us to consider why violence against life is normalized while violence against those who destroy life is unacceptable? Reading Jensen's books are like having an ice pick tapping against your forehead echoing tick-tock on the planetary clock. It is blow-by-blow reading where every indictment against our civilization is supported with a factual account of atrocity after atrocity. It is a real dilemma. The violence is real, and the costs are real, and our inaction is real. He goads us:

> We have the best excuse in the world to not act. The momentum of civilization is fierce. The acculturation deep. Those in power will imprison us if we effectively resist. Or they will torture us. Or they will kill us. There are so many of them,

> and they have weapons. They have the law.... Because of all of this, there really is nothing we can do. We may as well admit that. (Jensen, 2006a, p. 178)

Then there is the guilt problem of our culpability in participating in civilization, a tactic Jensen is quick to point out is designed to put the onus on us and not those in power. Jensen's rebuttal is that we can be forgiven for having to live in the world, "because we did not create the system, and because our choices have been systematically eliminated..." (2006a, p. 178). We become culpable when we do not exercise our agency, when we do not "stop them with any means necessary. For not doing that we are infinitely more culpable than most of us-myself definitely included-will ever be able to comprehend" (Jensen, 2006a, p. 178). This is the sheer power of Jensen's relentless rant; we are responsible for what happens to life on this planet. Yes, we use the technologies of this civilization, but we did not create the system that has eliminated our choice. But our knowledge of the destructive nature of these technologies demands that we once again assert our choice, our volition to end our servitude and complicity to the destruction of our land base. It will not happen without exercising our agency, our birthright to feel and think as our hearts and brains tell us.

How does Jensen respond to those who tell him that he is great at tearing down civilization but ask him what is the alternative? To this Jensen replies "I do not provide alternatives because there is no need. The alternatives already exist, and they have existed—and worked—for thousands and tens of thousands of years" (2006a, p. 889). To many this is simply a cop out. Ten thousand years ago there were not 7.1 billion people, and you cannot just let them die by shutting down the machine. In defense of Jensen (not that he needs it), the crisis of peak oil is predicted to cause more than a billion deaths alone, not to mention all the other crises that peak oil will create. One way or the other, we are headed for planetary ecological collapse. The four words most often uttered from the lips of most environmentalists are "we are all fucked." It is not easy to go back into denial after reading almost 900 pages of Jensen. He is absolutely right in saying that our culture is sick and destructive and needs to be destroyed. I am less sanguine than Jensen about the likelihood of 10,000 years of civilization sickness being wiped out by the actions of even a million dedicated eco-warriors. The cultural inertia of 10,000 years will need more than a human push. We are destined to undergo collapse and it is going to be an unfathomable experience for those who will have to bear witness. Human and nonhuman life is going to be decimated as catastrophic collapse implies.

Does Jensen's prescriptions help aid the sense of nihilism that many acutely feel? As the old Emiliano Zapata quote says, "it is better to die on your feet than to live on your knees" (as cited in Jensen, 2006a). To this end,

Jensen's anti-nihilism campaign is invaluable at least to those who still feel a sense of dignity and compassion for the living world. We must fight, but as Jensen himself admits he is not a killer. Nor am I, nor are most people. That is not to say that people do not care, but most people—myself included—do not think my going to jail for dismantling some apparatus of the machine is going to make much of a dent. I may feel like a martyr for the first day in jail, but after that I will just be in an even more restrictive cage, denied relationship with all that I love. There are many out there who would willingly give their lives—myself included—if we thought it would stop the rape, pillage, and genocide of our current culture. I attend many public meetings on environmental issues and listen to the depth of the sickness as it drones on out of the spokespeople that represent industry and our government. You could take one out, but there are going to be another 100 standing in line to take their place, and that can be said about every position of power in this society, all the way down to midlevel management. For this reason I know Jensen will continue to write, and I will continue to read and act. Embracing meaninglessness is not an option in a universe filled with life. To let those in power deny you access to life is not now nor is it ever acceptable. It is our duty and responsibility to resist the culture of nihilism and for this I harbor a sense of gratitude toward Jensen and all who engage in the struggle against the nihilist bind.

Conclusion

> *When one accepts nihilism as "just the way things are," it ceases to be a potential weapon against corrupt and decaying modes of thought.... The possibility of any kind of ethical, religious, or political transformation is de facto ruled out and the perpetuation of the status quo is covertly promoted. Any disagreements that do exist deteriorate ultimately into contests of power.*
>
> (Carr, 1992, 3p. 140)

The crisis of nihilism that pervades the environmental movement would be entertaining if the consequences from inaction were not so dire. At the core of this problem is a flawed way of living that simply is not sustainable. We are exceeding the carrying capacity of the earth, and its various ecological systems are beginning to collapse. This is a fact that no literate person can deny. The question remains: what are we going to do about it?

Given the unequal distribution of world resources, which has the richest 20% of the world's population consuming about 86% of all resources and the poorest 20% consuming less than 2%, the scope of the problem is beyond just political and economic solutions but requires a deeper look into moral and ethical agency. To believe that the political and economic systems of the richest 20% of the world's population are going to undergo a voluntary transformation to a sustainable life is to be uselessly idealistic. Change is not likely to come voluntarily from the top. We can all rest assured that change is coming, it is just a matter of whether change is going to be guided by moral and ethical agency or thrust upon us from natural forces. At this point, change will probably consist of natural catastrophes, social collapse, and the unleashing of human rage at being forced to live a meaningless existence for thousands of years.

The rage that is mounting in people all over the world at having their lives stolen from them is beginning to escalate—witness how governments all over the world are more frequently criminalizing dissent. We see this in the United States where we now classify property damage with a political emphasis as an act of terrorism. Those in power are scared—as they should be. As the younger generations come of age realizing they have little in the way of any future, it is doubtful that they will be contented with false promises and choices. Humans can understand emptiness, but in almost all circumstance they reject it in favor of a meaningful existence, and when the culture cannot provide meaning they will create it. Nihilism is an unbearable condition and extremely dangerous when fueled by desperation. It is for this reason that we must fight nihilism in our personal and public lives, replacing resignation with passion, alienation with connection, and inaction with action.

References

Assadourian, E., & Starke, E. (2009). Global Economic growth continues at expense of ecological systems. In E. Starke (Ed.), *Vital signs 2009: The trends that are shaping our future* (pp. 66–68). Washington, DC: World Watch Institute.

Carr, K. L. (1992). *The Banalization of nihilism: Twentieth-century responses to meaninglessness.* Albany, NY: State University of New York Press.

Everden, N. (1992). *The social creation of nature.* Baltimore, MD: The John Hopkins University Press.

Forbes, J. (2008). *Columbus and other Cannibals.* New York, NY: Seven Stories Press.

Hasbrouck, J. (2005). *Primitive dissidents: Earth Liberation Front and the making of a radical anthropology* (Doctoral dissertation). Retrieved from the University of southern California digital library.

Jensen, D. (2006a). *Endgame volume 1: The problem of civilization.* New York, NY: Seven Stories Press.

Jensen, D. (2006b). *Endgame volume 2: Resistance.* New York, NY: Seven Stories Press.

Nietzsche, F. (1968). *The will to power.* New York, NY: Vintage Books.

Pratt, A. (2009). Nihilism. In *Internet Encyclopedia of Philosophy*. Retrieved from http://www.iep.utm.edu/nihilism/#H1

Shellenberger, M., & Nordhaus, T. (2005, January 13). The death of environmentalism: Global warming politics in a post-environmental world. *Grist*.

Wedekind, J. (2009, March 1). The greenhouse gas lobby: Behind the lines. *Multinational Monitor*. Retrieved from https://www.highbeam.com/doc/1G1-200251900.html

Part II: Classic Writings on Ecoterrorism Rhetoric

4. *Ecoterrorism? Countering Dominant Narratives of Securitization: A Critical, Quantitative History of the Earth Liberation Front (1996–2009)*

MICHAEL LOADENTHAL

Introduction

In the wake of the 9/11 attacks, the often-linked fields of Terrorism Studies and Security Studies have witnessed a boom, accompanied by the more general rise in university studies directed at Islam, political Islam, terrorism and Middle Eastern politics (Kurzman & Ernst, 2009; Miller & Mills, 2009; Ranstorp, 2007; Richard, 2007; Shepherd, 2007; Silke, 2009; Suleiman & Shihadeh, 2007). Subsequently, new approaches have been developed within a host of "critical" fields, including *Critical* Terrorism Studies and *Critical* Security Studies (Brecher, Devenney, & Winter, 2010; Horup, 2012; Jackson, Murphy, & Poynting, 2010; Poynting & Whyte, 2012; Stump & Dixit, 2013). These attempt to problematize and clarify a methodology for those seeking to investigate political violence and responses to it through a non-orthodox, non-realist lens in an attempt to move beyond a purely securitization focus (Bellamy, 2004; Floyd, 2006, 2007; Salter & Mutlu, 2012; Shepard, 2013; Vaughan-Williams & Peoples, 2010). These approaches, while offering a host of new points of concern and criticism, often continue to base their study on raw data produced by state institutions. Thus, while such studies may critically examine findings, new scholarship is needed that draws its conclusions from the wealth of data offered by the social movements themselves.

Within this post-9/11 era of terrorism scholarship, a new class of "terrorism experts" emerged, poised to corner the academic market, often at the service of law enforcement, state-centric think tanks and a wider statecraft of securitization. The present study is not meant to serve as yet another quantitative tool for criminal and behavioural profiling (Greenwald, 2012). Instead it is meant to act as an example of a methodological break, wherein one surveys the difficult data offered by the practitioners of political violence or their supporters themselves. The analysis contained herein is not meant to be a tool for law enforcement but rather to serve as a counter-balance to the statist narrative concerning tactical trends and their relation to the criminalization of dissent. Overblown, inaccurate and fearmongering depictions of bomb-throwing masked vigilantes occupies much of the discussion of "ecoterrorism". In response, scholars have been careful to begin developing counter-narratives to discuss these movements within a more accurately nuanced language. Activist-aligned journalists and academics have also begun to offer critiques of the terrorist framing of these movements in an attempt to offer an alternative explanation to state rhetoric (Lovitz, 2010; Potter, 2011). Within state rhetoric, we see the Earth Liberation Front (ELF) and its sister entity the *Animal* Liberation Front (ALF) termed "ecoterrorists" since around 2002, when Federal Bureau of Investigations (FBI) Domestic Terrorism Chief James Jarboe invoked the label twelve times in a speech entitled "The threat of Eco-Terrorism". The same year, Dale Watson from the FBI's counterterrorism and counterintelligence division reported to the US Senate that the ALF/ELF represented a serious terrorist threat, characterizing them as the most active extremist elements in the United States.

The ELF has been active in the United States since 1996 and, through the use of decentralized, self-contained, underground cells of activists, has managed to not only carry out scores of attacks on property, but remain relatively immune from arrest. Activists inspired and motivated by the politics of the ELF are free to carry out acts of property destruction and claim them via the ELF moniker, provided they meet the movement's guidelines. According to widely circulated guidelines, ELF actions must economically harm the adversary, aim to educate the public and avoid harming both human and (non-human) animal life. Therefore, if an individual or a small group of activists agree with these three simple points, they are encouraged to act independently and to claim their attack through the ELF moniker. This has most often been done through written communiqués bearing the ELF name. In those rare instances when a communiqué is not issued, the letters E.L.F. have appeared in paint at the site of the incident. The various actors and cells that constitute the ELF should therefore be understood as an ideologically aligned network of autonomous, decentralized nodes, who share a strategic and tactical vision. They are not a unified movement in the traditional sense, nor are they a membership-based organization. They are a tactical, strategic and praxis-informed tendency supported by a similarly decentralized, ad hoc network. As a result, scholarship that insists on understanding such groupings as nothing more than radical splinters of traditional social movements (e.g. Earth First!) will continue to be inherently flawed.

In exploring these networks of clandestine eco-saboteurs and arsonists, one is often tempted to construct a definition of *terrorism* and, following that, present one's case comparatively to that set of parameters. The goal then becomes to decide if the evidence presented qualifies the object for inclusion within the *terrorist* designation. Conversely, this study seeks to present evidence which can then be held up against a variety of definitions of terrorism that have in common a focus on deliberate attacks against unarmed human beings in order to intimidate, coerce or otherwise influence a larger audience. Therefore the "dominant narratives" this study seeks to challenge are those that present the ELF not as a strategic social movement utilizing targeted property destruction, but as a violent terroristic threat to the nation state. It is not the main intention of this study to refute the FBI's classification of the ELF as domestic "ecoterrorists", but rather to discuss how data is represented through a divergent lens, and subsequently used to embolden such claims for the purposes of securitization. The intention of this study is to provide quantitative evidence to public, above-ground activists and scholars who seek to offer support in the creation of counter-narratives: explanations of an emergent social movement not based in state-centric terrorism rhetoric.

The following study seeks to determine the tactical, targeting, messaging and associated behavioural characteristics of the ELF through a dataset drawn from the movement's self-constructed mouthpiece. This analysis will draw from the movement's thirteen-year (1996–2009) history of global attacks in order to answer the question: What does a typical ELF attack look like, and secondly, how often are *atypical* incidents claimed under the ELF moniker? In order to develop such a behavioural profile, a series of statistical findings will be reviewed. These findings are drawn from an analysis of the movement's attack history as presented via their above-ground support structure, the North American Earth Liberation Front Press Office (NAELFPO) (BBM, n.d.; NAALPO, n.d.). All data analysed was gleaned from pubic (i.e. non-classified) sources and, as such, provides little to no utility for law enforcement as such entities compile their own incident databases from a host of clandestine (e.g. Law Enforcement Sensitive, Classified, etc.) sources. This attack chronology, documented by the NAELFPO, was used to develop a database of 707 events, each coded for 11 variables. Each attack was coded through a standardized decision tree based on the description provided by the

NAELFPO, as well as communications issued by the ELF cell directly. The data was then split into six datasets and analysed. These six distinct datasets were developed to account for the presence or absence of repeating events (e.g. one cell breaks the windows of four banks, claimed in one communiqué) and commonly occurring, distinct national locations. Throughout the discussion contained herein, the findings have been compared to studies presented in academic journals, as well as government reports, in an attempt to evaluate the ELF's attack history in the light of assertions made about the movement's behaviours.

Methodology: Process and Limitations

The database utilized throughout this analysis was created from the "diary of actions" hosted on the NAELFPO's website. According to the NAELFPO, "The actions contained on the pages below comprise a complete history of ELF actions in North America and globally" (NAALPO, 2009a). From the "diary of actions", a 707-entry database was created, each entry representing one ELF-linked attack. These 707 entries were comprised of 211 distinct events and 496 repeating attacks (e.g. a single cell vandalizing multiple, distinct targets in one outing claimed through a single communiqué) occurring between 14 October 1996 and 23 November 2009. The attacks were carried out across 14 countries, including 28 US states, while there were as well four attacks without a clearly discernable location. The data was coded manually and was used to create a database via the "SPSS Statistics 17.0" software suite. Each event was assigned distinct values within 11 variable fields. Many of the attack descriptions and cell communiqués are exceedingly descriptive regarding the tactics utilized and target selection, though in some cases this descriptive richness was lacking. Occasionally, attacks were recorded in the NAELFPO diary with only a single descriptive sentence, making the process of coding for 11 variables difficult. The variable categories were developed with such limitations in mind, and thus, "the goal in developing the [coding] taxonomy was to build a set of classes broad enough to capture the range of terrorist behaviour, but still simple enough to use, given the limitations in the descriptive data available on each individual terrorist incident" (Jackson & Frelinger, 2008, p. 564).

In cases where elements of the description necessary for coding were absent, attempts were made to estimate a reasonable scenario and to describe it through the most accurate terms available. For example, if a description stated that a laboratory was "attacked", "trashed" or "monkey wrenched", the attack was recorded as an act of "sabotage/vandalism/graffiti", as the broad nature of this tactic category was developed to allow for the coding of such events, events where the exact nature of the damage and tactics is unclear. If the description stated that the target was "covered in slogans", "paint bombed", "tagged" or used similar language, the tactic was recorded as "graffiti", despite the fact that such a term was not included in the communiqué text. Throughout the coding process, attention was paid to the stated motivation for attacking a target. For example, when a Wal-Mart or Nike shop was attacked and criticized for its *global* policies, it was recorded as an attack targeting a "multinational corporation", whereas the office of a regional energy company was recorded as a "business property". Similarly, two attacks, both targeting automobiles, could be coded differently depending on the owner's position vis-à-vis the larger ELF policy. The vandalism of a sports utility vehicle (SUV) in a dealership was recorded as targeting a "SUV/automobile", whereas similar vandalism of a specific individual's (e.g. CEO of targeted company, researcher engaged in controversial experimentation) car, targeted because it belonged to that individual, was recorded as the targeting of "personal property". In coding the data, the aim was to use as little interpretation as possible and to transparently decipher the language of the description and/or communiqué into the coding values through a standardized decision tree.

In order to accurately represent the scale of some attacks, some single events are recorded as multiple entries. For example, if four SUVs are firebombed, the events were recorded as four acts of arson because four targets were attacked. Conversely, the breaking of four windows of *one* office/SUV/home/etc., was counted as a single attack. However, if one window was broken on each of four separate offices, this was recorded as four attacks since multiple targets serve as the determining factor. Occasionally, the exact number of targets attacked was unclear. If the description stated that "numerous vehicles" or "a row of homes" was attacked, that incident was recorded as two entries despite the possibility that many more targets were attacked. Lastly, if an attack utilized two distinctly different tactics, the event was recorded as two incidents (Jackson & Frelinger, 2008, p. 602). This was done when both tactics fell outside of the "sabotage/vandalism/graffiti" category, such as in the case of an "animal liberation" that also involved the arson of the building. In this example, the event would be recorded as one "distinct incident" and one "multiple entry" (Jackson & Frelinger, 2008, p. 567). Because of the tendency for such a coding procedure to artificially inflate the appearance of some attacks, calculations were conducted separately within multiple datasets, one wherein multiple entries are *included* and another wherein only distinct (non-recurring) attacks are included. In this second, distinct incident dataset, multiple entries were condensed to single attacks. For example, if saboteurs were to slash the tires of four vehicles and claim it in a single communiqué, this would be recorded as *one* tire-slashing incident in the *distinct incident* dataset and recorded as *four* tire-slashing incidents in the *multiple-entry* dataset. For the purposes of analysis, the sample was split into three location-based categories. Each of the three datasets was then split into subsets (multiple entries and distinct incidents). All numerical findings were rounded to the nearest whole number when presented in the in-text data tables, and in doing so, some total to more than 100%.

In rare cases, the NAELFPO's data included attacks that were carried out by a known group that was not affiliated with the ELF. For example, between 10 October 2008 and 31 October 2008, four attacks were carried out in Canada targeting the EnCana Corporation (NAALPO, 2009c). These attacks were not claimed by an ELF cell despite the presence of a communiqué hosted by NAELFPO, and thus, these attacks were excluded from the sample. Occasionally a cell adopted the ELF name *after* the initiation of an attack campaign. For example, starting in June 2009, attacks in Mexico ceased to be claimed by the ELF and were instead claimed by "Eco-Arsonists for the Liberation of the Earth" (EpLT) (NAALPO, 2009b). Thus attacks claimed by the EpLT were excluded from the study sample, and only those prior attacks signed with the ELF name were included. Around the same time, additional attacks in Mexico were being carried out by "Luddites Against the Domestication of Wild Nature" (LADWN); these were similarly excluded as LADWN represented a distinctly new, non-ELF moniker. However, on 20 July 2009, LADWN announced in a communiqué that it now "form[ed] part of a cell of the Frente de Liberación de la Tierra" (NAALPO, 2009d), thus from that date onwards, the group's actions were recorded as attacks of the ELF (NAALPO, 2009e). Finally, when a group used the ELF name but specified a distinct unit within the movement, this attack was simply recorded as being carried out by the ELF. For example, on 5 March 2001, a graffiti attack was claimed by the "ELF Night Action Kids" (NAAPLO, 2009e) and was recorded in the database as being carried out by the ELF. These methodological decisions were made to allow for focus on the deployment of the ELF *moniker*—not the wider constituency make-ups. For example, despite the ideologically shared proclivities of the ELF and the EpLT, since the latter chooses an explicitly non-ELF moniker to claim its actions, it is excluded despite tactical, strategic and ideological similarities.

In coding for the "communiqué" variable, the presence of a communiqué linked from the NAELFPO website was recorded as such, just as the lack of a communiqué present on the website was recorded as "no communiqué issued", despite the possibility that a communiqué was available in another source. In order for an attack to be marked as ELF-linked without the presence of a communiqué, it was required that the letters "E.L.F." were let at the scene of the attack through graffiti, a banner, note or similar visual/written communication. For example, on 11 October 2004, a "package with the letters 'ELF'" (NAAPLO, 2009f) was let on a road in Philadelphia and treated as a possible improvised explosive device, though the box turned out to be harmless. Thus in the database, the attack was credited to the ELF as a "bomb threat".

Limitations exist in the data acquisition and categorization methodology employed. Of particular importance are concerns regarding the validity of the NAELFPO's "diary of actions" as the data was provided by an organization with vested interests in promoting the best image of the ELF. Despite other databases available, such as those created by the Foundation for Medical Research, the North American Animal Liberation Front Press Office reports and numerous scholarly articles, this study sought to utilize a single data source, thus eliminating the need to synthesize conflicting information (Foundation for Biomedical Research, 2009; NAALPO, 2002). In an attempt to remove the judgment of the researcher from the acquisition of a data sample, only the NAELFPO "diary of actions" was used despite the understanding that such a source may contain inherent bias. As previously discussed, the lack of descriptive detail present in accounts of some attacks led to the development of coding categories that were more broadly defined than would have been necessary if complete incident descriptions were present for all attacks (Helios Global, 2008; Leader & Probst, 2003). To correct for this tendency, the variable coding categories were defined broadly enough to be inclusive of the uncertainty present in the data, while attempting to maintain distinctions. The categories were designed so that a single event could *only* be classified within one category. Despite these limitations, the NAELFPO dataset provides a singular and complete source for analysis while avoiding the need for the researcher to decide which sources are legitimate and which are to be excluded. At its best, the "diary of actions" represents an accurate, well-researched history of the ELF and affiliated movements. At its most limited, this study analyses the manner in which the ELF's press office presents the movement to a wider audience: how the press office frames the cells' actions via their intended messaging.

Findings and Discussion: Targeting

This first section will analyze the targeting pattern present in the ELF data. Target data was coded within 24 targeting types ranging from the common (e.g. 208 attacks on automobiles) to the obscure (e.g. one attack on an advertisement). The results from the data analysis concerning targeting vary dependent on the portion of the sample utilized. When the complete 707-entry dataset (DB1) is analyzed, the following findings emerge as the eight most commonly attacked target types:

- SUV/automobile: 29%
- Phone booths: 17%
- Homes under construction/model homes: 12%
- Company vehicles: 11%
- Construction/industrial equipment: 10%
- Business property: 5%
- Farm/ranch/breeders: 2%
- McDonalds restaurants: 2%

The other 16 target types each account for less than 2% of the total attacks and collectively comprise only 12% of the total targets.

When the global dataset *excludes* the multiple entries (DB2), the predominance of attacks on automobiles and phone booths is reduced, as these targets are typically attacked in groups. In the 211-entry DB2 dataset, the 12 most commonly attacked targets are:

- Construction/industrial equipment: 14%
- Home under construction/model homes: 13%
- Business property: 12%
- SUV/automobile: 10%
- Phone booth: 8%
- Business vehicle: 8%
- McDonalds: 5%
- Farm/ranch/breeder: 5%
- GMO experiment/research: 4%
- Government property: 3%
- Trees: 3%
- Government vehicle: 2%

The other 11 target types each account for less than 2% of the total attacks and collectively comprise only 14% of the total targets.

When these findings are compared to that of prior scholarship, points of congruency and disagreement can be seen. Though no alternative study could be located using the exact sample data source, in an article based on a dataset consisting of "database of [109] ELF attacks", occurring between 1996 and 2001, Leader and Probst (2003, p. 43) report that the most commonly attacked ELF targets are:

> "corporations": 33 (36%) "Urban sprawl/development": 30 (33%) "logging & related": 18 (20%) "genetic engineering/biotech research facilities": 14 "facilities that threaten animals": 6 (7%) "government facilities": 5 (5%) "symbols of global economy": 3 (3%)

As one can see, a close comparison becomes quite difficult in the two studies as they adopt different frameworks for categorization; one based in the nature and identity of the targeted object and the other in a broader business type category.

Similarly, a US Department of Homeland Security (DHS) commissioned study, carried out by Helios Global Inc., reported comparable, though more generic results. The Helios study focuses on a conflated history of the ELF and the ALF over a longer timeline, from 1981 to 2005 (Helios Global, 2008). Because their sample incorporates the ELF as well as the ALF, and their timeframe predates the ELF's founding by 15 years, an exact comparison is not possible. Regardless of such limitations, according to the Helios study (2008, p. 7), the most frequently attacked "primary targets" of the ALF/ELF are:

- "Commercial enterprises and/or individuals engaged in housing and urban development
- Commercial enterprises and/or individuals involved in the logging industry
- Sports utility vehicle (SUV) dealerships
- Commercial enterprises and/or individuals involved in the production, sale, and distribution of animal products (leather and fur producers, sellers, and distributors; restaurants; and meat, poultry, and fish producers)
- Animal research facilities and personnel
- Commercial enterprises and universities involved in genetic engineering..."

From the Helios study, though different terminology is employed, the targeting findings are quite similar to those contained in the present study. Target category 1 overlaps this study's category termed, "house under construction/

mobile home", whereas the remaining five categories similarly overlap this study's use of categorical terms such as: "SUV/automobile", "business property", "farm/ranch/breeder", "laboratory" and "GMO crops or research", respectively.

Returning to the data presented herein, in this ELF-specific study, one can now examine the remainder of the datasets concerning targeting typologies. When the dataset is further reduced to only attacks carried out in the United States, including multiple entries (DB3), the results are largely the same, with the same target types occupying the higher echelons. Notable changes include the exclusion of attacks on "phone booths", as all such attacks occurred in Mexico, as well as the rising presence of the targeting of "trees" (via tree spiking) as the eighth most common target type, comprising 2% of the total attack pool. In the 462-entry DB3 dataset, the most commonly attacked targets are "SUV/automobile", "House under construction/model home", "Business property", "Construction/industrial equipment", "Business vehicle", "McDonalds", "Farm/ranch/breeder" and "Trees: 2%". The other 13 target types each account for less than 2% of the total attacks and collectively comprise only 12% of the total targets. When this dataset excludes multiple entries (DB4), the results are largely the same. With the exclusion of multiple entries, the targeting types are more evenly distributed, with 15/20 target types accounting for 2% or more of the total pool. Comparison between these two datasets is displayed below: the other 11 target types each account for less than 5% of the total attacks and collectively comprise 21% of the total targets.

Table 4.1. Most Commonly Attacked Targets. (Source: Author)

Target Type	**Dataset #**	
	DB3 %	**DB4 %**
SUV/automobile	35	14
House under construction/model home	18	17
Business property	14	7
Construction/industrial equipment	8	11
Farm/ranch/breeder	3	6
Trees	2	5
Business property	N/A	7
GMO crops or research	N/A	5
Business vehicle	4	N/A
McDonalds	3	N/A

When targeting type is further reduced to the pool of attacks carried out only in Mexico, including the incorporation of multiple attacks (DB5), the exceedingly high proportion of attacks on "phone booths" is visible. In this sample, 78% of all ELF attacks in Mexico targeted a Telmex phone booth; this is by far the most singularly focused targeting seen in any of the datasets. In the DB5 dataset, the second most commonly attacked target type is "business property", comprising only 7%. When multiple attacks are excluded from the dataset (DB6), attacks on phone booths still remain the most common target type, but only account for 39% of the total attacks. Attacks on business property similarly remain the second most commonly attacked target type, now accounting for 24%. Throughout the findings in regard to targeting, there is a pattern of ELF cells attacking unguarded, "sot target" sites such as vehicles, phone booths and construction sites. In general, such properties would be located in public areas with little or no security. In contrast, target types such as laboratories, ski resorts, banks and government property consistently occupy the lower levels of target selection, possibly because such areas would more commonly employ electronic surveillance systems or human guards. Regardless of the dataset examined, and independent of the inclusion or exclusion of multiple attacks, business properties and construction sites are routinely targeted. In the US datasets, there is a dominant pattern of targeting homes under construction and model homes, though this pattern is not seen in other national settings. This is likely a reflection of the growing "sprawl" critique as seen in anti-globalization, anti-gentrification, anti-capitalist and anarcho-primitivist movements in the United States, a site where many ELF activists find their ideological groundings (Rosebraugh, 2004, pp. 121–126). In other nations, such as Mexico, there is no record of the targeting of "sprawl" sites, as Mexican cells have focused their attention on attacking phone booths as part of a larger campaign against Telmex, a company described in communiqués as "earth destroying" and guilty of "biocide".

Throughout the data collection and coding process, attention was paid to determining if the ELF attack carried out was part of a larger stated campaign, thus leading to the specific location being targeted. In the United States, two campaigns were identified. The most prominent was the anti-sprawl campaign in Long Island, NY, comprised of eight distinct attacks (42 multiple entry attacks), occurring from September 2000 to January 2001 (Nocella & Best, 2006, 415; Ziner, 2001). Such attacks accounted for approximately 8% of the total attacks carried out in the United States. Craig Rosebraugh, former ELF spokesman, notes the prominence of the Long Island anti-sprawl campaign, writing that it "constituted the most focused and intensified campaign the ELF had ever undertaken" (Rosebraugh, 2004, p. 157), consisting of 11 "major" attacks including five arsons of homes and condominiums under construction (Rosebraugh, 2004, p. 161). The second largest campaign targeted aliases of animal research supplier Huntingdon Life Sciences (HLS). Six distinct anti-HLS attacks were carried out in the United States, comprising less than 2% of the overall attacks. Such campaigns do not appear to be prominent in the ELF's targeting system as approximately 91% of all US attacks were not part of a stated campaign. In Mexico, this trend dramatically changes, as over 81% of all multiple entry attacks and over 51% of all distinct entry attacks are part of the campaign targeting Telmex. Mexico also was the venue for one anti-HLS attack.

In examining atypical events, some of the data that comprises the rare incidents, the outliers, though not statistically significant, are deserving of brief discussion. In only one *distinct* attack a ski resort was targeted, yet despite its rarity, the Vail, Colorado arson is often the most commonly heard of ELF attack, possibly because it caused approximately $26 million in damages. At the time the datasets were being constructed, a photograph of the Vail fire was featured on the ELF's main page, cataloguing the movement's "diary of actions" (NAALPO, 2009d). A second atypical targeting discovery focuses on cells' decisions to target human life and not solely property. Throughout the movement's thirteen-year history, only one attack directly targeted a human being. On 3 June 2009, an Australian cell of the ELF "hand delivered" a written threat to the home of a Hazelwood Power Station CEO, located in Melbourne (NAALPO, 2006). The note threatened the individual's *property* not his person, but because the threat was addressed towards a specific person, the incident was recorded as an attack targeting an *individual* not their property. Finally, in the targeting of fast food establishments, 15 attacks were directed at McDonald's restaurants, while in only one attack, a Burger King was targeted. This particular action was taken in 2002 by an ELF cell in the US city of Richmond, Virginia (NAALPO, 2009g).

Findings and Discussion: Tactics

The findings related to the tactics employed by ELF cells show little variation when the different datasets are examined comparatively. In every dataset, regardless of national location or the inclusion of "multiple entries", the top three tactical choices are: "sabotage/vandalism/graffiti", "arson" and "graffiti". When viewed across datasets, the proportionality of these tactics changes, as does their usages vis-à-vis one another, but regardless of these variations, these three tactical choices consistently occupy the top three positions in the tactical tool belt. This cross-dataset trend can be viewed in the comparison chart in Table 4.2 wherein the frequency of the three most common tactical typologies is compared across DB1–DB6:

Table 4.2. Comparison of 1st, 2nd, 3rd Most Commonly Utilized Tactics. (Source: Author)

Tactic	**Dataset #**					
	DB1 %	**DB2 %**	**DB3 %**	**DB4 %**	**DB5 %**	**DB6 %**
SVG	44	40	34	37	65	37
Arson	28	35	31	37	31	49
Graffiti	13	8	18	9	3	7

In five out of six datasets, "sabotage/vandalism/graffiti" is the most commonly employed tactic, followed by "arson" and finally "graffiti". In only one instance does "arson" calculate as the most commonly employed tactic. After the first three most commonly employed tactics, the breakdown across the various datasets begins to show greater diversity. Table 4.3 gives a comparison of the fourth, fifth and sixth most commonly utilized tactics within the six datasets:

Table 4.3. Comparison of 4th, 5th, 6th Most Commonly Utilized Tactics. (Source: Author)

Commonality of Tactic	**Dataset #**					
	DB1	**DB2**	**DB3**	**DB4**	**DB5**	**DB6**
4th most common	Window	Animal lib	Window	TS	IED	IED
5th most common	At ar, IID	TS	At ar, IID	At ar, IID	b threat	b threat
6th most common	Tire	At ar, IID	TS	Animal lib		

When examined together, these tactical trends show some similarity, with the breaking of windows, tree spiking (TS), attempted arsons (abbreviated as "At ar") and animal liberations proving common in the first four datasets, while the Mexican datasets (DB5–6) show identical results.

The findings concerning tactics of this study can be compared to similar attempts in the scholarly literature. The Helios study summarizes the tactical choices of "ecoterrorists", a broad category including but not limited to the ELF, and represents the totality of such attacks within five tactical typologies.

According to the Helios study (Helios Global, 2008, p. 14), the tactical breakdown of "eco-militant tactics carried out between 1981 and 2005" can be summarized as:

- "Vandalism": 45%
- "het": 23%
- "Harassment": 15%
- "Arson": 10%
- "Bombing": 7%

The Helios results share some findings with this study, as both agree that "vandalism" (termed "sabotage/vandalism/graffiti" in this study) as broadly defined is the most commonly employed tactic, and "bombing[s]" (termed "IED" in this study), is the least commonly used tactic. Both studies also agree that within these extremes, "eco-militants" use other tactics including arson and theft. Although not detailed in the study, one can assume that "het" for Helios encompasses the removal (or release) of live animals from slaughterhouses, breeding facilities, laboratories, etc. The Leader and Probst (2003) article reports similar findings but utilizes smaller categorical groupings. Based on 92 attacks, according to the article (Leader & Probst, 2003, p. 41), the three most common tactic types are:

1. "Vandalism": 36 (33%) (Leader & Probst, 2003, p. 41)
2. "Arson": 32 (29%) (Leader & Probst, 2003, p. 41)
3. "Sabotage": 19 (17%) (Leader & Probst, 2003, p. 41)

Once again, such findings support those of this study, in that both report the most commonly utilized tactics combine sabotage, vandalism, graffiti, arson and attempted arson. In attempting to identify inaccuracies within the literature—such as those positioned to embolden security debates—the tactical descriptions of the ELF are likely the most important areas to examine. The Leader and Probst article asserts "ELF's prime weapon is arson" (2003, p. 41), though this claim is not supported by their own data nor the research presented herein, as more general sabotage and vandalism tactics generally show a higher predominance, as they do in datasets DB1–DB5, excluding DB6, where arson does surface as the most commonly utilized tactic in "distinct incident" attacks carried out in Mexico. In Leader and Probst's own findings, vandalism occurs slightly more commonly than arson, and thus the statement that the movement's "prime weapon is arson" appears hyperbolic for the sake of rhetoric. Linked to the tactics chosen for attack are issues of lethality and threat to life. Casualty data was collected for every incident in the datasets. Throughout the 707, multinational, all incident dataset (DB1), and thus all secondary datasets, no ELF attack is reported to have caused any injuries or fatalities to human beings. This finding is supported by the scholarly literature (Ackerman, 2003, p. 162; Borum & Tilby, 2005; Bron, 1998, p. 3, 8; Leader & Probst, 2003, p. 44) in every example surveyed and places a big question mark as to the rationale for the categorization of ELF activities under "terrorism".

Findings and Discussion: Claims of Responsibility

The presence of a communiqué documenting an attack is common throughout the different attacking cells. In cases where a formal communiqué is not issued, attackers sometimes leave ELF "calling cards" such as the group's name scrawled in graffiti, notes or banners. In Table 4.4, the comparison of communiqués and "calling cards" is shown across the six datasets:

Table 4.4. Comparison of the Presence of ELF Communiqué or "calling card". (Source: Author)

Communiqué	**Dataset #**					
	DB1 %	**DB2 %**	**DB3 %**	**DB4 %**	**DB 5%**	**DB6 %**
Communiqué present	58	68	41	56	100	100
No communiqué present	26	17	35	22		
"ELF" let at scene	16	15	24	22		

When compared across the six datasets, the trend is relatively uniform. In all cases, communiqués are more commonly issued than not, though their existence has varying degrees of regularity. In the first four datasets (DB1–4), communiqués are issued for more than half of all attacks (56% on average), and when a communiqué is not issued an ELF "calling card" is present in approximately 19% of all cases. In the Mexican datasets, communiqués are issued 100% of the time.

Although this study has focused on attacks carried out by cells self-identifying as members of the ELF, occasionally, attacks are claimed in the name of a cooperative endeavour by the ELF and ALF through either a communiqué or the NAELFPO's description of the attack. Table 4.5 shows the proportion of attacks claimed solely by the ELF, as compared to those claimed mutually by the ELF/ALF.

Table 4.5. Comparison of Group Claims. (Source: Author)

Group Claiming Attack	**Dataset #**					
	DB1 %	**DB2 %**	**DB3 %**	**DB4 %**	**DB5 %**	**DB6 %**
ELF only	94	88	95	90	99	98
ELF/ALF	6	12	5	10	1	2

This cross-dataset comparison shows that the claiming of ELF/ALF joint actions is marginally more common in the non-US, non-Mexican arena. In the Mexican-only datasets, such cooperative claims of responsibility are exceedingly rare, accounting for only one attack in the entire country's history.

Findings and Discussion: Location

According to the data collected, the ELF is active in fourteen countries across the continents of North America, South America, Europe and Australia. The highest concentration exists in the English-speaking, "Western" world of North America and Western Europe, though the presence of active cells in Mexico appears to be increasing (Ross, 2009, 2010). In DB1, the countries with the highest rates of attacks are the United States, Mexico, United Kingdom and Canada. The other ten national locations account for less than 2% of the total attacks each and collectively comprise only 7% of the total events. Making up this collective 7% are attacks carried out in Spain, Italy, the Netherlands, Sweden, Iceland, Russia, Australia, New Zealand, Chile and Colombia. When the dataset is reduced, examining only "distinct incidents", the results are the same, both in the nations identified and their general ranking vis-à-vis one another (Molland, 2006, p. 69). The most common countries remain the United States, Mexico, Canada and UK and the other ten nations account for less than 1% of the total attacks each and together comprise less than 3% of the total events. A comparison of these two similar findings is displayed in Table 4.6.

Table 4.6. Comparison of Location. (Source: Author)

Country	**Dataset #**	
	DB1 %	DB2 %
United States	64	65
Mexico	19	21
Canada	5	7
United Kingdom	5	4

These findings are similar to those reported in the Helios study wherein the authors state, "despite their global presence...acts of terrorism appear to be most prevalent in North America and Western Europe". Helios' exclusion of Mexico as a target of ELF attacks is expected, as the study was published in 2008, the year when Mexico began to experience activity from ELF cells.

The United States is overwhelmingly the focus of the ELF's international campaign, despite the fact that the movement as it exists today emerged in England (Helios Global, 2008, p. 11). The availability of information relating to the location of ELF attacks occurring in the United States lends itself well to analysis, as there is no need for the researcher to equitable develop categories or to extrapolate variable labels from attack narratives. Data on location as it pertains to state was available for all but one "distinct incident", and regional divisions were determined based on mapping provided by the US census report. In Table 4.7, attacks within the United States (DB3–DB4) will be compared in regard to region and regional division.

Table 4.7. Comparison of US Attack Location Regionally. (Source: Author)

Region	**Dataset #**	
	DB3 %	DB4 %
WEST	**52**	**46**
Pacific	42	35
Mountain	10	11
SOUTH	**19**	**10**
West South Central	0.2	1
East South Central	4	2
South Atlantic	15	7
NORTHEAST	**18**	**23**
Middle Atlantic	15	21
New England	3	2
MIDWEST	**10**	**18**
West North Central	2	5
East North Central	8	13

From this data, one can see that the Western region (specifically the five state Pacific division) has been a particular centre of activity, concentrated in the states of California, Oregon and Washington. Similarly, the Middle Atlantic three state division within the Northeast region has been particularly active, as numerous attacks have been carried out in the states of New York and Pennsylvania. The least active region appears to be the Midwest despite its large geographic area. The least active regional division is the West South Central, four state grouping of the southern region.

In surveying the literature concerning the ELF, only the Helios study includes a detailed discussion of location as it pertains to regions throughout the United States. Although the Helios sample is different from the one employed herein, the findings are similar. The Helios study concludes "eco-terrorists" are "particularly active in the Western and West Coast states. In particular…Oregon, California and Washington…the Midwest and East Coast have a smaller percentage of eco-terrorist incidents" (Helios Global, 2008, p. 11). Certainly, the data contained herein supports the Helios claim that such movements are especially active in the "Western and West Coast states", and state-specific data supports the claim that high levels of activity are seen in Oregon, California and Washington. The claim, that the Pacific Northwestern region "is the most prominent environmental hot spot in the nation", is also offered by former ELF spokesmen Craig Rosebraugh (2004, p. 76). Also, both the Helios study and this study agree that the Northeast (called East Coast in the Helios study) and Midwest occupy the lower regions for ELF activity.

Conclusion

This chapter draws on the case study of the ELF to demonstrate the analytical potential of conducting a quantitative tactical analysis of activism of social movement groups in order to debunk hyperbolic tropes of "terrorism". For example, methodological decisions related to categorization, coding and data sourcing can be used to skew data towards hyperbole, fearmongering and securitization or they can be used to approach greater accuracy, nuance and balance. The preceding dataset challenges the framing of radical environmental groups as terrorist threats to the nation state. This rhetorical framing—especially that dealing with tactics and targeting—supports the increased government repression of leftist movements through targeted legislation such as the Animal Enterprise Terrorism Act of 2006 and the larger atmosphere of the "Green Scare" (Potter, 2011). Future research can challenge limits to dissent through quantifying movement group actions and calling into question government tropes about *radical* movements. With such political and methodological concerns in mind, the data presented can allow scholars to develop an incident-driven history of the ELF beyond broad state framings as bomb-throwing "terrorists" and "arsonists".

The ELF, a transnational movement of direct action eco-saboteurs, follows a definitive targeting and tactical pattern, focusing its attacks on unguarded properties associated largely with commercial and residential construction, automobiles (especially SUVs) and various regional, national and multinational business interests. ELF cells target such entities clandestinely, and with low-tech tactics, often striking multiple sites within one target type in rapid succession. For example, it is common for one cell to vandalize dozens of SUVs in one outing. The targeting patterns follow regional indicators concerning campaigns developed through attack histories in that locale. In the United States, such attacks have focused on targets associated with "sprawl" and residential development, SUV sales and ownership, and construction sites. In Mexico, attacks have focused on a campaign targeting Telmex phone booths and other affiliated properties. The majority of ELF attacks are not part of larger attack campaigns, though in about 5% of US attacks and 34% of Mexican attacks, the communiqué stated that the target was chosen as part of a long-term campaign, focusing strikes on a specific set of entities linked thematically.

Tactically, ELF cells tend to rely on varied combinations of vandalism (including graffiti), sabotage and arson. Throughout all of the data, a combining of vandalism and sabotage has dominated the tactical history, with arson occurring as the second most commonly employed tactic. In extremely rare instances (six attacks out of 707 equalling 0.85%), cells have used tactics that direct violence against humans or present the threat of violence against humans through the use of improvised explosive devices, bomb hoaxes and written threats.

Through a combined analysis examining targeting as it compares to tactics, one witnesses the defining modus operandi of ELF cells. When distinct attacks are examined globally, the arson of residential units, as well as the sabotage and vandalism of construction vehicles and other business properties emerges as the most dominant attack patterns. When the United States is examined separately, one sees the same pattern of homes being targeted through arson, business properties targeted through sabotage and vandalism, and SUVs targeted with graffiti. In the purely Mexican context, targeting and tactics collide at the vandalism, sabotage and arson of Telmex phone booths and, to a far lesser extent, the arson of construction and industrial equipment.

Concerning cells' claims of responsibility for attacks, in the global context, attacks are claimed via a formal communiqué sent to either an aboveground press office or other media outlet more than 70% of the time. In other instances, the ELF name is let at the scene of the crime to indicate the movement's claim of responsibility. In only approximately 17% of ELF attacks is the incident linked to the movement, but not formerly claimed via a communiqué or other form of communication. The ELF movement rarely reports that its cells have jointly carried out attacks cooperatively with cells affiliated with the ALF. The ELF/ALF cooperative moniker is seen globally in approximately 6% of cases, with a greater frequency seen in attacks occurring outside of North America.

Trends in attack location indicate that the ELF, while focused primarily in the United States, is having an expanded sphere of activity in Mexico. Other sustained areas of activity include Canada—centred in Ontario province—and the United Kingdom—especially within England. Within the United States, attacks have focused on the eastern and western coastal areas, centred in the states of California, Oregon, Washington, New York and Pennsylvania. Outside of North America and the UK, sparse attacks have been documented in the continents of Europe (especially Western Europe and Scandinavia), South America and Australia. At the present time, there are no reports of ELF attacks within Antarctica or the African or Asian continents.

Finally, the preceding analysis has attempted to diagram the attack history of the ELF through a transparent methodology. In doing so, one is able to comparatively evaluate its findings alongside that of other, more opaquely authored studies. While it is true that the preceding findings were constructed around an a priori agenda—namely providing defensible data for furthering nuanced and well-informed debates regarding emergent social movements—this is no different than scholarship that came before or will likely come after. If the critical analysis of state and academic scholarship is seen as "having an agenda" could one not say the same thing of well-circulated papers built around an a priori agenda of securitization? For example, the 2013 DHS-funded report authored by the National Consortium for the Study of Terrorism and Responses to Terrorism (START), while academically rigorous, cannot be described as politically impartial. The report explicitly describes its mission in the opening pages stating:

> This report is part of a series in support of the Prevent/Deter program. The goal of this program is to sponsor research that will aid the intelligence and law enforcement communities in assessing potential terrorist threats and support policymakers in developing prevention efforts. (NCSTRT, 2013, p. 1)

Studies such as those conducted by START or Helios are funded by, and produced explicitly for, the policing of dissent, as they are identified as projects of the DHS. The collection and publication of such data is into a larger American, post-9/11 shift in domestic policy—a shift from *policing to national security*. Scholarship of this nature can be as rigorous (or manipulative) as any academic pursuit, but to contend that it does not possess a pre-existent ideological framework and political agenda is to misunderstand statecraft as a neutral endeavour.

The data suggests that the label "ecoterrorism" has been misapplied to a form of political militancy that falls short of what can reasonably be called "terrorism" since there have been practically no deliberate deadly attacks on civilians that would warrant the use of such a loaded term. While the actions of the perpetrators are often unlawful since these tend to involve acts of vandalism, arson or sabotage, and while these acts are meant to convey a message to a wider audience, that is still a far cry from the bloody terrorism of, for example, salafist jihadists. The terrorism label loses its potency if it is stretched beyond credibility. It should be used sparingly, rather than loosely and be limited to certain categories of gross violations of human rights and international humanitarian law—roughly the peacetime equivalent of war crimes (Alex & Schmid, 2004).

References

Ackerman, G. (2003). Beyond arson? A threat assessment of the Earth Liberation Front. *Terrorism and Political Violence*, *15*(4), 162.

Alex, C. F., & Schmid, P. (2004). Frameworks for conceptualising terrorism. *Terrorism and Political Violence*, *16*(2), 197–221.

Bellamy, A. (2004). *Critical security studies, in international society and its critics*. Oxford: Oxford University Press.

Bite Back Magazine. (n.d.). Retrieved from http://www.directaction.info/

Borum, R., & Tilby, C. (2005). Anarchist direct actions: A challenge for law enforcement. *Studies in Conflict & Terrorism*, *28*(3), 201–223.

Brecher, B., Devenney, M., & Winter, A. (2010). *Discourses and practices of terrorism: Interrogating terror*. New York, NY: Routledge.

Bron, T. (1998). Religion, violence and radical environmentalism: From Earth First! to the Unabomber to the Earth Liberation Front. *Terrorism and Political Violence*, *10*(4), 1–42.

Floyd, R. (2006). *When Foucault met security studies: A critique of the "Paris School" of security studies*. Poster session presented at the British International Studies Association Annual Conference, University of Cork, Ireland.

Floyd, R. (2007). Towards a consequentialist evaluation of security: Bringing together the Copenhagen and the welsh schools of security studies. *Review of International Studies*, *33*(2), 327–350.

Foundation for Biomedical Research. (2009). Illegal Incidents (1997–Present). *Foundation for Biomedical Research*. Retrieved from http://www.bresearch.org/Media/MediaRoom/Backgrounder/IllegalIncidentsMap/tabid/960/Default.aspx

Greenwald, G. (2012, August 15). The Sham "terrorism expert" industry. *The Salon.* Retrieved from http://www.salon.com/2012/08/15/the_sham_terrorism_expert_industry/

Helios Global. (2008). *Ecoterrorism: Environmental and animal-rights militants in the United States.* Washington, DC: Helios Global Inc./U.S. Department of Homeland Security.

Horup, M. (2012). *An intellectual history of terror: War, violence and the state.* New York, NY: Routledge.

Jackson, B., & Frelinger, D. (2008). Riling through the terrorists' arsenal: Exploring groups' weapon choices and technology strategies. *Studies in Conflict & Terrorism, 31*(7), 538–604.

Jackson, R., Murphy, E., & Poynting, S. (2010). *Contemporary state terrorism: Theory and practice.* New York, NY: Routledge.

Kurzman, C., & Ernst, C. W. (2009). *Islamic studies in U.S. universities.* Presented at the Social Sciences Research Council workshop on "The Production of Knowledge on World Regions: The Middle East." Chapel Hill, NC: University of North Carolina.

Leader, S. L., & Probst, P. (2003). The Earth Liberation Front and environmental terrorism. *Terrorism and Political Violence, 15*(4), 37–58.

Lovitz, D. (2010). *Muzzling a movement: The effects of anti-terrorism law, money, and politics on animal activism.* Herndon: Lantern Books.

Miller, D., & Mills, T. (2009). The terror experts and the mainstream media: The expert nexus and its dominance in the news media. *Critical Studies on Terrorism, 2*(3), 414–437.

Molland, N. (2006). A spark that ignited a flame: The evolution of the Earth Liberation Front. In A. J. Nocella II & S. Best (Eds.), *Igniting a revolution: Voices in defense of the earth* (pp. 47–58). Oakland, CA: AK Press.

National Consortium for the Study of Terrorism and Responses to Terrorism (NCSTRT). (2013). *Overview of bombing and arson attacks by environmental and animal rights extremists in the United States, 1995–2010.* College Park, MD: U.S. Department of Homeland Security Science and Technology Center of Excellence.

Nocella, A. J., II, & Best, S. (2006). *Igniting a revolution: Voices in defense of the earth.* Oakland, CA: AK Press.

North American Earth Liberation Front Press Office (NAALPO). (n.d.). Retrieved from http://animalliberationpressoffice.org/NAALPO/

North American Earth Liberation Front Press Office (NAALPO). (2002). Year-end direct action report. *North American Animal Liberation Front Press Office.* Retrieved from www.tao.ca/~naalfpo/2001_Direct_Action_Report.pdf

North American Earth Liberation Front Press Office (NAALPO). (2006). Communiqué: 06.03.09 Earth Liberation Front leaves message for Hazelwood CEO Graeme York in Australia. *ELF Press Office.* Retrieved from http://www.elfpressoffice.org/comm060309.html

North American Earth Liberation Front Press Office (NAALPO). (2008). Year-end report on animal liberation activities in North America. *North American Animal Liberation Front Press Office*. Retrieved from http://www.animalliberationpressoffice.org/pdf/year_in_review1.pdf

North American Earth Liberation Front Press Office (NAALPO). (2009a). Earth Liberation Front Diary of Actions. *Elf Press Office*. Retrieved from http://www.elfpressoffice.org/doa.html

North American Earth Liberation Front Press Office (NAALPO). (2009b). Earth Liberation Front diary of actions—2009. *Elf Press Office*. Retrieved from http://www.elfpressoffice.org/actions2009.html

North American Earth Liberation Front Press Office (NAALPO). (2009c). Earth Liberation Front diary of actions—2008. *Elf Press Office*. Retrieved from http://www.elfpressoffice.org/actions2008.html

North American Earth Liberation Front Press Office (NAALPO). (2009d). Earth Liberation Front communiqué 07.20.09: ELF targets Telmex branch office in Mexico. *Elf Press Office*. Retrieved from http://www.elfpressoffice.org/comm072009.html

North American Earth Liberation Front Press Office (NAALPO). (2009e). Earth Liberation Front diary of actions—2001. *Elf Press Office*. Retrieved from http://www.elfpressoffice.org/actions2001.html

North American Earth Liberation Front Press Office (NAALPO). (2009f). Earth Liberation Front diary of actions—2004. Earth Liberation Front press office, 2009. Retrieved from http://www. elfpressoffice.org/actions2004.html

North American Earth Liberation Front Press Office (NAALPO). (2009g). Earth Liberation Front diary of actions—2002. *Elf Press Office*. Retrieved from http://www.elfpressoffice.org/actions2002.html

Potter, W. (2011). *Green is the new red: An insider's account of a social movement under siege*. San Francisco, CA: City Lights Publishers.

Poynting, S., & Whyte, D. (2012). *Counter-terrorism and state political violence: The "war on terror" as terror*. New York, NY: Routledge.

Ranstorp, M. (2007). Introduction: Mapping terrorism research. In M. Ranstrop (Eds.), *Mapping terrorism research: State of the art, gaps, and future directions* (pp. 1–28). London: Routledge.

Richard, J. (2007). Constructing enemies: Islamic terrorism in political and academic discourse. *Government and opposition: An international journal of comparative politics*, *42*(3), 394–426.

Ross, J. (2009, October 6). Our fires illuminate the night: Wave of anarchist bombings strikes Mexico. *Counterpunch*. Retrieved http://www.counterpunch.org/ross10062009.html

Ross, J. (2010, January 11). The Mexican Revolution at 100: Mexico welcomes 2010 with bombs and riots. *Counterpunch*. Retrieved from http://www.counterpunch.org/ross01112010.html

Rosebraugh, C. (2004). *Burning rage of a dying planet: Speaking for the Earth Liberation Front.* Herndon: Lantern Books.

Salter, M., & Mutlu, C. (2012). *Research methods in critical security studies: An introduction.* New York, NY: Routledge.

Shepard, L. J. (2013). *Critical approaches to security: An introduction to theories and methods.* New York, NY: Routledge.

Shepherd, J. (2007, July 3). The rise and rise of terrorism studies. *The Guardian.* Retrieved from http://www.guardian.co.uk/education/2007/jul/03/highereducation.research

Silke, A. (2009). Contemporary terrorism studies issues in research. In R. Jackson, M. B. Smyth, & J. Gunning (Eds.), *Critical terrorism studies: A new research agenda* (pp. 34–48). New York, NY: Routledge.

Stump, J. L., & Dixit, P. (2013). *Critical terrorism studies: An introduction to research methods.* New York, NY: Routledge.

Suleiman, Y., & Shihadeh, A. (2007). Islam on campus: Teaching Islamic Studies at higher education institutions in the UK. *Journal of Beliefs & Values, 28*(3), 309–329.

Vaughan-Williams, N., & Peoples, C. (2010). *Critical security studies: An introduction.* New York, NY: Taylor & Francis.

Ziner, K. (2001, May). Burning down the houses–Earth Liberation Front tries to stop urban growth in Rural Long Island. *E: The Environmental Magazine.* Retrieved from http://indarticles.com/p/articles/mi_m1594/is_3_12/ai_74628715/

5. *Activism as Terrorism: The Green Scare, Radical Environmentalism and Governmentality*

COLIN SALTER

The events of September 11, 2001 (herein 9/11) have had a profound and often underestimated, State-mobilized and intentional, impact on dissent and counterhegemonic ideas in contemporary Western societies. Within six weeks of 9/11 the term "domestic terrorism" came into law in the USA PATRIOT Act, with similar legislation passed in a number of countries across the globe.[1] Alongside increasing penalties and other sweeping legislative changes, including "enhanced" surveillance powers with limited judicial review, use of the term *terrorism* continues to lend itself to specific discursive functioning. The frame-bridging of actively criticizing the State and "terrorism" has sought to construct a social boundary around what is acceptable democratic dissent, manifested in and by the asymmetrical targeting of sources of dissent (i.e., social-environmental justice activists), as opposed to specific tactics of dissent (i.e., causing physical injury). By dissent I am referring to public discourse that challenges the State and corporate interests in the sociopolitical arena, in the forms of speech and collective action (i.e., it is explicitly performative)—be this printed or electronic media, and broader participatory activities such as marches and other forms of protest.[2] Sources of dissent considered here are situated counter to hegemonic discourse (pre- and) post-9/11, which is manifested in signifiers of a renewed patriotic (shared, nationalist) identity and what were constructed as normative notions of (national) (in)security and freedom that were the foundations of sweeping legislation including the USA PATRIOT Act (Nabers, 2009). Hegemony, as used here, drawing from Rose and Miller's influential analysis of political power and the State, "is not so much a matter of imposing constraints upon citizens as of 'making up' citizens capable of bearing a kind of regulated freedom" (Rose & Miller, 2010, p. 272).[3] A key element and characteristic of State hegemony post-9/11 is action at a distance: the management of dissent without always needing direct, overt forms of influence. The outcome is a perception of autonomy, which is betrayed by self-regulation (at the individual level in line with the social context).[4] It is here that the perception of action at a distance rests. The foundations of self-regulation are located in (contested) hegemonic discourse (Rose & Miller, 2010, pp. 277–279).

The active, and increasingly pre-emptive, repression of dissent in the West after 9/11 has significant and far-reaching implications. Directly visible effects include a reduced willingness of some to criticize the State and (openly) participate in certain types of political activity. Less visible implications include one's own reflections on what ideas and actions are suitable in a normative and socially acceptable sense. To put it simply, the social costs of dissent have increased (Gillham & Edwards, 2003). Notions of democratic tolerance in a Marcusean sense, directly influenced by the context of 9/11 and bourgeois capitalist hegemony, continue to influence what is positioned as socially acceptable.[5] How tolerance is manifested is twofold. Broadly, perspectives on the protection of certain liberties have changed, with an increasing number of people willing to accommodate a number of restrictions under the guise and rhetoric of increased safety and security (Schneiderman & Cossman, 2001, p. 173). The aim of producing a pliable, disciplined populace, Foucault's docile bodies, continues to be broadly achieved in this sense (Foucault, 1995).

Paralleling such willingness to sacrifice civil liberties, certain types of dissent, specifically those that challenge normative ideals in a radical sense, are relationally positioned as deviant and socially unacceptable. Those who seek to expose and undermine the exploitation of all animals, laying the foundations for "a revolutionary society based on critiques of the multiple fronts of systemic oppression," find themselves routinely positioned as domestic terrorists, despite not having harmed a single person and having a stated tenet of eschewing physical harm (Kahn, 2005, p. 2).[6] For example, those who take such actions under the banner of the Animal Liberation Front (ALF) and the Earth Liberation Front (ELF) have been labeled as the "most serious domestic terrorism threats" in the United States, after more than a decade of lobbying by corporate agribusiness seeking such an outcome, yet not a single person has been harmed by anyone from these groups (Lewis, 2005).[7] Actions under these banners are criminalized based on a direct threat to "animal capital" (Sanbonmatsu, 2011, p. 26). Steven Best describes "the animal liberation movement…[as] one of the most dynamic and important political forces on the planet," with its importance to radical social movements emerging from an anarchist politics of total liberation in which all forms of oppression are targeted, keeping "radical resistance alive" (Best, 2009, p. 19).

This chapter explores the increased State and corporate focus on those who take actions seeking to foster an essential and critical dialogue in exposing and challenging the exploitation of all animals, focusing on specific aspects of the far-reaching implications of targeted and pre-emptive repression. Beyond exposing asymmetrical targeting, this chapter reflects on manifestations and implications of self-censorship in individuals and the broader collective of those taking action in a post-9/11 context. The State and corporate interests' subversive and repressive tolerance of reformist organizations (i.e., regulated freedom), such as the Humane Society of the United States (HSUS), have constructed a false dualism in which certain reforms are tolerated as "good" dissent while direct action is demonized as "bad" (directly paralleling nationalist discourse of good and evil to justify the Bush war on terror). Following past approaches to the suppression and repression of dissent, a social boundary has been erected between what are positioned as acceptable mainstream reform-based organizations and the more radical grassroots ideals of those who directly challenge the State and corporate interests (Gibson, 2010). State attempts to demarcate and conflate differences between radical grassroots and mainstream reformist organizations are obvious forms of wedge politics. Indicative of the efficacy of this approach and building on forms of strategic ignorance, the "terrible new menace" of those exposing and challenging the exploitation of all animals has faced attack from both the Left and the Right (Sorenson, 2011, p. 220). Attacks from the Left are rooted in an inability to escape human chauvinist and speciesist attitudes toward nonhuman animals, "from Kropotkin and Marx, to Bookchin and beyond" (Best, 2009, pp. 190–193).[8,9] The wedge politics have manifested in forms of self-regulation, shaped in part by the politics of fear in the post-9/11 climate, with the rhetoric of radical activism demarcated as terrorism in the Manichean worldview of the Bush Administration (see Debrix & Barder, 2009). Self-censorship is panoptic, in the Foucaultian sense, with opinions withheld, falling along a scale of risk: threat of persecution, being labeled a terrorist, and the social boundary between what are framed normative-democratic speech acts and those marginalized as unacceptable.

Contemporary approaches to the suppression of dissent in post-9/11 Western societies have historical precedent. The politically motivated nature of current McCarthy-esque attempts to criminalize "nearly every form of dissent" can be traced back to the period known as the second "Red Scare" of the 1940s and 1950s, and subsequent covert State apparatuses including the

Federal Bureau of Investigation's (FBI) "actions program" COINTELPRO, which provided the foundations for covert activities and legislation aimed at suppressing certain dissenters (Best, 2004, p. 305). The N30 mass demonstrations in Seattle in November 1999 provided a more recent visible example of increasing attempts to manage and manipulate public opinion in the West. The broadly supported protest actions were organized to coincide with a World Trade Organization (WTO) meeting taking place in the downtown area, and "went beyond demanding change..., rather using protest to delegitimate capital itself" (Gordon, 2009, p. 253). The overtly coercive actions of the police, alongside a number of acts of property damage by protestors, provided numerous images in which those demonstrating were framed as violent, anti-social and *un-American*.

In Seattle, and in the post-9/11 period, police preparedness and responses were positioned as justified, and provided a foundation for future events and actions, based on the constructed image of the threat posed (Churchill, 2001). What 9/11 enabled, and had been sought for some time by those specifically promoting corporate interests, was "frame bridging": the linking of specific ideas behind dissent with a master narrative of the threat of terrorism (Fernandez, 2008; Panitch, 2002). Simply, the potential threat (to the State, cultural hegemony and the status quo) of radical and revolutionary ideas was intentionally equated with perceived threats emergent in feelings of fear and insecurity in the wake of 9/11: revolutionary ideas comprise a threat to freedom, liberty and the American way of life. Building on earlier attempts to position specific activist movements as terrorist-like, the alter-globalization movement became a convenient target for corporate interests, in part given the potential threat posed by a mass reorientation of values and associated actions, and laid the foundations to focus more specifically on ideas that were starting to have an impact.[10] For example, ALF and ELF, considered as at the forefront of the radical environmental movement, are considered *domestic terrorist* organizations by the FBI.[11]

The suppression of certain dissenting voices was and is asymmetrical and inconsistent. The foci continue to be specific individuals, movements and groupings based on ideological constructions of threat to the cultural hegemony of neoliberalism, mobilized within the context of a politics of fear and insecurity. Drawing on the similarities with the Red Scare, the term "Green Scare" has been adopted to describe targeted suppression of radical ecosocial activists (Potter, 2011; Rosenfeld, 2006; Smith, 2008). An awareness of the manifestations and implications of the Red Scare and selected targeting of groups such as the Black Panthers facilitates a greater strategic understanding of the types of attacks being waged currently and a greater ability to effectively respond. This chapter seeks to move beyond an exposure of the visible impacts of (pre-emptive) repression on those considering and offering radical critiques of the State and neoliberalism, highlighting how this has manifested in self-censorship and self-regulation of certain behaviors beyond direct physical intervention from the State.

Chilling Effects: Self-Censorship and Self-Regulation

Attempts to facilitate self-regulation can be traced back to Jeremy Bentham's 18th-century architectural model for a prison, the panopticon, in which those incarcerated can be observed without knowing if they are being observed. The ability to observe as a one-way relationship is specifically designed into the structure of the prison itself. The aim is to foster a form of self-social regulation built on the potential of being surveilled: the idea is that those surveilled would alter their behavior as the odds of being caught—specifically the perception of this, the apparent risks—appear too high. Panoptic, as it is used here, refers to the self-regulation of one's own behavior in what Foucault describes as disciplinary society: societies in which observance and judgment are normalized (Foucault, 1995).[12] The self-regulatory nature of one's own behavior directly locates power as existing in relationships, a pluralistic and decentered conception of power.[13]

Acts of self-regulation are everywhere in society and not inherently negative. We make choices every day, from the micro- to the macro-level, and often against our own self-interest.[14] Visible regulatory mechanisms consist of responses to active suppression through State crackdowns on public dissent: the specific targeting of the promoting of certain ideas, including through targeted legislative change. Individuals self-regulate based on the potential sanctions faced through forms of incarceration and the implications of being labeled in a certain way (i.e., social boundaries). Examples here include the proliferation of the asymmetric use of ideologically and politically orchestrated terms such as *domestic terrorist.* For example, the broad application of the term ecoterrorist, without any accompanying interrogation of ecocidal practices in industrial capitalism, provides a means to facilitate public support for legal prosecution of specific individuals and groups through the mobilization of post-9/11 security discourses.[15] The targeted segmentation of populations, or "social sorting," is illustrated in the wealth of "graduated forms of surveillance" in contemporary society. Whereas segmentation is designed to facilitate differential treatment, potentially including socially constructive ends, post-9/11 security discourse is a clear example of surveillance as governance (Henman, 2004, pp. 174–177). Governance is used here, drawing on Foucault's ideas, as action, as (ways of) acting, relationally: "the invention and assemblage of particular apparatuses and devices for exercising power and intervening on particular problems" (Rose, 1999, p. 19). Not solely the domain of the State, governance exists where there are relations of (political) power, which, in turn, only exist where there are resisting activities. Used as a rhetorical device, the terrorist label associates the promotion of certain ideas as socially and ethically deviant. The term is utilized to segment ideas, without engagement with the sociopolitical basis for the actions of those targeted. As an extension, the implication of this is (State) governance at a distance. Techniques of governance are mobilized by individuals effectively acting, in part, as agents of the State (i.e., self-regulation), in line with the achievement of certain ends (Rose & Miller, 2010, p. 279).

The threat of being labeled in such a way, directly or through association, continues to have an impact on self-censorship. Drawing from Noelle-Neumann's (1974, 1993) influential work on the "spiral of silence," Hayes, Glynn, and Shanahan (2005) distinguish self-censorship, the withholding of one's opinion based on active consideration of normative discourse from those perceived to disagree, with "opinion expression inhibition," a general reticence to express one's opinion publicly. Central to this distinction is the intersection of one's willingness to withhold an opinion, to self-censor and to resist, and how this is differentiated between different people.[16] Self-censorship can be difficult to identify. Research on the spiral of silence continues to grapple with this challenge, often explored in reference to the use of hypothetical situations to gather quantitative data (see Hayes et al., 2005, p. 453). In the wake of 9/11, the crackdown on dissent continues to foster individual and collective self-regulation and self-censorship. Self-regulatory mechanisms are the mechanisms through which governance is manifested (Rose, 1999, p. 17).

The self-questioning of one's (potential) actions is a form of self-censorship (repression) in a pre-emptive sense, paralleling the focus on pre-emption which dominates post-9/11 State rhetoric and discourse. What is important here is more than surveillance. It is the discourse in which this surveillance is situated (i.e., governance). By way of a simple example, we can see this in the fragmentation of the electoral–political "left."[17] There were a number of vocal opponents to military action including the invasions of Iraq and Afghanistan, as well as tentative supporters (some who later spoke out against the actions taken). Along similar lines, some spoke out against legislative changes that diminished or repealed civil liberties. Many others were caught between concern for the use and implications of legislation such as the USA PATRIOT Act, and in not wanting to appear to support actions in which (Western) civilians were targeted. The full title of this legislation—*Uniting and Strengthening America by Providing Appropriate Tools Required to Intercept and Obstruct Terrorism* Act—and the acronym more directly were designed to invoke specific discursive–ideological notions. In the wake of 9/11 and the spectacular mass killing often far removed from everyday life in the West, the "war on terror" entered popular discourse alongside normative notions of what was right and just (i.e., hegemonic notions of freedom and liberty).

What emerges is a perception that to speak out against any military action or laws that infringe on civil liberties, both of which are framed as being about (national) security, has the implication of being positioned as opposing that which is right and just. One example of the former is the notion of saving Muslim women, which was mobilized as a means to justify intervention in Afghanistan. Framed as a (perceived) social good, any critical discussion of the culturally imperialist implications of saving the racialized *other* (i.e., saving someone *from* something requires saving them *to* something, in this example based on notions of *West is best*) was effectively sidelined (see Abu-Lughod, 2002). An example of the latter is the terminology used in an American Civil Liberties Union (ACLU) report which attempted to expose the crackdown of dissent post-9/11: "In separate but related attempts to squelch dissent, the government has attacked the patriotism of its critics…" (emphasis added, Anthony D. Romero, Executive Director, ACLU, in Unknown, 2003, p. i). What is positioned as important is one's *patriotism*, without any engagement of the implications of the term for dissenting voices, specifically how it has been mobilized in a post-9/11 context (i.e., *with us or against us*).

Historical Precedents: The Red Scare and COINTELPRO

There is a direct parallel between the use of the term *domestic terrorism* today with the Red Scare of the 1940s and 1950s. The term *terrorist* has today replaced *communist* as the foremost rhetorical wedge for contemporary Western states.[18] The selective and asymmetric targeting of individuals, groups and groupings indicates that "green" has effectively become the new "red" (Potter, 2011). Publications of the Ayn Rand Institute and right-wing Christian organizations express this clearly:

> Did you ever wonder what happened to the left wing "intelligentsia" following the humiliating collapse of the Soviet Union and its Communist puppet states? Well, they are alive and well, and they are continuing to promote the Communist ideals of state control over resources. The only things that have changed are the terminology they use, and the names of the organizations they belong to. Roll

> over Marx and Lenin! Today's trendy and leftist causes are animal rights and radical environmentalism. (Dave Matheson, quoted in Sorenson, 2011, p. 223)

U.S. Senator Joseph McCarthy is widely considered as being the spearhead and figurehead of the Red Scare, an ideological attack on progressive ideals in which thousands of people in the United States were subjected to intense public scrutiny and paraded before extrajudicial panels and hearings for aggressive questioning of their activities. The most famous of these were the hearings, not directly linked to McCarthy, conducted by the House Committee on Un-American Activities (HUAC) in the late 1950s. During this period, progressive individuals who exposed and challenged normative ideals were labeled as *communists* or *communist sympathizers*, with the implication that they were disloyal, subversive or treasonous (i.e., unpatriotic) in their actions. These politically loaded terms were utilized to position individuals as a threat to what were rhetorically identified as common sense: the positioning of capitalist and bourgeoisie values as socially desirable and normative (Gramsci & Buttigieg, 1992).

The overall intent was to discard any criticism (valid or not) of capitalism and imperialism in the wake of the influence of the Communist Party of the United States of America (CPUSA) and to foster self-regulation based on the threat of surveillance and risk of exposure as being un-American. Policing would be undertaken by the populace in the sense of the social boundary created and by individuals based on perceptions of risk and not wanting to be labeled as unpatriotic. In essence, the status quo and the interests served were rendered unquestioned, with the actions of individuals, as unpatriotic, becoming the focus of public debate.[19] The potential for discussions of overt and more structurally exploitative State actions was sidelined. By structural exploitation, I am referring to the exploitation embedded in the very fabric of a society, where systems, institutions, policies or cultural beliefs can and do meet the needs and rights of some at the expense of others (Schirch, 2004).[20]

Alongside the public actions of Senator McCarthy, emanating from his February 1950 speech in which he referred to a blacklist of *communist sympathizers* working in the State Department, were those of FBI director J. Edgar Hoover. Hoover was a fervent anti-communist and drew on the broad reach of the FBI to implement his agenda. His approach to prosecuting, under the guise of investigating, those accused of being *communists* or *communist sympathizers* included keeping the identity of the accuser secret. Being labeled as a *communist* or a *sympathizer* was a de facto conviction in the public arena. Once labeled, one was guilty until proven otherwise. The striking parallels between the Red Scare and the mass hysteria surrounding the 1692–1693 Salem witch trials were most notably exposed by Arthur Miller, himself a target of Senator McCarthy's campaign, in his 1953 play *The Crucible.* The arbitrary ability to convict based on little or no reputable evidence was considered too limiting for Hoover and led to the establishment of the FBI's covert and at times illegal counterintelligence program known as COINTELPRO.

COINTELPRO, an umbrella for covert actions and other programs targeting domestic groups, was established in the 1950s. Moving beyond the use of the *communist* label to imply individuals were disloyal, subversive or treasonous in their actions, this "actions program" sought to *disrupt* and *neutralize* target groups and individuals (Blackstock, 1976; Hilliard, 2007). Documents from a 1975–1976 United States Senate Select Committee to Study Governmental Operations, chaired by Senator Frank Church (the "Church Committee"), provide details of attempts by the FBI to foster friction between different leftist groups, often seeking to encourage violent acts between them and a subsequent spiral into retributive violence. Individuals and groups considered "subversive" were specifically targeted. Those covertly, and often illegally, surveilled included Martin Luther King Jr., Malcolm X, Fred Hampton and Bill Ayers, alongside groups such as the National Association for the Advancement of Colored People (NAACP), the American Indian Movement (AIM), the Black Panther Party, the Student Nonviolent Coordinating Committee (SNCC) and Students for a Democratic Society (SDS). King and others were targeted based on their potential to "unify and electrify" movements for peace and justice, which were seen to threaten certain ideological and corporate interests. In seeking out such an aim, the FBI program sought to prevent targeted individuals and groups from achieving "respectability" in any societal circles (Bloom & Breines, 2003, pp. 319–324).[21] Some of those targeted, such as Dr King, are seen today as pioneers of justice—including by many who supported their prosecution.

The activities undertaken under the COINTELPRO banner were exposed after the "Citizen's Committee to Expose the FBI" seized over 1000 classified documents from a Pennsylvania field office. Following publication of these documents, exposing the directives and actions undertaken, COINTELPRO officially ceased to exist in 1971. The Church Committee, following a year-long investigation, proposed specific legislation to set limits on FBI surveillance of political activities protected by the First Amendment to the US Constitution. Congress did not pass the legislation. Internal guidelines were established by Attorney General (AG) Edward Levi in 1976, and subsequently watered down by successive AGs (see Chang, 2002, pp. 30–37; Chomsky, 1999). The watering down of the guidelines reflects the continuation and reemergence of tactics adopted under the COINTELPRO banner.

(More Than) Rhetorical Criminalization

A rise in the positioning of grass roots activism as a threat within, drawing on the lessons of COINTELPRO, became broadly visible in the criminalization of dissent surrounding social movements of the mid-late 1990s including Reclaim the Streets Festivals, Carnivals Against Capitalism and what has since become known as the alter-globalization movement. The turning point was the preparation for, and direct response to, the N30 demonstrations in Seattle coinciding with the WTO meeting.[22] The selective mass media portrayals of protestor violence during the Seattle and other demonstrations have been routinely, posthumously, used to justify the actions of the police in the use of chemical and other weaponry (i.e., capsicum spray, tear gas, concussion grenades and rubber bullets), overt physical force and mass arrest. Ward Churchill has clearly identified this in referring to the weapons technologies mobilized by the Seattle Police Department:

> All of a sudden the Police Chief and the Mayor...ran out and bought themselves a SWAT [Special Weapons And Tactics] team, a couple Armored Personal Carriers, a whole inventory of tear gas. Got everybody trained and equipped and coordinated to get out there in the street. That all happened in about 28 minutes.... (Churchill, 2001)

Churchill's sarcastic comments draw attention to the preparedness of the police forces well before the WTO meeting and demonstrations. The preparedness, specifically the possession of such weaponry and the completion of training required for their use, contrasts directly with circularly justified arguments of such weaponry as being necessary as a result of the protest.

The framing of protests against the World Economic Forum (WEF) in Melbourne, Australia, on September 11, 2000, a year after the N30 events in Seattle and days prior to the international spectacle of the Summer Olympic Games in Sydney, enabled the passing of specific legislation. The *Defence Legislation Amendment (Aid to Civilian Authorities) Bill 2000* was introduced three months prior with the stated and very broad purpose of "establishing a regime for the use of Defence Forces to protect the States and self-governing Territories and Commonwealth interests from 'domestic violence,' expanding upon a more limited existing regime in the *Defence Act 1903*." Selectively citing an act of political violence more than 20 years prior, the amendments reduced restrictions on the deployment of the Australian Military domestically and removed the need to consult with State government requests. The Bill explicitly provided increased powers of search, seizure and detention without a warrant or formal arrest, including the use of deadly force against civilians.[23] Of specific note was the absence of the terminology of terrorism, in a pre-9/11 context.

Post-9/11 it is the potential, constructed or otherwise, of perceived "threats" to the State-capitalist order that justify the mobilization of large numbers of police and anti-personnel weapons against civilians. The pre-emption of the *Defence Legislation Amendment (Aid to Civilian Authorities) Bill* has effectively become the standard response of capitalist states around the world, with circular reasoning mobilized to justify such an approach—*ex post facto*.[24] Such an approach was adopted during the June 2010 G20 March for Justice in Toronto, Canada, with the juxtaposition of a handful of people smashing windows and damaging vehicles in Toronto to the actions of the 10,000-strong police contingent, which included arbitrary kettling (also known as corralling) of anyone on the street into a confined space surrounded by armed police for hours (including those in the designated "protest zone"), beatings, snatch squads and mass arrests.[25] The passing of laws (including misinformation) designed to restrict civil liberties and mass mobilization of State resources from several provinces, with *security* cost estimates of over $1 billion prior to the scheduled meetings and mass demonstrations, required political justification (Freeze, 2010). Some have argued *agents provocateur* were used to incite acts of violence as a precursor to the police state tactics used during the protest and to legitimize the pre-emptive actions of the State.[26] Pre-emptive action (arrests, curtails of civil liberties) prior to summit protests acutely resemble the tactics mobilized in the "war on terror" (see Fernandez, 2008, p. 149).

Constructing the Green Menace

The US *Animal Enterprise Protection Act* (AEPA) was passed in 1992. The Bill created the crime of animal enterprise terrorism, seeking to label those who acted under the banner of the ALF as *terrorists* (Black & Black, 2004). Established in England in the 1970s, ALF is based on anarchist and anarcha-feminist ideals of decentralization, without leadership, and comprised of autonomous and anonymous collectives, or cells (Jones, 2004).[27] Actions that fall within the guidelines of ALF abide by principles including: the liberation of animals, the exposure and infliction of economic damage on exploitative industries and the operational dedication to do no harm.

The constructed need for the legislation did not materialize into prosecutions. Soon after its passage, front groups who had lobbied for its passing sought to expand its scope and penalties (see Potter, 2011, pp. 122–124). The events of 9/11 were specifically seized upon, seeking to utilize the emergent and promoted fear to serve neoconservative interests. A 24-page, 2003 report prepared by the American Legislative Exchange Council (ALEC), titled *Animal & Ecological Terrorism in America*, specifically positioned "radical" ecological–environmental and animal rights activists alongside "other terrorist groups like al-Qaeda" (*Animal & Ecological Terrorism in America*, 2003, p. 4). The report also suggested the roots of such "domestic terrorism" emerged and "migrated from the personal quarters and inquisitive considerations of collegiate academia…[who] are hell-bent on revolutionizing a system of perceived abuse into one that abides by deeply rooted philosophies of fundamental animal equity and environmental preservation" (*Animal & Ecological Terrorism in America*, 2003, p. 5). This shift from an anthropocentric and human chauvinist notion of welfare toward one of "rights" for other species is specifically considered a threat to the corporate interests ALEC was founded to promote and protect. The roots of changing perceptions away from a Descartian notion of animals as machines were identified as founded in Darwin's (1871) *The Descent of Man* (incorrectly cited in the ALEC report as published in 1859). A very selective chronological history of legislation seeking to reduce the suffering of animals exploited for human use and protect endangered species, academic debates and publications, alongside specific actions of animal rescues and property damage, is presented as a timeline of "sustained struggle" to support the claimed need for a crackdown.

The ALEC report broadened the threat to include ELF, which emerged in the early 1990s and is modeled on the decentralized and leaderless ALF, as another source of domestic terrorism. The report links formal registered organizations such as People for the Ethical Treatment of Animals (PETA) with ALF. The clear intent was to drive a wedge between more mainstream and broadly supported organizations (the good) and radical grassroots activists (the bad), seeking to ferment disagreement on one level, and movement-splintering on another. What we can locate here is an attempt to manage dissent, to regulate freedom in the sense of entertaining certain dissent (i.e., reformist ideas) and not others (radical and revolutionary thought) through the use of the rhetoric of terrorism, specifically, the management of (acceptable) ideas in a post-9/11 context as an element of the complex assemblage through which governance is mobilized. A specific definition of terrorism was developed to represent a particular reality and facilitate the promotion of certain interests (Rose, 1999, p. 280). These interests were then directly tied to precise notions (i.e., patriotism) in seeking to align public choices (in the way of actions) with State and corporate interests (Rose, 1999, p. 286). There are direct parallels to the COINTELPRO approaches in seeking to undermine the efficacy and effectiveness of solidarity among groups. The dualistic construction of good/bad forms of dissent, the construction of social boundaries, intentionally positions those who challenge State and corporate interests as separate from "acceptable" reformist organizations. Such boundary work comes with incentives and risks for the more mainstream groups to distance themselves, based on perceptions of self-protection, from the more radical groups (i.e., those labeled as *terrorist* or *terrorist-like*) (see Gibson, 2010, p. 10).

In providing a selective history of the ALF and ELF, constructing a path between Darwin's research and what is framed as the inevitable violent turn of activists, the ALEC report explicitly and dishonestly neglects to mention the principle of *do no harm*, a cornerstone of activities that fall under ALF and ELF banners. Building on this selective misrepresentation, it is implied that "cutting throats" of anyone who exploits nonhuman animal and the environment is the logical evolution of actions taken, directly constructing a (false) *terrorist-like* menace in the post-9/11 context. There is no mention that not a single person has been injured in an action attributed to an ALF or ELF underground cell. ALF and ELF target property used to directly facilitate the exploitation of the environment and nonhuman animals, which is a direct threat to corporate agribusiness, pharmaceutical and related industries. The aim is to create economic costs for continued exploitation of nonhuman animals, not to physically harm any being.

The (repressive) tolerance mobilized in positioning reform-based ideals as part of a democratic society (see Kahn, 2006, pp. 397–398), with more radical critiques seeking revolutionary change positioned outside acceptable social boundaries, has produced some of the outcomes desired by State and corporate interests. For example, mainstream groups such as the HSUS have gone beyond self-censorship, explicitly speaking out against grassroots activism and seeking to distance themselves from being labeled as a supporter of anyone positioned as domestic *terrorists* by neoconservative groups such as the ALEC. We could consider this example as the pinnacle manifestation of State governance facilitating "action at a distance" (Rose & Miller, 2010, p. 278). The foreseen sanctions (i.e., being publicly associated with those labeled *terrorist-like*) were enough to facilitate the bearing of a form of regulated freedom. A broader, and intentional, implication has been individuals and groups—including those promoting veganism and an end to all exploitation—becoming unwilling to be seen associating with those surveilled or investigated under The USA PATRIOT Act, AETA and other provisions (Potter, 2011, pp. 198–199). Here we see the manifestation of the overall intent of self-regulatory governmentality. The perception is that risks are too high, and those previously willing to speak out against the State and corporate interests shift toward Foucault's docile bodies, of citizens being molded into a pliable form "that may be subjected, used, transformed, and improved" (Foucault, 1995, pp. 135–136). Such pliability is evidenced in the self-censorship of actions in line with the interests of the neoliberal State: in this instance, of being unwilling to be seen as associating with those positioned as outside social boundaries of acceptable State critique and forms of democratic dissent. The mobilization of political power, in which such regulation is policed among the populace, is a key feature of the contemporary attempts to manage, suppress and pre-emptively repress dissent (Rose & Miller, 2010, p. 272).

Living and Resisting in a Repressive Society

Will Potter's detailed journalistic account of the persecution of grassroots activists under the banner of *domestic terrorism* provides a clear indication of the self-regulatory effects of the Green Scare. He was visited by the FBI in 2003, after being associated with activists being monitored. After explicit threats were made (including being added to the *domestic terrorists watch list*), he was reflexively aware of the implications of this visit, of how he would be perceived by others at this workplace, the newsroom of the *Chicago Tribune*. In 2006, he was invited to give a presentation to a *Congress Subcommittee on Crime, Terrorism and Homeland Security* hearing on the AETA. After initially seeing the invitation as recognition of his investigative journalism, Potter began to question the implications of testifying, of his describing the AETA as reminiscent of the Red Scare, with *terrorist* replacing *communist* as the most powerful political term: "Would I be smeared as an 'animal rights terrorist'?" (Potter, 2011, p. 117). The fear instilled by the initial FBI visit, the rhetoric and discourse surrounding the PATRIOT Act, and the politics of fear mobilized in a post-9/11 environment to serve specific ends had stayed with him. The intended implications are clear. Not only would he question his actions, his decisions would also be shaped (disciplined) by the memory of the FBI visit and the threats, both actual and perceived. His willingness to report on radical activities that challenge the State and corporate interests under the banner of journalistic freedom was shaken.

The implications were clearer in the actions of HSUS. Whereas there was an awareness of the erosion of civil liberties (HSUS had privately raised concerns about the content of the AETA and how this would impact on its own work), the organization declined to publicly present their concerns as they did not want to be seen as opposing a bill about terrorism. This is governance in the sense that the actions of HSUS (and others, as sought) were "reshaped…in a space of regulated freedom" (Rose, 1999, p. 22). HSUS waited until after the bill passed before sending out an e-mail to supporters indicating some concerns. In this way the organization could safely express concerns about the implications of the AETA without being seen to oppose passage of the Bill—or be associated with groups and individuals labeled as *terrorists* or supporters of *terrorist* organizations. The concerns raised by HSUS were explicitly juxtaposed with condemnations of direct action tactics, and unidentified individuals and groups, by adopting the same ideological terminology and rhetoric mobilized by front groups such as ALEC (Rimmer, 2006). HSUS acted as a docile body in the Foucaultian sense, and in a tolerable way, by not rocking the boat (too hard). The broad implication was to assist in the frame-bridging engaged in by State and corporate interests: the linking of certain ideas behind dissent within a master narrative of the threat of terrorism.

The actions of HSUS and Will Potter's reflections are indicative of the broader implications of pre-emptive repression mobilized in the wake of the events of 9/11. The politics of fear, (in)security and the proliferation of "an aggressive right-wing patriotic correctness" continue to foster self-censorship and self-regulation in the interest of the State (Giroux, 2010, p. 661). In reflecting on the manifestations of contemporary suppression of dissent, we can draw from historical examples such as the Red Scare and COINTELPRO eras, and note that current approaches are not as far-reaching. The perceived and real threats of militant groups that target human life are quite different from those of the Cold War period.

Constructing a pretense of fear requires sustained ideological and politic rhetoric to ferment insecurity and mobilize signifiers of a specific, constructed, patriotic identity. We have already witnessed a litany of distortions utilized to justify the "war on terror," the crackdown on civil liberties and the asymmetrical targeting of radical ideas that challenge State and corporate interests. In such contexts, there is also potential for radical and revolutionary change. Delegitimation, alongside direct action and networking, is a key element of anarchist praxis. Revolutionary and radical struggle pose a real *threat* to State and corporate interests. For example, actions that fall under the banner of ALF (and ELF) directly challenge the legitimacy of capitalism, in that the property status of animals is rejected (rather than a focus on the treatment, such as those of reform-based organizations).[28] This is how the State, in line with corporate lobby groups pushing for the criminalization of such dissent, has set out to "defend animal capital" through sustained and pre-emptive approaches to repression (Sanbonmatsu, 2011, p. 26).

It is in this context of renewed attempts to repress and suppress that the potential for challenge is also visible. At times, those seeking to manipulate discourse and debate (not always intentionally) show their hands. This can take the form of the extent of political donations being directly linked to policy and more overt statements (see Berry, 2011). While seeking the Republican nomination for the 2012 US Presidential Election, Newt Gingrich indicated the ideological aims of a backlash on critical pedagogy in reference to the attack on outspoken academic Ward Churchill: "We are going to nail this guy and send the dominos tumbling. And everybody who has an opinion out there and entire disciplines like ethnic studies and women's studies and cultural studies and queer studies that we don't like won't be there anymore" (cited in Giroux, 2010, p. 102). Critical Animal Studies, Peace Studies and other disciplines that critically challenge neoliberal ideas and ideology are similarly targeted.[29]

Being aware and prepared for the overt, alongside the more subtle, approaches of the State and their implications is a foundational element of effectively living and resisting in a repressive society. Another effective approach is to build and maintain strong networks (Martin, 2005). There are numerous examples of effective resistance during the COINTELPRO era.[30] The maintenance of strong networks undermines attempts to foment differences and disagreements, such as the (at times successful) targeted, murderous attacks on the Black Panthers and others including MOVE. Such linkages are potentially more possible today, given the rise of new media technologies. This does not mean that there should be complete agreement across the spectrum of ideas. Rather, attempts to pit individuals, groupings and organizations against each other, to link radical and revolutionary direct action with *terrorism*, should be seen as a fundamental tool utilized to reduce the effectiveness of dissent and facilitate suppression.

In the wake of increasing repression and attempts to stifle dissent, there are also positives. Will Potter chose not to hold back at the Congressional Hearing. Some of those prosecuted under the AETA have not succumbed to legal and extrajudicial threat, refusing to cooperate and testify against others (Potter, 2011, pp. 197–198). Some have chosen not to appear before Grand Juries. Many have made direct criticisms of the mass mobilization of police, such as those during the 2010 G20 Summit in Toronto. Such actions indicate that repression can be resisted in a number of different ways. Politically expedient reincarnations including the Green Scare can be challenged and delegitimized. Resisting and actively undermining the rhetoric of terrorism, the associating and framing of dissent as *terrorist-like*, limits the ability of the State and corporate interests to label activists in such ways, enabling dissent and the renewal of revolutionary efforts aimed at moving society toward an existence free from multiple fronts of systemic oppression: a society embracing total liberation and a true liberatory politics. Such actions are an essential element of living and resisting, justly, in a repressive society.

Notes

1. The term *domestic terrorism* was defined by the FBI's Terrorist Research and Analytical Center as early as 1994 and was widely used in a 1996 report on domestic incidents (Unknown, 1996). The PATRIOT Act "created a new definition of 'domestic terrorism,' in order to correspond to the existing definition of 'international terrorism'" (Unknown, 2004, p. 31). The "ecoterrorist" label is accredited to Ron Arnold, used in a 1983 article in Reason magazine. Arnold has held the position of vice president of Center for the Defense of Free Enterprise (CDFE) since 1984. CDFE is a right-wing think tank which pioneered the "wise use" movement and has attacked the environmental movement since its inception in 1974. In June 1998, a Congressional subcommittee was convened under the title of *Acts of Ecoterrorism by Radical Environmental Organizations*, in which Ron Arnold testified (see Smith, 2008).
2. Drawing from Rose and Miller, engagement with political discourse enables illumination of "systems of thought," and "systems of action" through which specific ideas of *reality* are mobilized, reinforced and perpetuated (Rose & Miller, 2010, pp. 275–279). Rose also refers to language constitutive of governance, making it *possible* (Rose, 1999, p. 29). See also Bleiker (2000, p. 35), for a discussion of agency, discourse and dissent.
3. Rose marks a distinction between "freedom as a formula of resistance from freedom as a formula of power." The former being that "deployed in contestation" and the latter defined/represented by the State (Rose, 1999, pp. 65, 96).
4. I am not implying overall effectiveness here. Dissent is a very clear locale of contention: what form (and content) of dissent is considered acceptable locates the exercise of power.
5. See Brown (2006) for an exploration of offensive uses of tolerance, which incorporate dissent, such as that (partly) illustrated in the actions of HSUS referred to herein.
6. See also Rosebraugh (2004) and Pickering (2002).
7. See also Best (2004).
8. Use of the term "human chauvinism" here refers to the groundbreaking ecological philosophy of Richard Routley (later Sylvan) and Val Routley (later Plumwood) (Routley, 1973; Routley & Routley, 1979). Simply, human chauvinism labels and identifies the socially and politically constructed notion of human separateness and superiority to the natural world. In many ways, the left has adopted a neo-Cartesian view, or what Best describes as a "Cartesian-Marxist mechanistic view of [nonhuman] animals" to rationalize continued nonconsideration. He goes further to describe "leftist theory and practice [a]s merely *Stalinism towards [nonhuman] animals*" (Best, 2009, p. 193). See also *Critical Theory and Animal Liberation* (2011), particularly chapters by Boggs and Benton. The latter directly explores the animal question in Marx's writings.

9. As early as 1906, direct linkages between the exploitation of workers and nonhuman animals, from a left perspective, were made. See Upton Sinclair's (2003) *The Jungle*, a portrayal of the life of immigrant workers in the United States, through the example of the Chicago Stockyards.
10. See Chalk, Hoffman, Reville, and Kapsuki for a predication that the alter-globalization movement would embrace radical social–environmental ideas potentially leading "to the emergence of a new radical left-wing fringe" (2005, p. 51).
11. A ski resort at Veil, Colorado, was destroyed by fire in October 1998, causing $12M damage. A communiqué released claimed responsibility by the ELF. A photograph of the fire adorned the cover of the joint U.S. Department of Justice and FBI *Terrorism in the United States 1998* report (Unknown, 1998). Photograph of another action claimed by the ELF was on the cover of the 1999 report. The 1996 report included specific reference to actions of the ALF, with the 2000–2001 report referring to the "first recorded ALF attack" taking place in April 1987 (Unknown, 1996, 2004).
12. See Simon (2005) for a broader reflection on the Foucault's concept of panopticism post-9/11.
13. Gene Sharp's (1973) work on nonviolent resistance provides an interesting exploration of power relations in the context of dissent.
14. Foucault explores responses to the plague as an early example of the discourse of social management, from which some positive self-regulatory measures, in the sense of sanitary practices, emerged.
15. Slavoj Žižek addressed the asymmetrical use of the term *terrorist* to describe Julian Assange and Wikileaks, in that the idea behind Wikileaks—the shift in power relations it facilitates, is a threat to the State (and corporate interests) in the way Gandhi was to the British Empire, and as such could be described as a *terrorist.* If this description was to be accepted, it would require an open and full embrace that the State routinely acts in a *terrorist* manner ("Julian Assange in Conversation with Slavoj Žižek Moderated by Democracy Now!'s Amy Goodman," 2011). The recording can be viewed online at http://wlcentral.org/node/1976
16. Class and social standing play specific roles in shaping self-censorship and willingness to speak out (which is very different to opinion expression inhibition), shaping differentiation beyond that explored by Hayes et al. (2005). The routinely nonconsidered role of strategic resisting practices, also differentiated by class and social standing, may be misinterpreted as self-censorship in a negative sense, or as opinion expression inhibition (see Hoagland, 2007).
17. The left also intentionally fragments itself, at times based on left ideologies. Women's equality was considered a secondary issue for some time, in much the same way as the exploitation of animals is today. Addressing these issues was and is inconvenient and threatens vested interests (Sorenson, 2011, p. 232). For the former, it was men's roles and the benefits afforded them by patriarchy. For the latter, it is strategic ignorance of one's own complicity at the whim of certain desires (*Race and Epistemologies of Ignorance*, 2007).
18. Whereas use of the term *domestic terrorist* is mobilized in similar ways to that of *communist*, it is important to note that its application is not as far-reaching.
19. See Ayres (2004) for a broader discussion of the centrality of meaning, the constructing of frames, to dissent.

20. Following Brian Martin, I use the term "structural exploitation" in place of Johan Galtung's notion of structural violence: "The main problem with the expression 'structural violence' is that it adds an enormous burden onto the term violence. Most people think of violence as direct physical violence. For much communication, terms such as exploitation and oppression may be clearer than 'structural violence'" (Martin, 1993, p. 43).
21. Hoover was also involved in enabling the violence perpetrated on the *Freedom Riders* in the early 1960s (see *Freedom Riders*, 2009).
22. The actions, and the police response, received widespread international mainstream media coverage, in substantive part based on the emergence of the Indymedia model of open source citizen journalism (see Miekle, 2002).
23. The new powers given to the military exceeded police powers and included the right to: shoot to kill "where necessary" without fear of prosecution; detain people without a formal arrest or charge; and seize and search persons, places, vehicles or personal belongings without a warrant.
24. See, for example, David Carlin describing that he need not provide proof of activists being *dangerous*, rather a feeling that their intent is (quoted in Sorenson, 2011, pp. 221–222).
25. These actions were mirrored in the response to student protests in opposition to funding for public education and other "austerity" measures in the UK in early 2011.
26. The use of agents provocateur is not without precedent in Canada. For example, Dave Coles, president of the Communications, Energy and Paperworkers Union, noted three undercover police officers inciting violence at the Security and Prosperity Partnership Summit in Montebello ("Quebec Police Admit They Went Undercover at Montebello Protest," 2007). This incident received mainstream press coverage as it was captured on video and uploaded to the internet, forcing a formal admission by Quebec Police (see Canadians Nanaimo, 2007).
27. Jones (2004) and Kheel (2006) engage with the need to reflect on the appeal and implications of the *heroic ideal* seen in some ALF actions, including the potential attractiveness of macho posturing, for young men seeking out destructive behavior. Kahn (2005) has similarly noted "a risk of (the ALF and ELF) devolving into both a sort of vanguard elitism and despondent nihilism without a stronger theoretical basis."
28. Gary Francione's (2000) criticism of animal *use* centers on the property status of animals.
29. We need not look further than the inane ramblings of David Horowitz for a wealth of examples.
30. The Black Panther Party initiated a number of social justice programs that continue (in some form) today, including the Free Breakfast for School Children Program. For an insider's broader perspective, see Acoli (1995).

References

Abu-Lughod, L. (2002). Do Muslim women really need saving? Anthropological reflections on cultural relativism and its others. *American Anthropologist*, *104*(3), 783–790.

Acoli, S. (1995). *A brief history of the Black Panther Party. Its place in the black liberation movement.* Sundiata Acoli Freedom Campaign. (Original work published April 2, 1985). Retrieved from http://www.hartford-hwp.com/archives/45a/004.html

Animal & Ecological Terrorism in America. (2003, September 1). Washington, DC: American Legislative Exchange Council. Order # 0329.

Ayres, J. M. (2004). Framing collective action against neoliberalism. *Journal of World Systems Research, 10*(1 Special Issue: Global Social Movements Before and After 9–11), 11–34.

Benton, T. (2011). Humanism = speciesism? Marx on humans and animals. In J. Sanbonmatsu (Ed.), *Critical theory and animal liberation* (pp. 99–135). Lanham, MA: Rowman & Littlefield.

Berry, P. (2011, May 27). Tobacco donations a risky business: AMA. *Ninemsn.* Retrieved from http://news.ninemsn.com.au/national/8254275/ashtray-award-to-arrive-in-plain-packaging

Best, S. (2004). It's war! The escalating battle between activists and the corporate-state complex. In S. Best & A. J. Nocella II (Eds.), *Terrorists or freedom fighters: Reflections on the liberation of animals* (pp. 300–339). New York, NY: Lantern Books.

Best, S. (2009). Rethinking revolution: Total liberation, alliance politics, and a prolegomena to resistance movements in the twenty-first century. In R. Amster, A. DeLeon, L. A. Fernandez, A. J. Nocella II, & D. Shannon (Eds.), *Contemporary anarchist studies: An introductory anthology of anarchy in the academy* (pp. 189–199). New York, NY: Routledge.

Black, J., & Black, J. (2004). The rhetorical "terrorist": Implications of the USA patriot act on animal liberation. In S. Best & A. J. Nocella II (Eds.), *Terrorists or freedom fighters: Reflections on the liberation of animals* (pp. 300–339). New York, NY: Lantern Books.

Blackstock, N. (1976). *COINTELPRO: The FBI's secret war on political freedom.* New York, NY: Vintage Books.

Bleiker, R. (2000). *Popular dissent, human agency, and global politics.* Cambridge: Cambridge University Press.

Bloom, A., & Breines, W. (2003). *"Takin' it to the streets": A sixties reader.* New York, NY: Oxford University Press.

Boggs, C. (2011). Corporate power, ecological crisis and animal rights. In J. Sanbonmatsu (Ed.), *Critical theory and animal liberation* (pp. 71–96). Lanham, MA: Rowman & Littlefield.

Brown, W. (2006). *Regulating aversion: Tolerance in the age of identity and empire.* Princeton, NJ: Princeton University Press.

Canadians Nanaimo. (2007). *Police provocateurs stopped by union leader at anti SPP protest.* Retrieved from http://www.youtube.com/watch?v=St1-WTc1kow

Chalk, P., Hoffman, B., Reville, R., & Kapsuki, A. B. (2005). *Trends in terrorism: Threats to the United States and the future of the terrorism risk insurance act.* Santa Monica, CA: RAND Corporation. Retrieved from http://www.rand.org/pubs/monographs/2005/RAND_MG393.pdf

Chang, N. (2002). *Silencing political dissent: How post-September 11 anti-terrorism measures threaten our civil liberties.* Open Media.

Chomsky, N. (1999). Domestic terrorism: Notes on the state system of oppression. *New Political Science, 21*(3), 303–324.

Churchill, W. (2001, November 16). *Ward Churchill—Pacifism and pathology in the American left* [CD]. AK Press.

Darwin, C. (1871). *The descent of man, and selection in relation to sex.* London: John Murray.

Debrix, F., & Barder, A. D. (2009). Nothing to fear but fear: Governmentality and the biopolitical production of terror. *International Political Sociology, 3*, 398–413.

Fernandez, L. (2008). *Policing dissent: Social control and the anti-globalization movement.* Piscataway, NJ: Rutgers University Press.

Foucault, M. (1995). *Discipline and punish: The birth of the prison.* New York, NY: Vintage Books.

Francione, G. L. (2000). *Introduction to animal rights: Your child or the dog?* Philadelphia, PA: Temple University Press.

Freedom Riders. (2009). [Videocassette]. American Experience.

Freeze, C. (2010, May 28). Billion-Dollar G20 security cost not a "blank cheque," security czar argues. *The globe and mail.* Retrieved from http://www.theglobeandmail.com/news/world/g8-g20/article1585355/

Gibson, A. (2010). State-Led social boundary change: Transnational environmental activism, "ecoterrorism" and September 11. In T. Oleson (Ed.), *Power and transnational activism* (pp. 130-150). London: Routledge.

Gillham, P., & Edwards, B. (2003). Global justice protesters respond to September 11th terrorist attacks: The impact of international disaster on demonstrations in Washington, DC. In J. L. Monday (Ed.), *Beyond September 11: An account of post-disaster research* (pp. 483-520). Boulder, CO: Natural Hazards Research and Applications Information Centre.

Giroux, H. (2010). Higher education after September 11th: The crisis of academic freedom and democracy. In A. J. Nocella II, S. Best, & P. McLaren (Eds.), *Academic repression: Reflections from the academic industrial complex* (pp. 92–111). Oakland, CA: AK Press.

Gordon, U. (2009). Dark tidings: Anarchist politics in the age of collapse. In R. Amster, A. DeLeon, L. A. Fernandez, A. J. Nocella II, & D. Shannon (Eds.), *Contemporary anarchist studies: An introductory anthology of anarchy in the academy* (pp. 249–258). New York, NY: Routledge.

Gramsci, A., & Buttigieg, J. A. (1992). *Prison notebooks.* New York, NY: Columbia University Press.

Hayes, A. F., Carroll, F. G., & Shanahan, J. (2005). Validating the willingness to self-censor scale: Individual differences in the effect of the climate of opinion on opinion expression. *International Journal of Public Opinion Research, 17*(4), 443–445.

Henman, P. (2004). Targeted! *International Sociology, 19*(2), 173.

Hilliard, D. (2007). *The Black Panther: Intercommunal news service.* New York, NY: Atria Books.

Hoagland, S. (2007). Denying relationality: Epistemology and ethics of ignorance. In S. Sullivan & N. Tuana (Eds.). *Race and epistemologies of ignorance* (pp. 95–118). Albany, NY: State University of New York Press.

Jones, P. (2004). Mothers with monkeywrenches: Feminist imperatives and the ALF. In S. Best & A. J. Nocella II (Eds.), *Terrorists or freedom fighters: Reflections on the liberation of animals* (pp. 137-156). New York, NY: Lantern Books.

Julian Assange in Conversation with Slavoj Žižek Moderated by Democracy Now!'s Amy Goodman. (2011, July 2). Retrieved from http://www.viddler.com/explore/front lineclub/videos/536/

Kahn, R. (2005). From Herbert Marcuse to the Earth Liberation Front: Considerations for revolutionary ecopedagogy. *Green Theory & Praxis: The Journal of Ecopedagogy, 1*(1), 77-99.

Kahn, R. (2006). Radical ecology, repressive tolerance, and zoöcide. In S. Best & A. J. Nocella II (Eds.), *Igniting a revolution: Voices in defense of the earth* (pp. 394–403). Oakland, CA: AK Press.

Kheel, M. (2006). Direct action and the heroic ideal: An ecofeminist critique. In S. Best & A. J. Nocella II (Eds.), *Igniting a revolution: Voices in defense of the earth* (pp. 306–318). Oakland, CA: AK Press.

Lewis, J. (2005, May 18). *Statement of John Lewis, Deputy Assistant Director, Federal Bureau of Investigation.* Oversight on eco-terrorism specifically examining the Earth Liberation Front ("ELF") and the Animal Liberation Front ("ALF"). U.S. Senate Committee on Environment & Public Works. Retrieved from http://epw.senate.gov/hearing_statements.cfm?id=237817

Martin, B. (1993). *Social defense, social change.* London: Freedom Press.

Martin, B. (2005, November 30). *Get ready for repression: Living and resisting in a repressive society.* Retrieved from http: //www.bmartin.cc/dissent/documents/rr/get ready.pdf

Miekle, G. (2002). *Future active: Media activism and the internet.* Annandale, NSW: Pluto Press.

Miller, A. (1953). *The crucible: A play in four acts.* New York, NY: Viking Press.

Nabers, D. (2009). Filling the void of meaning: Identity construction n U.S. foreign policy after September 11, 2001. *Foreign Policy Analysis, 5*, 191–214.

Noelle-Neumann, E. (1974). The spiral of silence: A theory of public opinion. *Journal of Communication, 24*(2), 43–51.

Noelle-Neumann, E. (1993). *The spiral of silence: Public opinion—Our social skin* (2nd ed.). Chicago, IL: University of Chicago Press.

Panitch, L. (2002, September). Violence as a tool of order and change the war on terrorism and the antiglobalization movement. *Options Politiques*, 40–44.

Pickering, L. J. (2002). *The Earth Liberation Front: 1997–2002.* South Wales, NY: Arissa Media Group.

Potter, W. (2011). *Green is the new red: An insider's account of a social movement under siege.* San Francisco, CA: City Lights Books.

Quebec Police Admit They Went Undercover at Montebello Protest. (2007, August 23). Retrieved from http://www.cbc.ca/news/canada/story/2007/08/23/police-montebello.html

Rimmer, T. (2006, November 19). Act passes but bill still stalled: Animal agency not happy with vote. *Associated content.* Yahoo. Retrieved from http://www.associatedcontent.com/article/89168/act_passes_but_bill_still_stalled.html?cat=9

Rose, N., & Miller, P. (2010). Political power beyond the state: Problematics of government. 1992. *The British Journal of Sociology*, *61*(Suppl. 1), 271–303.

Rose, N. S. (1999). *Powers of freedom: Reframing political thought.* Cambridge & New York, NY: Cambridge University Press.

Rosebraugh, C. (2004). *Burning rage of a dying planet: Speaking for the Earth Liberation Front.* New York, NY: Lantern Books.

Rosenfeld, B. (2006, March 10). The great green scare and the fed's "case" against Rod Coronado. *Counterpunch.* Retrieved from http://counterpunch.org/rosenfeld03102006.html

Routley, R. (1973). Is there a need for a new, an environmental ethic? In Organising Committee (Ed.), *Proceedings of the XII world congress of philosophy* (pp. 205-210). Varna: Sophia Press.

Routley, R., & Routley, V. (1979). Against the inevitability of human chauvinism. In K. E. Goodpaster & K. M. Sayre (Eds.), *Ethics and problems of the 21st century* (pp. 36-59). Notre Dame, IN: University of Notre Dame Press.

Sanbonmatsu, J. (2011). Introduction. In J. Sanbonmatsu (Ed.), *Critical theory and animal liberation* (pp. 1–32). Lanham, MA: Rowman & Littlefield.

Schirch, L. (2004). *The little book of strategic peacebuilding.* Intercourse, PA: Good Books.

Schneiderman, D., & Cossman, B. (2001). Political association and the anti-terrorism bill. In R. J. Daniels, P. Macklem, & K. Roach (Eds.), *The security of freedom: Essays on Canada's anti-terrorism bill* (pp. 172–194). Toronto, ON: University of Toronto Press.

Sharp, G. (1973). *Power and struggle.* Boston, MA: P. Sargent.

Simon, B. (2005). The return of panopticism: Supervision, subjugation and the new surveillance. *Surveillance & Society*, *3*(1), 1–20.

Sinclair, U. (2003). *The jungle.* Tucson, AZ: See Sharp Press. (Original work published February 28, 1906)

Smith, R. K. (2008). "Ecoterrorism"?: A critical analysis of the vilification of radical environmental activists as terrorists. *Environmental Law*, *38*, 537–576.

Sorenson, J. (2011). Constructing extremists, rejecting compassion: Ideological attacks on animal advocacy from left and right. In J. Sanbonmatsu (Ed.), *Critical theory and animal liberation* (pp. 219–247). Lanham, MA: Rowman & Littlefield.

Unknown. (1996). *Terrorism in the United States 1996.* Washington, DC: Counterterrorism Threat Assessment and Warning Unit, National Security Division, Federal Bureau of Investigation.

Unknown. (1998). *Terrorism in the United States 1998.* Washington, DC: Counterterrorism Threat Assessment and Warning Unit, National Security Division, U.S. Department of Justice/Federal Bureau of Investigation.

Unknown. (2003). *Freedom under fire: Dissent in post-9/11 America.* New York, NY: American Civil Liberties Union.

Unknown. (2004). *Terrorism 2000/2001.* Washington, DC: U.S. Department of Justice/Federal Bureau of Investigation.

6. The Myth of "Animal Rights Terrorism"

John Sorenson

Addressing Serious Issues

The prevention of terrorism is a serious matter, especially if we look at the original use of the term to describe violence conducted by states against civilians, as in the Reign of Terror during the French Revolution. There is no shortage of significant examples from recent times: aerial bombing of cities and civilians in World War II, notably the use of atomic bombs on Hiroshima and Nagasaki; the US campaign against Cuba from 1959 to the present, US-backed coups in Guatemala in 1954, Brazil in 1964, Chile in 1973; and support for dictators and death squads throughout Latin America, US efforts to depose the Sandinista government in Nicaragua, including mobilization of a proxy army, assistance with anti-Communist massacres in Indonesia in 1965 and support for other murderous regimes elsewhere. If we limit the scope to terrorist activities by non-state actors and focus on those committed by groups opposed to the foreign policies of Western states, we still find many serious crimes from recent years: the July 2010 bombings at the Kyadondo Rugby Club and the Ethiopian Village restaurant in Kampala, the 2002 and 2005 Bali bombings, the 2004 Madrid train bombings, the 2001 World Trade Center bombing, all attributed to Islamist groups such as al-Qaeda, Jemaah Islamiyah and al Shabab. In all of these cases, we see that terrorism involves the murder of innocent victims. Indeed, Chalecki points out that "the violent death of unsuspecting people" in events such as hijacked airliners crashing into the World Trade Center is what comes to most people's minds in association with the word "terrorism" (2001, p. 3). Similarly, Schmid cites the communicative aspect of murderous violence as the central aspect of terrorism (2005, p. 138). The US Department of State's Office of the Coordinator for Counterterrorism emphasizes "life threatening attacks" in its definition of terrorism as "politically-motivated violence perpetrated against noncombatant targets by subnational groups or clandestine agents" (NCC, 2007). However, if we turn from these serious cases to the topic at hand, concerning "animal rights terrorism," we find nothing comparable. Indeed, I suggest that "animal rights terrorism" is the creation of industries that profit from the exploitation of animals.

Animal Exploitation Industries

These industries are responsible for hideous suffering and premature deaths of billions of individual beings. The scale and the degree of suffering endured by nonhuman animals at our hands are scarcely conceivable. These industries not only inflict agonies and death on individual animals but devastate the environment, destroying the habitat of other animals, leading to extinction of entire species, as well as endangering the future of human life, through pollution of air, soil and water, global warming, production of new pathogens and global pandemics (in addition to epidemic levels of obesity and a host of serious health threats among those who consume animals as food). Animal activists raise serious questions about these issues and in doing so challenge the financial interests of these industries.

To meet these challenges, industries mobilize to delegitimize, marginalize and demonize their critics. Those who profit from exploitation of animals construct narratives to justify their power over others and to make exploitation of other beings seem natural, normal and acceptable. Anti-animal rights propaganda draws on a wide range of support including farmers, hunters, ranchers, the pet industry, circuses, rodeos and other forms of animal-exploitation-based entertainment, the fashion industry and dealers in fur and leather, restaurants and grocery chains, and major corporations in the agribusiness, biomedical, pharmaceutical and vivisection industries, as well as the military. Despite their varied interests, all these voices agree on what they consider their right to exploit animals and create a chorus of anti-animal rights propaganda.

In order to make animal exploitation seem acceptable rather than cruel, foolish and murderous, industries must present the violence they conduct against animals in a better light. Corporations and their hired public relations experts work to recreate reality by shaping public discourse. Industry propaganda normalizes exploitation while presenting critics as dangerous and irrational. One propaganda tactic is to portray industry as the real protectors of animals through a discourse of "animal welfare." Another tactic is to portray activists as extremists, fanatics and, increasingly, as terrorists. In order to present their critics as irrational, industry propaganda depicts animal activists as "anti-human," despite the fact that the animal rights movement historically has been associated with concern for other social justice issues of direct concern to humans, such as the movement for women's rights, anti-slavery and human rights generally. Corporate managers understand that depicting activists as terrorists is preferable to recognizing them as serious critics with identifiable goals supported by rational arguments that would have to be acknowledged and answered.

Animal Rights "Violence"

Propaganda against animal rights focuses on "violence" committed by activists. This masks far more extensive violence conducted by animal exploitation industries on a massive scale: intensive breeding, warehousing and killing of animals in factory farms and slaughterhouses, under appalling conditions, and horrifying torture in vivisection laboratories. These violent practices are normalized, accepted as industry standards and legally permitted. Just as powerful states engage with impunity in actions that are condemned as terrorism and punished with military force when committed by others, so do we accept the most hideous atrocities when the victims are animals; it is simply prejudice to denounce violence only when it affects humans.

Industry propaganda consistently describes activists as "extremists" who use violence to achieve their ends. Unlike terrorists who deliberately target innocent people, such as in the 2005 London public transit bombings that killed 52 and injured 700, most animal activists engage in legal activities such as leafleting, demonstrations and vegan potluck dinner events. Activists use various strategies, including vegetarian advocacy, humane education, boycotts, media campaigns, protests, undercover investigations of factory farms, slaughterhouses and laboratories and open rescues in which activists do not conceal their identities while removing animals from horrifying conditions. Only a few activists engage in illegal actions and many of those acts are not "violent" at all but consist of offences such as trespassing. Even when activists engage in illegal practices, much of this consists of rescuing animals from situations where they will be harmed or killed. Most people agree that animals should not be subjected to unnecessary suffering and consider it praiseworthy to rescue them from such situations, no matter who it is that inflicts such suffering. We feel instinctive sympathy for animals in pain and can empathize with those who rescue them. Indeed, when Sarah Whitehead and three other activists were arrested in 2006 for rescuing birds, dogs, rabbits and rodents from Philip Porter's pet-shop breeding operation in Sussex, Judge John Sessions refused to sentence them to prison or to order compensation to Porter because he recognized that the conditions in which the animals were kept were appalling (Payne, 2006). Many felt it was unjust when another judge sentenced Whitehead to a two-year prison sentence in 2008 for rescuing an abused dog from a garden where he was kept muzzled in a cage and beaten regularly.

Some activists have damaged property but much of this is minor, such as breaking locks or windows to gain entry to rescue animals. Other forms of property damage consist of such things as gluing locks or spray-painting slogans. In an even smaller number of cases, activists have damaged equipment used to harm or kill animals. However, it is arguable that this does not constitute violence. Philosopher Mark Rowlands, for example, argues that one cannot be violent toward inanimate objects, only toward living beings (2002, p. 188). Even if one thinks property destruction constitutes violence, it seems that the ends are commendable: the prevention of suffering.

While animal activists have not directed violence against humans, some have engaged in intimidation. Some of these activities are illegal and undoubtedly have been unpleasant for those who were targeted. However, it does not seem appropriate to equate activities such as sending of black faxes to companies involved in vivisection or even demonstrations at vivisectors' homes with the deliberate mass murder of innocent people at a restaurant or sports bar, or setting off car bombs in crowded markets with the intent to kill as many passersby as possible. With very few exceptions, animal activists have not engaged in violence against humans. Even if we follow the lead of industry lobbyists and propagandists and ignore the vast majority of actions taken by animal activists and focus on a small number of illegal actions, we still find little to justify efforts to brand these as "terrorism." Unlike Al-Qaeda or white supremacist militias, which deliberately target and kill humans, the Animal Liberation Front (ALF; considered one of the most extreme expressions of animal advocacy) holds as one of its key principles that no harm should be done to animals, including humans, in the course of ALF actions. These principles contrast with the attitudes of actual terrorists such as, for example, white supremacist Timothy McVeigh who detonated his truck bomb at the Alfred P. Murrah Federal Building in Oklahoma City in 1995, killing 168 people, including young children at a day care, and wounding hundreds more; afterward, McVeigh dismissed the deaths as "collateral damage."

The Terrorism Label

Propagandists use the language of "terrorism" to compare those who act on behalf of animals to those who commit mass murder and thus to automatically delegitimize them. No matter how noble their motives, how rational their arguments or how much their actions are congruent with what many people claim to believe, applying the "terrorist" label to activists places them outside acceptable moral boundaries. This also serves to make excessive force and diversion of resources to address the threat seem necessary.

Industry propagandists, their lobbyists and right wing think tanks feel free to make the most outrageous claims about the animal rights movement, unburdened by any need to provide evidence. For example, after the 2001 destruction of the World Trade Center, industry propagandists made frequent comparisons between animal activists and al-Qaeda; the anti-environmentalist American Policy Center even suggested that Islamists and ecoterrorists were collaborating to destroy America. Industry propagandists exaggerate violence committed by activists and increasingly use the term "terrorism" to demonize them. In its "Report to Congress on the Extent and Effects of Domestic and International Terrorism in Animal Enterprises," *The Physiologist* uses a very "broad, inclusive" definition of the term. The authors note the FBI's definition already quite broad:

> the unlawful use of force or violence against persons or property to intimidate or coerce a government, the civilian population, or any segment thereof, in furtherance of political or social objectives. (Unknown, 1993, pp. 207, 247–259)

The Animal Enterprise Protection Act (AEPA) "characterizes terrorism as physical disruption caused to the functioning of an animal enterprise" (Unknown, 1993, p. 247). However, even these definitions are not broad enough for *The Physiologist*'s authors, who consider "a wider range of activities than is covered by either the Act or FBI's definition of terrorism" (Unknown, 1993). These definitions contrast with those of ALF supporters, however, who do not consider the rescue of animals from dangerous conditions or property damage to be violence, let alone terrorism. They also reject the idea that animals are property, owned by the enterprises that exploit them, arguing that animals have their own interests and right to life.

The Physiologist emphasizes the dangers of animal rights "terrorism." One way of doing this is by omitting key aspects of the ALF guidelines, which they quote only in part:

> to liberate animals from places of abuse, i.e. laboratories, factory farms, fur farms, etc, and place them in good homes where they may live out their natural lives,

> free from suffering to inflict economic damage to those who profit from the misery and exploitation of animals; and to reveal the horror and atrocities committed against animals behind locked doors. (Unknown, 1993, pp. 248–249)

In fact, the ALF guidelines are as follows:

- To liberate animals from places of abuse, i.e. laboratories, factory farms, fur farms, etc, and place them in good homes where they may live out their natural lives, free from suffering.
- To inflict economic damage to those who profit from the misery and exploitation of animals.
- To reveal the horror and atrocities committed against animals behind locked doors, by performing non-violent direct actions and liberations.
- To take all necessary precautions against harming any animal, human and non-human. (ALF, n.d.)

Clearly, *The Physiologist* deliberately distorts the ALF's philosophy, tactics and goals by deleting the words "by performing non-violent direct actions and liberations" from the third guideline and by omitting the fourth guideline entirely (Unknown, 1993). This distorts the ALF's approach in very significant ways, characterizing it as a violent group ready and willing to harm humans to help other animals. The intended effect is to promote a negative image of animal activism by concealing its non-violent principles.

Although the anonymous authors of *The Physiologist* emphasize the danger of "terrorism" their own data suggest that this is an exaggeration. Reviewing 313 actions from 1977 to 1993, they note that the "most common" of these only constitute "minor vandalism" such as graffiti, broken windows and glued locks. These account for almost half the actions (Unknown, 1993, p. 253). The second most common form of "extremist incident" is the "theft or release of animals" (Unknown, 1993). *The Physiologist* acknowledges that "most extremist animal rights-related acts continue to be small-scale and fairly haphazard" (Unknown, 1993, p. 251). Only 26 of the 313 incidents are designated as "major vandalism" that the FBI categorizes as "domestic terrorism" (Unknown, 1993, p. 253). Furthermore, *The Physiologist* finds "no evidence…[of]…any operational, logistical or financial connections" of ALF groups internationally (Unknown, 1993, p. 257). *The Physiologist* also notes considerable public sympathy for animal rights, which they describe as a "mainstream" movement with hundreds of thousands of supporters and at least 7,000 organizations in the United States alone (Unknown, 1993, p. 248). Typically in discussions of so-called "ecoterrorism" actions of animal activists and environmentalists are lumped together in chronologies of events. This overlooks philosophical disagreements between animal and environmental activists. Most animal activists probably have concern for the environment, recognizing that protection of habitat is necessary to protect endangered species. However, many environmentalist groups do not endorse animal rights and in fact denounce animal activism as a sentimental concern of misguided urban types. Opponents of both animal and environmental activists, however, find it very convenient to merge these movements together, since it helps to present their opponents as larger and more monolithic.

Corporate Front Groups in the United States

Nevertheless, corporate propagandists and lobby groups pushed the eco-terrorism label and used it to demand stronger laws to specifically protect animal exploitation industries. These industries channel millions of dollars to public relations firms, lobbyists and front groups to create and disseminate anti-animal rights propaganda and to influence government to create laws to silence their critics (as well as giving money directly to government officials to purchase their services). For example, in 2010 pharmaceutical corporations spent at least $135 million and agribusiness $35 million on lobbying in the United States (DDD, 2010).

One influential corporate front group is the Center for Consumer Freedom (CCF). The CCF is a vigorous campaigner for repressive laws against animal activists, while asserting that it is defending consumer choice and promoting common sense about the use of animals. The CCF originated as the Guest Choice Network, established in 1985 by Richard Berman with funding from Philip Morris tobacco company with the objective of uniting tobacco, food and restaurant industries against anti-smoking, anti-drinking and anti-meat campaigns designed to improve public health. The GCN became the CCF in 2002, with Berman claiming that so-called anti-consumer activists were escalating their assault on personal freedom and that a more militant proconsumer approach was needed. Financial backing expanded to include other major corporations, particularly those in the food, alcohol and restaurant industries, such as Anheuser-Busch, Brinker International, Cargill, Coca-Cola, HMSHost Corp, Monsanto, Pilgrim's Pride, RTM Restaurant Group, Smithfield Foods, Tyson Foods and Wendy's, among others. The CCF is a front group for these corporations and runs negative campaigns against their critics. Thus, the CCF opposed unionization, minimum wage legislation, anti-drunk driving legislation, smoking bans and warning labels on food while rejecting health concerns about alcohol, antibiotic use for livestock, genetic engineering, mad cow disease, meat, mercury levels in fish, obesity, pesticides, salmonella poisoning and tobacco. The CCF opposes "Big Brother government" and claims to promote individual choice. Berman's widely quoted strategy is to "shoot the messenger" and the CCF and its related organizations and websites produce attacks on groups such as the Centers for Disease Control, Greenpeace, the Humane Society of the United States, Mothers Against Drunk Driving and PETA. Especially in the case of animal welfare groups, the CCF alleges that while these groups claim to act within the law they are, in reality, supporters of terrorists. For example, the CCF ran a print advertising campaign (archived on its website) denouncing "PETA's Fiery Links to Arsonists." The advertisement features a large photograph of a burning building and asserts that PETA has given over $100,000 to "convicted arsonists and other violent criminals" and, thus, is "not as warm and cuddly as you thought" (CCF, n.d.b).

Berman runs over a dozen industry-funded, tax-exempt front groups and holds various positions within all of them. He shifts funds between various organizations he has created, hiring his own public relations and lobbying firm to do research and channelling the money into his own pocket, which led Citizens for Responsibility and Ethics in Washington to call upon the Internal Revenue Service to revoke the CCF's tax-exempt status. In addition to working on behalf of specific corporations, Berman provides propaganda for capitalism itself through the Center for Union Facts, the Employment Freedom Action Committee, the Employment Policies Institute and the First Jobs Institute.

Corporate funders benefit from various CCF campaigns. For example, the CCF's anti-union campaigns are welcomed by Smithfield Foods, which strongly opposed unionization in its plants, but Smithfield is also the world's largest "producer" of pig-flesh, as well as "producing" significant quantities of cow-flesh, so it also benefits from the CCF's attacks on animal activists. Smithfield is notorious for its appalling environmental record, especially concerning its storage of millions of gallons of untreated fecal waste in holding lagoons in North Carolina. Humans living in the vicinity of these lagoons experienced serious health problems and complained of the overpowering stench that kept them inside their homes. In the late 1990s, Smithfield was fined $12.6 million for violations of the Clean Water Act. Although this was a comparatively large fine, it was only a miniscule fraction of Smithfield's profits, less than 1% of annual sales (Toetz, 2007). In 1999 when hurricanes hit North Carolina, these lagoons overflowed and polluted rivers and waterways throughout the region. Smithfield's operations in Mexico were cited as a likely source of the 2009 swine flu epidemic. Residents living near Smithfield operations complained of health problems similar to the ones experienced in North Carolina as well as about swarms of flies at the lagoons; these flies were suspected as a vector of the disease.

As well as its opposition to unionization, lack of concern for human health and disregard for the environment, Smithfield is notorious for opposition to even minor modifications to the treatment of the animals it kills. Only after a long campaign by the Humane Society of the United States did Smithfield agree to gradual phasing out of gestation crates for pigs. For most of their lives, female pigs are confined in crates that do not allow them to turn around; they are trapped in a continuous cycle of artificial impregnation, gestation and farrowing until their litter sizes decrease and they are killed and replaced by other victims. In 2007 Smithfield touted its grudging agreement to slowly decrease the use of these crates as concern for what the industry calls "animal welfare" and HSUS hailed this as a major step forward by the industry; however, in 2009 Smithfield announced with less fanfare that it would no longer comply with this plan (HSUS, 2010).

The CCF promotes the idea that wealthy animal rights groups are dictating what ordinary hard-working people can do and limiting their personal choices. For example, in its television advertisement "Food police smashing your choices?", the CCF plays on a sense of entitlement and resentment about feelings of loss of personal freedom:

> Everywhere you turn, someone's telling us what we can't eat. It's getting harder just to enjoy a beer on a night out. Do you always feel like you are being told what to do. (CCF, n.d.b)

"Personal choice" is the final resort of those who cannot respond to logical arguments about why they should not eat animals. Insistence that "it's a personal choice" is intended to halt further discussion, invoking sacred freedoms that must not be restricted in any way. Yet our personal choices are constantly regulated and it would be impossible to live in a situation where there were absolutely no restrictions on personal freedom. We have accepted that the slave-owner's personal choice is an insufficient justification for him to force others to work on his behalf. Similarly, the claim that one's personal choice justifies the murder of animals overlooks the personal choice those animals would make to remain alive. The CCF manipulates feelings of resentment and personal powerlessness to claim that animal activists are forcing an extremist agenda on ordinary folks. They present advocacy as intimidation and brand this terrorism, claiming that mainstream groups such as PETA are funding violent attacks.

Another corporate front group is the National Animal Interest Alliance (NAIA). The NAIA says its mission is "to promote the welfare of animals, to strengthen the human-animal bond, and safeguard the rights of responsible animal owners" (NAIA, n.d.a). However, the NAIA does nothing "to promote the welfare of animals" and, indeed, actually works against the interests of animals (NAIA, n.d.b).

Whereas the NAIA claims to advocate for animals, its Board members come from animal exploitation industries including circuses, rodeos, vivisection industry, dog breeders, the racing industry and agribusiness. Their advocacy is for continued exploitation of animals, not for the animals themselves. For example, NAIA's president, Larry S. Katz, associate professor and chairman of the Animal Sciences Department at Rutgers University, works in wildlife management and "sits on the board of directors of the Foundation for Animal Use and Education. He is an outspoken advocate for biomedical research in print and broadcast outlets across the US, and his effectiveness in these appearances has made him a frequent target of animal rights harassment" (Sourcewatch, 2011b). Bob Speth, pharmacy professor at Nova Southeastern University, "has written widely in support of the use of animals in biomedical research" (Sourcewatch, 2011b). Professor John Richard Schrock of the Biological Sciences department at Emporia State University "defends appropriate animal use in education" (Sourcewatch, 2011b). NAIA's national director, Patti Strand co-authored *The Hijacking of the Humane Movement: Animal Extremism* in 1993, advertised as "the first US book exposing the extremism of the animal rights movement" (Sourcewatch, 2011b).

Cindy Schonholtz, NAIA vice president who is director of Industry Outreach for the Professional Rodeo Cowboys Association, "handles government relations for the PRCA relating to animal issues leading to the defeat of numerous bans on rodeo" and works with "many other animal use industries…to educate the public on animal welfare issues" (Sourcewatch, 2011b). She also works for Friends of Rodeo and operates the Animal Welfare Council, both supporters of the Animal Enterprise Terrorism Act (AETA, see below). The AWC represents the rodeo industry but also promotes ranching, the premarin industry, horse slaughter, the carriage horse industry and circuses. Member organizations include various rodeo and cowboy associations, carriage horse operators and circus groups such as Feld Entertainment but also Americans for Medical Progress, a pro-vivisection lobby group. Obviously, "medical progress" is even less likely to be served through rodeos than through vivisection but the AMP's willingness to join with these entertainment industries shows the convergence of interests in denouncing animal rights. AMP also collaborates with the Fur Council USA, another organization unlikely to advance medical progress, but one willing to embrace AMP's propaganda efforts to link animal activists with al-Qaeda (Ward, 2001).

AMP was also a strong supporter of the AETA. AMP is a front group for the vivisection industry, with a Board of Directors that includes top executives from pharmaceutical and vivisection companies such as Abbott Laboratories, AstraZeneca, Charles River Laboratories, GlaxoSmithKline, Pfizer and Wyeth. These corporations have a record of violations of even the few animal welfare laws that do exist and have been the object of campaigns by animal advocates as well as various public health and consumer groups. For example, Charles River Laboratories is the world's main laboratory animal supply company. The pharmaceutical corporations that belong to AMP have been clients of Huntingdon Life Sciences, target of a major animal rights campaign. Wyeth was the subject of specific campaigns about abuse of horses in the Premarin industry (PETA, n.d.). Four universities (Harvard, Oregon Health and Science University, University of North Carolina Chapel Hall and Tulane) on the AMP Board were cited by PETA as being among the "Ten Worst Laboratories" (Sourcewatch, 2010c), and Oregon National Primate Research Center (ONPRC) at Oregon Health and Science University was criticized by the Humane Society of the United States, In Defense of Animals, PETA and Stop Animal Exploitation Now! for abusing primates in alcohol, nicotine, maternal deprivation and obesity studies (Sourcewatch, 2011b). Despite the fact that other primates do not develop HIV/AIDS as humans do and that animal models are widely criticized, the ONPRC continued to use animals in these studies as well. Whistleblowers, undercover investigations and even a 2001 report by Dr. Carol Shively, professor of pathology and psychology at Wake Forest University Medical School, who had been hired by ONPRC itself to assess the psychological condition of the Center's primate prisoners, revealed ghastly abuse of these animals, by poorly trained technicians (Sourcewatch, 2011b).

Another NAIA board member, Gene Gregory, is president and CEO of United Egg Producers, which represents 97% of US egg production. Paul Mundell, National Director of Canine Programs for Canine Companions for Independence, is a consultant for the United States Marine Corps, helping them train dogs for military use. Board member Kenneth A. Marden is a dog breeder, former president of the American Kennel Club and "a lifetime hunter and fly fisherman…[who]…actively opposes unfair dog legislation and laws proposed by animal rights fanatics in their attempts to restrict hunting, fishing and trapping [and]…has a deep understanding of the negative consequences of animal extremism and terrorism on the lives of farmers and ranchers" (NAIA, n.d.d). NAIA's advisory board includes Sheila Lehrke from the

International Professional Rodeo Association, Michael Manning, a Roman Catholic priest who "devotes much of his pastoral time defending the unique sanctity of human life from those who would place all living beings on the same spiritual plane" and retired Lt. Col. Dennis Foster, executive director of Master of Foxhounds Association, "an avid horseman, [and] an internationally recognized expert in the tactics of the animal rights movement" (NAIA, n.d.d).

While claiming to promote "animal welfare," the NAIA does everything it can to undermine it by opposing animal rights, promoting anti-environmental messages, campaigning against spay and neuter programs and fighting legislation against horse slaughter and the Prevention of Farm Animal Cruelty Act. In contrast, it supports hunting, vivisection, use of animals as entertainers by "circuses, zoos, wild animal parks, aquariums, and private entertainers and foundations" (NAIA, n.d.c). It also supports "husbandry" practices involving mutilation of animals such as "dehorning,…ear cropping, tail docking, and debarking of dogs, and removing the claws of cats" and endorses the breeding and raising of animals for food, fibre and draft as well as the fur industry (NAIA, n.d.b). In short, there is virtually no abuse of animals that the NAIA does not endorse and promote. Seemingly, for the NAIA, "animal welfare" is synonymous with "animal exploitation."

The NAIA warns against those who do protect animals:

> Animal rights and environmental extremists do more than demonstrate and push radical legislation. They also use physical assaults, intimidation, vandalism, harassment, theft, property destruction and terrorism. (NAIA, 2010)

The NAIA maintains what it calls "the most complete chronology of animal rights and eco-criminal acts on the Internet" (2010). No sources are cited for any incidents listed. Far from being a record of "terrorism" many of the incidents described are nonviolent, such as releasing animals from their prisons. Other incidents involve minimal damage to property, as in an example from the UK in March 2010 in which "Hunt saboteurs claim to have removed signs advertising hunt point-to-points and paint-stripped hunters' cars" (NAIA, 2010). Other minor acts of vandalism include spray-painting graffiti, gluing locks and breaking windows. Some incidents are unlikely to be the work of animal or environmental activists:

> February 27, 2010 Monza, Italy: Oil was released into the Po river, after tanks at an abandoned refinery were tampered with. Valves were opened, and several tanks were ruptured. Authorities called the sabotage an act of environmental terrorism. (NAIA, 2010)

Environmentalists would be unlikely to open valves and rupture tanks to release oil into rivers. The NAIA's description suggests this was more likely an act of thoughtless vandalism undertaken for its own sake rather than a political act. But the NAIA is determined to include every act of destruction that it can characterize as the work of its opponents.

Other corporate front groups created to combat animal advocacy are the Foundation for Biomedical Research, the National Association for Biomedical Research (NABR) and Policy Directions Inc. All work from the same address in Washington DC and were created by Frankie Trull to serve the vivisection industry. Trull has lobbied against even minor amendments to the Animal Welfare Act, such as a 1985 provision to provide caged dogs periodic exercise. Opposing this, Trull argued:

> There are no scientific data which say any minimum exercise per day, or per week, is physiologically better. You just sleep better at night because you think if exercise is good for you, it must be good for the dog. (Sourcewatch, 2011a)

When the Alternative Research and Development Foundation tried to amend the Animal Welfare Act to include some consideration of mice and rats used in laboratories Trull persuaded Senator Jesse Helms to amend a farm subsidy bill so that "animal" would be defined to exclude rats, mice and birds (Sourcewatch, 2011a). On its website, the NABR warns:

> In the past 20 years, the animal rights movement (ARM) has successfully manufactured a climate of public opposition to research involving animals. (NABD, n.d.)

To combat public opinion, the NABR emphasizes terrorism:

> According to the Federal Bureau of Investigation (FBI), the Earth Liberation Front (ELF) and its sister organization the Animal Liberation Front (ALF), were responsible for the vast majority of terrorist acts committed in the United States in the 1990s. (NABD, n.d)

In addition to opposing legislation to protect animals, Trull was instrumental in having repressive laws passed to specifically target animal activists and to have them designated as terrorists. Trull boasts:

> Two of NABR's accomplishments of which I am most proud are the passage of the 1992 Animal Enterprise Protection Act and the 2006 Animal Enterprise Terrorism Act. (2009)

The Animal Enterprise Terrorism Act

Influential animal exploitation industries pushed for stronger legislation to stop animal activism. In 2005 Senator James Inhofe organized and chaired the Senate Committee on Environment and Public Works hearing on "Oversight on Eco-Terrorism Specifically Examining the Earth Liberation Front ("ELF") and the Animal Liberation Front ("ALF")." In his opening statement, Inhofe called the ALF and ELF "terrorists by definition" [who used] "intimidation, threats, acts of violence, and property destruction to force their opinions...upon society" and held them accountable for damages over $110 million in over a thousand "acts of terrorism" (USSCEOW, 2005). Inhofe treats the ALF and ELF interchangeably: although it was the ELF that claimed responsibility for arson at Garden Communities' condominium construction site in San Diego, California, Inhofe calls it "the largest ALF attack in history" (USSCEOW, n.d., p. 2). Inhofe compared the ALF and ELF to al-Qaeda and claimed that, like the latter, these "terrorist" groups draw money from "mainstream activists" including PETA. Admitting that "although they have not killed anyone to date" Inhofe asserts "it is only a matter of time" until they do (USSCEOW, n.d., p. 3). Most speakers continued in the same vein. Louisiana Senator David Vitter, later notorious for his use of prostitutes, acceptance of major financial contributions from the oil industry and efforts to block the Senate from forcing BP to accept full responsibility for the cleanup of the massive oil spill in 2010, applauded Inhofe's description of the ALF as "terrorists" and described two ALF actions at Louisiana State University (USSCEOW, n.d., p. 3). In the first, in 2003, ALF activists entered a toxicology laboratory, spray-painted slogans and damaged equipment; in 2005, ALF activists rescued ten mice, painted slogans and damaged equipment. Vitter said these incidents caused "psychological harm" to vivisectors and, like Inhofe, warned "it is only a matter of time" before the ALF kills humans (USSCEOW, n.d., p. 8).

John Lewis, deputy assistant director of the FBI's Counterterrorism Division, told the Committee that "the No. 1 domestic terrorism threat is the eco-terrorism animal rights movement," identified the ALF, ELF and SHAC as "today's most serious domestic threats," stated the FBI "certainly shares your opinion that these individuals are most certainly domestic terrorists" and identified this as one of the FBI's top priorities, calling for expanded federal laws to allow them to "dismantle these movements" (USSCEOW, n.d., p. 12). Like Inhofe and Vitter, Lewis acknowledges that these "terrorists" have never killed a human but predicts it will happen, citing an "escalation in violent rhetoric" (USSCEOW, n.d.). Comparing these groups to anti-abortionists, the KKK and right wing extremists, Lewis said the ALF and ELF "are way out in front in terms of the damage they are causing" (USSCEOW, n.d., p. 15).

Senator Frank Lautenberg briefly questioned Lewis about more serious violence from anti-abortionists and anti-gay activists before Inhofe forced him to stop. Lewis denied that anti-abortionist groups could be defined as terrorists, despite the fact that they use violence to "force their opinions on society" (USSCEOW, n.d., p. 18). Anti-abortionists had murdered at least seven people in the United States and seriously wounded at least twelve others in shootings, arsons, acid attacks and bombings prior to the time of the 2005 Hearings and another murder, along with additional attacks, followed (NAF, 2009). Many "pro-life" groups endorsed this (Army of God, n.d.). Comparing acts of "extremist violence" by animal rights and anti-abortion groups from 1977 to 1993, and including acts against people (murder, attempted murder, kidnapping, acid attack, assault and threats) and acts against property (bombings, arson, attempted bombings or arson, major and minor damage, theft, bomb hoaxes and kidnapping) Johnson finds a total of 1,079 incidents committed by anti-abortion activists as opposed to only 337 by animal activists (Johnson, 2008). Of the actions against property, the second-highest number (below "minor property damage") of actions committed by animal activists is in the category of "thefts" (whereas it is in the category of "arson" for anti-abortionists) and likely refers to the rescue of animals from vivisection laboratories or fur farms where they are subjected to close confinement, painful procedures and prematurely killed. As noted, in other circumstances, rescuing animals from danger is regarded as praiseworthy. Of actions against people included in the table, animal activists are responsible for only threats and have committed no murder, kidnapping, acid attacks or assaults. One incident of attempted murder is noted but Johnson says no FBI information on the incident is included; possibly it was the case of Fran Trutt[1] (Johnson, 2008).

In contrast to non-violent actions of animal activists, anti-abortion activists were responsible for violence against humans and property at a rate of three to one (Johnson, 2008). Johnson points out that more murders, attempted murders, acid attacks, bombings, arsons and death threats were conducted by anti-abortionists after 1993 and that the FBI steadfastly refused to categorize this as terrorism. Clearly, Lewis's statements about animal activists being "way out of front" are inaccurate.

Nevertheless, the Hearing was a prelude to establishing the Animal Enterprise Terrorism Act (AETA), passed by Congress and signed into law by President George W. Bush on November 27, 2006, replacing the 1992 AEPA.

The AEPA had been crafted by the NABR and created the term "animal enterprise terrorism." Other instigators of the AETA included influential agribusiness and biomedical industry lobby groups such as the Animal Enterprise Protection Coalition, American Legislative Exchange Council, Foundation for Biomedical Research (FBR), NABR and the Fur Commission. However, scores of other animal exploitation groups endorsed the Act. Animal exploitation industries guided the legislation through to its passage, assisted by politicians such as Inhofe whose services they had purchased through financial contributions and who had personal investments in industries the legislation would affect.

Inhofe has substantial personal investments in oil and gas industries, has received hundreds of thousands of dollars from chemical and forestry industries, oil and gas companies, the nuclear energy industry and their political action committees and is one of the major recipients of funding from these sources (Sourcewatch, n.d.). He consistently voted against environmental and public health safety regulations that would affect these industries (Sourcewatch, n.d.). On September 23, 2009, on C-Span's *Washington Journal*, Inhofe said he would fly to the UN Climate Change Conference in Copenhagen to campaign against the international consensus of scientific experts. Calling himself a "one-man truth squad," Inhofe said climate change was a "hoax" perpetrated by the UN and "the Hollywood elite" (Shepherd, 2009). In his 2010 Minority Report, Inhofe named seventeen leading scientists associated with the Intergovernmental Panel on Climate Change and the United States Global Change Research Program as "key players" in an international conspiracy; he said their actions violated basic ethical principles concerning publicly funded research and federal laws and called for prosecution by the US Justice Department (USSCEPW, 2010, p. 1). Facing pressure for corporate accountability after the 2010 BP oil spill, Inhofe blocked a bill to increase liability of oil companies responsible for spills that pollute the environment, kill thousands of animals and destroy human livelihoods. Inhofe also tried to limit the Environmental Protection Agency from regulating emissions from power plants and refineries and rolled back rules on increased fuel efficiency for automobiles manufactured from 2012 to 2016 (Broder, 2001). Inhofe is a staunch defender of the cruel practices of factory farming. In turn, these industries have lauded Inhofe's services to them:

> in 2008, the Oklahoma Independent Petroleum Association honoured Inhofe for "voting consistently in the 110th Congress to protect the interests of the oil and gas industry" and in 2004 the National Association of Chemical Distributors named him "legislator of the Year" while the American Farm Bureau designates him as an official "Friend of Farm Bureau," the Oklahoma Farm Bureau gave

> him a "Lifetime Achievement Award" and the Oklahoma Pork Council recognized his efforts on their behalf with a "Distinguished Service Award." (Lovitz, 2010, pp. 85–86)

Senator Dianne Feinstein co-sponsored the bill. Although she is not directly funded by animal exploitation industries, her husband, Richard Blum, is chairman of the board of CB Richard Ellis Group, a real estate firm that serves the vivisection industry and associated major pharmaceutical companies: American Pharmaceutical Partners, Astra Zeneca, Bayer Pharmaceuticals, Chiron, DuPont, Eli Lilly and Company, Johnson & Johnson, Merck, Novartis, Pfizer, Schering Plough and Wyeth (Lovitz, 2010, pp. 84–85). Co-sponsors of the AETA in the House also had financial interests in the industries served by the legislation. Representative Tom Petri is funded by the dairy industry and headed the Badger Fund, a political action committee funded by American Foods Group, owner of slaughterhouses; Representative Robert Scott has investments in Johnson & Johnson, Procter & Gamble and Yum! Brands (Lovitz, 2010, p. 85).

In 2006, the AETA also received a warm welcome from the Chair of the Committee on the Judiciary, Representative James Sensenbrenner, who owns stock in major pharmaceutical, chemical, petroleum and defence industries (Lovitz, 2010, p. 87). Sensenbrenner also blocked the Animal Fighting Prohibition Act, intended to increase penalties for those who engage in animal fighting activities, despite the fact it unanimously passed the Senate and had hundreds of co-sponsors (the Act was finally passed in 2007) (Cochran, 2006). Like Inhofe, Sensenbrenner is a climate change denier. After *Rolling Stone* magazine voted him one of the planet's worst enemies in a cover story on climate change deniers, entitled "You Idiots!", Sensenbrenner complained "I should have been No. 1, not No. 7" (Myers, 2010). In December 2009, as a member of the House Select Committee on Energy Independence and Global Warming, he wrote to Dr. Rajendra Pachauri, Chair of the Intergovernmental Panel on Climate Change, demanding that scientists named in e-mails stolen from the UK Climatic Research Unit be blacklisted as participants, contributors or reviewers of the IPCC's upcoming Fifth Assessment Report (Piltz, 2009). In his statement to the Committee at the Hearing on "The State of Climate Change Science", Sensenbrenner charged that the scientists were engaged in a "massive international scientific fraud" and "scientific fascism" (Fox news, 2009). Sensenbrenner was a vigorous advocate for the AETA, claiming that existing laws were inadequate and that animal activists had carried out over a thousand terrorist actions, causing millions in damages (Potter, 2009, p. 682).

Animal exploitation industries cheered the new legislation they had created. The NAIA "applaud[ed] the passage" of the AETA and National director Patti Strand said:

> We are grateful to Senators Inhofe and Feinstein, and Representative Petri for introducing companion bills in the Senate and House recognizing the threat to our country posed by animal-rights extremists. (NAIA, 2006)

However, animal activists and civil rights advocates said the AETA was too broad and vague, and that it did not even clearly define "animal enterprise" so that the law could be applied to any business that involves animals in some way. Penalties imposed by the AETA are out of proportion to actions covered, imposing longer sentences for nonviolent actions that cause a loss of profit to animal enterprises than for actions that cause direct harm to people. Opponents also said the AETA would have a chilling effect on legal protest in general because activists would fear being charged as terrorists (Potter, 2009).

Other Proponents of Animal Rights "Terrorism"

In addition to those who profit directly from exploitation of animals, others have an interest in exaggerating "ecoterrorism." As Herman and O'Sullivan pointed out two decades ago, there is a large network of experts and institutions organized to produce politically useful analyses, definitions and understandings of terrorism (Herman & O'Sullivan, 1989). Since then, that industry has grown substantially. The *Washington Post* reported that since 2001 the United States has developed a defence and intelligence bureaucracy "that has become so large, so unwieldy, and so secretive that no one knows how much money it costs, how many people it employs, or whether it is making the United States safer" (Priest & Arkin, 2010). At least 1,271 government organizations and 1,931 private companies are engaged in secret counterterrorism, homeland security and intelligence programs in over 10,000 locations, with 33 new building complexes to house these bureaucracies in Washington DC alone (Priest & Arkin, 2010). At least 850,000 personnel with Top Secret clearances work for these agencies but thousands of other jobs are associated with them, through provision of technology, services, construction and so on (Priest & Arkin, 2010). Clearly, these security and intelligence operations involve significant amounts of money. However, "major problems" include "lack of coordination between agencies" as well as "redundancy and overlap" (Priest & Arkin, 2010).

Just as the military requires constant production of new enemies to justify its existence and the continued inflow of public funds, so do counterterrorism operations need a constant supply of terrorists. Promoting the menace of animal rights terrorism provides income for "experts" who advise business on ecoterrorism and sell security technology. Hundreds of millions of dollars are spent on counterterrorism research and can provide a financial boost to universities. For example, in 2004 New Jersey Institute of Technology under Director Donald J. Sebastian was designated as the site of the New Jersey Homeland Security Technology Systems Center. Counterterrorism funding, along with military and biotechnology research brought in "$100M in 2010, placing NJIT in the top 10 engineering universities in the nation" (NJIT, 2010). Security organizations have a vested interest in portraying illegal actions against animal exploitation industries as terrorism rather than ordinary crimes. For example, writing for Stratfor Global Intelligence, Fred Burton says direct actions should be "categorized as terrorism because of their political motive" (2007). By portraying this as terrorism, the threat to business is made to seem greater (and certainly the penalties imposed by the courts are heavier), so the role of the "expert" is made to seem more vital. Thus, headlines about direct actions typically claim, without proof, that these actions are growing more serious. For example, Scott Stewart in Stratfor's *Security Weekly* warns against "Escalating Violence From the Animal Liberation Front" but provides no data to show that destruction of a business selling sheep-skin products in Colorado and a leather factory and a restaurant in Utah actually represents an "escalation" (Stewart, 2010). "Animal rights terrorism" is a growth area for security firms and organizations such as the Inkerman Group produce reports such as "The War on Eco-Terror" for industry clients. The "experts" frequently called upon by mainstream media to explain "ecoterrorism" are those such as Ron Arnold, from the Center for the Defense of Free Enterprise, a virulent critic of environmentalism and founder of the Wise Use movement, a corporate-backed anti-environmental coalition organized under the idea of property rights, and linked to militia and anti-government groups. Arnold is widely quoted on his strategy toward environmentalists, "We're out to kill the fuckers. We're simply trying to eliminate them. Our goal is to destroy environmentalism once and for all" (Helvarg, 2004, p. 7).

Influence on Government in the UK

We can clearly see the influence of animal exploitation industries on government policy in the case of the British pharmaceutical industry. The Association of the British Pharmaceutical Industry (ABPI) is the major pharmaceutical industry lobby group in the UK, representing at least 75 companies that supply most of the drugs prescribed by the National Health Service (NHS). The ABPI represents industry interests and works to shape laws to benefit these companies. The pharmaceutical industry is the biggest export industry in the UK, after North Sea oil and the ABPI is one of the most powerful lobby groups in the country, exerting strong influence over policy. Also, many government officials and industry regulators have significant financial interests in the industry (Corporate Watch, 2003a, 2003b).

Despite industry propaganda about dedication to saving lives, the main concern is profit. The ABPI lobbied for support to biotechnology, looser regulation of advertising of drugs, including direct marketing to consumers, opposed calls for disclosure of research data, despite the fact that this information would be of use to academics, consumer groups and public safety advocates, opposed plans to lower drug prices and opposed South Africa's plan to provide affordable AIDS medicines (Corporate Watch, 2003a, 2003b). ABRI also advocates for vivisection and stated its concerns about animal activists who challenge these practices (Corporate Watch, 2003a, 2003b).

Although the Liberal Party campaigned on promises to reduce and eventually end vivisection and to establish a Royal Commission to investigate the actual need for animal research, those promises were abandoned after they were elected (SPEAK, n.d.). Instead they began closer cooperation with vivisection, biotechnology and pharmaceutical industries and stated they would support the industry by making regulation more "flexible" (Unknown, 2006). In November 1999 British Prime Minister Tony Blair joined CEOs of giant pharmaceutical corporations of Astra Zeneca, Glaxo Wellcome and Smith-Kline Beecham on the Pharmaceutical Industry Competitiveness Task Force (PICTF) "to retain and strengthen the competitiveness of the UK business environment for the innovative pharmaceutical industry" (DHABPI, 2001a). The PICTF appointed Lord Sainsbury as chair of a working group to cut vivisection regulations. No animal advocacy groups were included.

Sainsbury is a billionaire, one of the UK's richest men, with investments in supermarkets and biotechnology, including the Sainsbury Laboratory. Sainsbury used his wealth to buy positions of influence within the government; he gave over 11 million pounds to the Liberal Party and was rewarded by Tony Blair with an appointment to the House of Lords as Lord Sainsbury of Turville and then appointed Parliamentary Under-Secretary of State for Science in the Department of Trade and Industry (Unknown, 2003). Sainsbury later resigned, following a police investigation of government corruption (Wilson, 2006). Sainsbury's group made drastic changes to policies, supplanting responsibilities of the Home Office, and weakening existing regulations on vivisection as measures to guarantee the pharmaceutical corporations' profits.

The PICTF also agreed on protection of patents and intellectual property, demonstrating support for industry's opposition to governments of underdeveloped countries' efforts to manufacture cheaper life-saving drugs. The PICTF proposed other important policy changes, including safety assessment of new drugs and drug purchases by the National Health System and advised that industry should be consulted on any new policy changes considered by the government. As pharmaceutical company executives gained unprecedented influence over government policy, corporate interests rather than public health became the key concern. In its 2001 Final Report, PICTF warned:

> the increasing complexity of the regulatory process involved in obtaining licences to carry out animal studies…and the possible implications of the new Freedom of Information Act, have meant that the UK is increasingly perceived by industry as an unfavourable environment in which to conduct research involving animals…[and] there is a danger that, as a result, research may be moved abroad. (DHABPI, 2001a)

Government heeded ABPI's warning about inconvenient regulations: in 2004, PICTF announced that the time required to obtain approval for vivisection had fallen to its lowest level (DHABPI, 2004). In 2006, Prime Minister Tony Blair expressed strong personal support to the industry, going as far as to write an article for *The Sunday Telegraph* explaining why he signed an online petition in support of vivisection (Blair, 2006). Blair's support for the vivisection industry included a proposed new law to exempt the industry from legal requirements to publish details of shareholders.

The PICTF also called for amendments to the Criminal Justice and Police Bill, the Malicious Communications Act and the Companies Act "to tackle harassment and intimidation by animal rights campaigners" and noted that "Amendments have subsequently been brought forth by the Government" (DHABPI, 2001a, pp. 55–56). In response to industry's demand for stronger legislation to silence critics, the 2001 Police and Criminal Justice Act penalized various forms of protest that were proving effective.

As well, the Association of Chief Police Officers Terrorism and Allied Matters unit created the National Coordinator Domestic Extremism (NCDE) to combat "domestic extremism" in England and Wales. The NCDE has an annual budget of 9 million pounds and a staff of 100 police officers (Evans, Taylor, Hirsch, & Lewis, 2011). It includes three units. The National Public Order Intelligence Unit (NPOIU) provides "intelligence" on thousands of animal activists gathered by police surveillance groups called Forward Intelligence Teams and Evidence Gatherers. These spies photograph activists at public meetings, rallies and protests and collect detailed information for entry into national computer databases. The National Domestic Extremism Team carries out secret investigations and the National Extremism Tactical Coordination Unit (NETCU) provides information to local police for political campaigns. Although organized specifically to spy on animal activists, the NCDE does not identify any groups that it considers domestic extremists, saying this would "compromise" its investigations (ACPO, n.d.). However, according to the NETCU's website:

> Domestic extremism is most commonly associated with "single-issue" protests, such as animal rights, environmentalism, anti-globalisation or anti-GM crops... We support industry, academia and other organisations that have been targeted or could be targeted by extremists.... (n.d.)

Although the website claims that NETCU and the police remain strictly impartial on the issues, this is clearly not true, as the website formerly listed several links to pro-vivisection groups such as the Coalition for Medical Progress, the Research Defense Society and the Victims of Animal Rights Extremism (these links have now been removed). However, no links to any organizations that present an anti-vivisection argument were ever listed so claims of impartiality are unconvincing and the pro-vivisection links and statements of support to industry indicate a clear bias. NETCU's main focus is animal activism and few other forms of "domestic extremism" are mentioned on their website. The NCDE says it:

> does not usually focus those who choose to protest peacefully and lawfully. The unit is mainly concerned with those who commit criminal offences in furtherance of their campaign. ... The units will have less interest in those who choose to sit down in the road or fasten themselves to gates to protest—we are mainly concerned with those who commit more serious offences. However, police forces will always need to deal with such incidents. (ACPO, n.d.)

While the NCDE says it does not "usually" focus on peaceful protest and is "mainly" concerned with criminals, it gives itself the mandate to monitor legal protests and suggests that significant links exist between those who protest legally and those who commit criminal acts. (This is a standard assertion by industry front groups. While regularly intoning belief in the right to dissent, they constantly strive to make dissent ineffective and to show that lawful protest hides criminal intent and associations.) Indeed, one regular task of the police is to monitor such protests, photograph activists and collect information on them as well as to infiltrate activist groups. The NCDE also monitors journalists at political events and demonstrations (Lewis & Evans, 2010).

Through the NCDE, police on a national basis have collected personal details of thousands of activists who have taken part in political events and protests and have stored these data on a secret network of intelligence databases, even if those activists have committed no crimes. Noting that "domestic extremism" is a term with no legal definition and has simply been invented by the police, *The Guardian* reports: "Senior officers say domestic extremism... can include activists suspected of minor public order offences such as peaceful direct action and civil disobedience" (Evans & Taylor, 2009). *The Guardian* notes that police acknowledged that crimes associated with animal rights had been decreasing and the NCDE was branching out to spy on "anti-war and environmental groups that have only ever engaged in peaceful direct action" (Evans & Taylor, 2009). Presumably, identifying other activists as terrorists is one means to justify the budget of organizations such as the NCDE but these groups are also considered as comparable threats to the interests of industry and as logical targets of the police mentality that sees the public as the enemy and the expression of dissent as a threat to "order" and "security" (Evans & Taylor, 2009).

"Terrorists" Apprehended

We may ask what sorts of terrorists have been apprehended through this legislation and increased police powers. In 2009 four activists—Adriana Stumpo, Nathan Pope, Joseph Buddenberg and Maryam Khajavi—were charged under the AETA for protesting at the homes of University of California vivisectors in 2007 and 2008. Police said they wore bandanas and wrote "Stop the Torture," "Bird Killer" and "Murder for Scientific Lies" on the pavement with chalk (Marris, 2010). In July 2010, a federal judge dismissed the indictment because it was too vague and because prosecutors could not specify how the activists had broken any laws (Marris, 2010).

In June 2010 in Britain, the NPOIU classified 85-year-old John Catt and his 50-year-old daughter, Linda Catt, as "domestic extremists" for attending legal anti-war demonstrations in a campaign against a Brighton weapons factory operated by US-owned EDO MBM Technology. John Catt's activities at these protests consisted of making sketches of scenes he observed. The Catts have no criminal records and only engaged in legal activities:

> "Our activities were totally legitimate—we were not interested in non-violent direct action," said Linda. "My dad likes to sketch and I will hold a banner and shout a few things. But I'm careful about what I say." (Lewis & Evans, 2010)

Although Canada has not enacted specific laws against animal rights activism, it provides an example of how devalued the term "terrorism" has become as those who oppose animal activists go to absurd lengths to demonize their enemies. In January 2010, Canadian Liberal MP Gerry Byrne called on the federal government to investigate US-based PETA under Canada's anti-terrorism laws after activist Emily McCoy pushed a tofu cream pie into the face of Fisheries Minister Gail Shea during a speech in Burlington, Ontario, as an act of protest against the government's support for the seal hunt. Byrne, an MP for Newfoundland and Labrador, said the incident met the legal definition of terrorism:

> When someone actually coaches or conducts criminal behaviour to impose a political agenda on each and every other citizen of Canada, that does seem to me to meet the test of a terrorist organization. I am calling on the Government of Canada to actually investigate whether or not this organization, PETA, is acting as a terrorist organization under the test that exists under Canadian law. (Lewandowski, 2010)

However, PETA's president told *The Canadian Press*:

> Mr. Byrne's reaction is a silly, chest-beating exercise... It is unlikely to impress anyone who has a heart for animals or who is bright enough to spot the difference between a bomb and a tofu cream pie. (Lewandowski, 2010)

For some closing insight on "animal rights terrorism," we may refer to the Statement of then-Senator, now-President Barack Obama at the 2005 Senate Committee Hearings on Ecoterrorism. Noting that there had been a "downward trend" in so-called ecoterrorist crimes, Obama suggested that those crimes should be seen in the context of the much greater number of hate crimes and environmental crimes committed by industry that resulted in worker endangerment, public health threats and environmental damage. Obama advised the Committee to "focus its attention on larger environmental threats, such as the dangerously high blood lead levels in hundreds of thousands of children" and that the Committee's time would be better spent on more serious issues (Obama, 2005). We would still do well to heed that advice.

Note

1. In 1988, Trutt was charged with attempted murder after trying to place a bomb near a parking spot used by Leon Hirsch, CEO of US Surgical Corporation, producer of biomedical tools. In fact, Trutt was incited to violence by Mary Lou Sapone, an undercover agent for Perceptions International, a security firm specializing in actions against the animal rights movement. Hirsh hired Sapone and other undercover agents to infiltrate animal rights groups and prod them to commit illegal activities. The plot to entrap Trutt was discussed at a meeting that included representatives of the Federal Bureau of Alcohol, Tobacco and Firearms, the Connecticut States Attorney's office,

US Surgical Corporation's security director and Perceptions International (Berlet, 1991). Sapone had approached numerous other activists, all of whom rejected her incitements. Perceptions International agents pretended to befriend Trutt, suggested the bombing, paid for the equipment and drove her to the US Surgical parking lot. Trutt was reluctant to continue and called another "friend" (also a Perceptions International agent), who encouraged her to carry out the operation.

References

ACPO. (n.d.). National coordinator domestic extreminism. *ACPO.* Retrieved from http://www.acpo.police.uk/NationalPolicing/NCDENationalCoordinatorDomesticExtremism/FAQ.aspx

Animal Liberation Front (ALF). (n.d.). The ALF and Guidelines. *Animal liberation front.* Retrieved from http://www.animalliberationfront.com/ALFront/alf_credo.html

Army of God. (n.d.). Defensive Action Statement. *Army of God.* Association of Chief Police Officers. Retrieved from http://www.armyofgod.com/defense.html

Berlet, C. (1991, August 22). Attacks on Greenpeace and other ecology groups. *ACPO.* Retrieved from http://www.acpo.police.uk/NCDE/FAQ.aspx

Blair, T. (2006, May 14). Time to Act Against Animal Rights Protestors. *The Telegraph.* Retrieved from http://www.telegraph.co.uk/news/uknews/1518328/Tony-Blair-Time-to-act-against-animal-rights-protesters.html

Broder, J. M. (2001, March 3). Inhofe and Upton: Just say no to the E.P.A. *New York Times.* Retrieved from http://green.blogs.nytimes.com/2011/03/03/inhofe-and-upton-just-say-no-tothe-e-p-a/

Burton, F. (2007, May 23). Direct action attacks: Terrorism by another name? *Stratfor Global Intelligence.* Retrieved from http://www.stratfor.com/direct_action_attacks_terrorism_another_name

Center for Consumer Freedom (CCF). (n.d.a). PETA's fiery links to arsonists. *Consumer Freedom.* Retrieved from http://www.consumerfreedom.com/advertisements_detail.cfm/ad/15

Center for Consumer Freedom (CCF). (n.d.b). Food Police Smashing Your Choices? *Consumer Freedom.* Retrieved from http://www.consumerfreedom.com/advertisements_detail.cfm/ad/38

Chalecki, E. L. (2001). *A new vigilance: Identifying and reducing the risks of environmental terrorism.* Oakland, CA: Pacific Institute for Studies in Development, Environment, and Security.

Cochran, J. (2006, November 14). House judiciary chairman blocks bill against animal fighting. *ABC News.* Retrieved from http://abcnews.go.com/Politics/story?id=2652909&page=1

Corporate Watch. (2003a). New science minister another industry crony. *Corporate Watch.* Retrieved from http://www.corporatewatch.org/?lid=2224

Corporate Watch. (2003b). The Association of the British Pharmaceutical Industry. *Corporate Watch.* Retrieved from http://www.corporatewatch.org/?lid=332

Department of Health and the Association of the British Pharmaceutical Industry (DHABPI). (2001a). *Pharmaceutical industry competitiveness task force, final report.*

Department of Health and the Association of the British Pharmaceutical Industry (DHABPI). (2004). *Pharmaceutical industry competitiveness task force competitiveness and performance indicators.*

Drug Discovery and Development (DDD). (2010, January 4). Growing pressure to stop antibiotics in agriculture. *DDD Magazine.* Retrieved from http://www.dddmag.com/news-Growing-Pressure-to-Stop-Antibiotics-In-Agriculture-10410.aspx

Evans, R., & Taylor, M. (2009, October 25). Police in £9m scheme to log "domestic extremists". *The Guardian.* Retrieved from http://www.guardian.co.uk/uk/2009/oct/25/police-domestic-extremists-database

Evans, R., Taylor, M., Hirsch, A., & Lewis, P. (2011, January 13). Rein in undercover police units, says former DPP. *The Guardian.* Retrieved from http://www.guardian.co.uk/environment/2011/jan/13/mark-kennedy-undercover-police-acpo

Fox news. (2009, December 9). Sensenbrenner to tell Copenhagen: No climate laws until "scientific fascism" ends. *Fox news.* Retrieved from http://www.foxnews.com/politics/2009/12/09/sensenbrenner-climate-fascism

Helvarg, D. (2004). *The war against the greens.* Boulder, CO: Johnson.

Herman, E. S., & O'Sullivan, G. (1989). *The terrorism industry.* New York, NY: Pantheon.

Humane Society of the United States (HSUS). (2010). HSUS exposes inhumane treatment of pigs at Smithfield. *Human Society.* Retrieved from http://www.humanesociety.org/

Johnson, D. E. (2008). Cages, clinics, and consequences: The chilling problems of controlling special-interest extremism. *Oregon Law Review, 86,* 249–294.

Lewandowski, J. (2010, January 26). Is a pie in the face a terrorist act? *Globe and Mail.* Retrieved from http://www.theglobeandmail.com/news/politics/is-a-pie-in-the-face-a-terrorist-act/article1444392/

Lewis, P., & Evans, R. (2010, June 25). Peace campaigner, 85, classified by police as "domestic extremist." *The Guardian.* Retrieved from http://www.guardian.co.uk/uk/2010/jun/25/peace-campaigner-classified-domestic-extremist

Lovitz, D. (2010). *Muzzling a movement.* New York, NY: Lantern.

Marris, E. (2010). Animal rights "terror" law challenged. *Nature,* 466, 424.

Myers, J. (2010, January 2010). Inhofe on "enemies" list. *Tulsa World.* Retrieved from http://www.tulsaworld.com/news/article.aspx?subjectid=16&articleid=20100113_16_A1_OlhmeJ156867&archive=yes

National Abortion Federation (NAF). (2009). NAF violence and disruption statistics. *Pro choice.* Retrieved from http://www.prochoice.org/pubs_research/publications/downloads/about_abortion/violenc e_stats.pdf

National Animal Interest Alliance (NAIA). (n.d.a). Mission statement. *NAIA* online. Retrieved from http://www.naiaonline.org/about/mission.htmhttp://www.naiaonline.org/about/mission.htm

National Animal Interest Alliance (NAIA). (n.d.b). Position statement: Animal husbandry practices. *NAIA* online. Retrieved from http://naiaonline.org/articles/archives/policy_husbandry.html

National Animal Interest Alliance (NAIA). (n.d.c). Policy statement: Animals in entertainment and zoos. *NAIA* online. Retrieved from http://naiaonline.org/articles/archives/policy_animent.html

National Animal Interest Alliance (NAIA). (n.d.d). NAIA board, advisory board, staff, and volunteer staff. *NAIA* online. Retrieved from http://www.naiaonline.org/about/board.htm

National Animal Interest Alliance (NAIA). (2006, November 20). Voices of reason win the day AETA awaits president's signature. *NAIA* online. Retrieved from http://www.naiaonline.org/library/AETA.htm

National Animal Interest Alliance (NAIA). (2010). Animal rights and environmental extremists use intimidation and violence to achieve their ends. *NAIA* online. Retrieved from http://www.naiaonline.org/body/articles/archives/arterror.html

National Association for Biomedical Research (NABR). (n.d.). Animal rights extremism. *NABR*. Retrieved from http://www.nabr.org/Animal_Activism/Animal_Rights_Extremism.aspx

National Extremism Tactical Coordination Unit (NETCU). (n.d.). About NECTU. *NETCU*. Retrieved from http://www.netcu.org.uk/about/about.jsp

New Jersey Institute of Technology (NJIT). (2010). Donald H. Sebastian. *NJIT*. Retrieved from http://www.njit.edu/news/experts/sebastian.php

Obama, B. (2005). Statement of Senator Barack Obama oversight on eco-terrorism specifically examining the Earth Liberation Front (ELF) and the Animal Liberation Front (ALF). *IWAR*. Retrieved from http://www.iwar.org.uk/cyberterror/resources/eco-terror/obama.htm

Payne, S. (2006, August 19). Animal activists freed over filthy breeding centre. *The Telegraph*. Retrieved from http://www.telegraph.co.uk/news/1526690/Animal-activists-freed-over-filthy-breeding-centre.html

People for the Ethical Treatment of Animals (PETA). (n.d.). Victory! Wyeth closes premarin plant. *PETA*. Retrieved from http://www.peta.org/features/wyeth-closes-premarin-plant.aspx

Piltz, R. (2009, December 9). Sensenbrenner IPCC witch-hunt: Attempt to blacklist climate scientists must be rejected. *Climate Science Watch*. Retrieved from http://www.climatesciencewatch.org/2009/12/09/sensenbrenner-ipcc-witch-hunt-attempt-to-blacklist-climate-scientists-must-be-rejected/

Potter, W. (2009). The green scare. *Vermont Law Review*, *33*, 671–689.

Priest, D., & Arkin, W. M. (2010, July 18). Top Secret America. *Washington Post*. Retrieved from http://projects.washingtonpost.com/top-secret-america/

Rowlands, M. (2002). *Animals like us.* London: Verso.

Schmid, A. P. (2005). Terrorism as psychological warfare. *Democracy and Security*, 1, 137–146.

Shepherd, K. (2009, December 17). James Inhofe's one-man truth squad. *Mother Jones.* Retrieved from http://motherjones.com/blue-marble/2009/12/inhofe-brings-one-man-truth-squad-0

Sourcewatch. (n.d.). James M. Inhofe. *Sourcewatch.* Retrieved from http://www.sourcewatch.org/index.php?title=James_M._Inhofe

Sourcewatch. (2011a). NABR & the Animal Welfare Act. *Sourcewatch.* Retrieved from http://www.sourcewatch.org/index.php?title=NABR_%26_the_Animal_Welfare_Act

Sourcewatch. (2011b). Oregon national primate research centre. *Sourcewatch.* Retrieved from http://www.sourcewatch.org/index.php?title=Oregon_National_Primate_Research_Center

Sourcewatch. (2011c). Ten worst laboratories. *Sourcewatch.*Retrieved from http://www.sourcewatch.org/index.php?title=Ten_Worst_Laboratories

SPEAK. (n.d.). Bad politics. *SPEAK.* Retrieved from http://speakcampaigns.org/sitepages.php?a=5

Stewart, S. (2010, July 29). Escalating violence from the Animal Liberation Front. *Security Weekly.* Retrieved from http://www.stratfor.com/weekly/20100728_escalating_violence_animal_liberation_front/content/congressional-hearing-administration-officials-respond-e-mail-controversy

Toetz, J. (2007, February 14). Boss hog. *Rolling Stone.* Retrieved from http://www.rollingstone.com/politics/story/12840743/porks_dirty_secret_the_nations_top_hog_producer_is_also_one_of_americas_worst_polluters

Trull, F. (2009). Thirty years…time flies! *National Association for Biomedical Research.* Retrieved from http://www.nabr.org/AboutNABR/AskFrankie/tabid/952/EntryId/10/Thirty-years-Time-Flies.aspx

Unknown. (1993). Report to Congress on the extent and effects of domestic and international terrorism in animal enterprises. *The Physiologist*, *36*(6), 247–259.

Unknown. (2003, April 1). £2.5m Labour donation attacked as "corruption." *The Guardian.* Retrieved from http://www.guardian.co.uk/politics/2003/apr/01/labour.uk

Unknown. (2006, November 10). Lord Sainsbury insists "cash for honours" not behind his resignation. *Daily Mail.* Retrieved from http://www.dailymail.co.uk/news/article-415740/Lord-Sainsbury-insists-cash-honours-resignation.html

U.S. Department of State Office of the Coordinator for Counterterrorism (NCC). (2007). *Annex of statistical information to country reports on terrorism.* Washington, DC: National Counterterrorism Center. Retrieved from http://www.state.gov/s/ct/rls/crt/2006/82739.htm

United States Senate Committee on Environment and Public Works (USSCEPW). (2005). *Eco-terrorism specifically examining the Earth Liberation Front ("ELF") and the Animal Liberation Front ("ALF").* (S. HRG. 109-947). Washington, DC: U.S. Government printing office.

United States Senate Committee on Environment and Public Works (USSCEPW). (2010). *Consensus' exposed: The CRU controversy.* Washington, DC: U.S. Government printing office.

Ward, S. (2001). Media link September 11 with ecoterror. *Fur Commission USA*. Retrieved from http://www.furcommission.com/news/newsF03p.htm

Wilson, G. (2006, November 11). Sainsbury quits "in anger over loan affair." *The Telegraph.* Retrieved from http://www.telegraph.co.uk/news/uknews/1533851/Sainsbury-quits-in-anger-at-loans-affair.html

7. *Leaderless Resistance and Ideological Inclusion: The Case of the Earth Liberation Front*

PAUL JOOSSE

April 19, 2005 marked the tenth anniversary of the bombing of the Murrah Federal Building in Oklahoma City, an act that some have described as being an example of "leaderless resistance" (Burghardt, 1995; Kaplan, 1997; Mitrovica, 2004). Leaderless resistance is a strategy of opposition that allows for and encourages individuals or small cells to engage in acts of political violence entirely independent of any hierarchy of leadership or network of support. Although Louis Beam, a Klansman with strong connections to the Aryan Nations, developed and popularized the concept of leaderless resistance in the hopes of mobilizing many acts of violence from the far right (Beam, 1983, 1992), such acts have been relatively rare. The notion of leaderless resistance may have inspired the bombings carried out by Timothy McVeigh and Eric Rudolph (Mitrovica, 2004), but it has thus far failed to take hold widely among adherents of the racist far right in the way that Beam envisioned (Beam, 1983, 1992).

Another social movement, however, has been employing the strategy of leaderless resistance with a much higher degree of success. The radical environmentalist movement—the Earth Liberation Front (ELF)[1] in particular—offers a contemporary example of leaderless resistance in action (Garfinkel, 2003; Leader & Probst, 2003, pp. 37–58; Pressman, 2003). Although the ELF's acts are less severe than those of Timothy McVeigh or Eric Rudolph,[2] they are far more numerous. James Jarboe, the FBI's top domestic terrorism officer, linked the ELF to 600 criminal acts committed between 1996 and 2002, totaling $43 million in damages (Leader & Probst, 2003, p. 38). Most destructive of these was the arson of a Vail, Colorado, ski resort resulting in $12 million in damages. The ELF communiqué claiming responsibility for the Vail fire was written "on behalf of the lynx," an endangered species threatened by Vail Inc.'s expansion plans, and further warned that "We will be back if this greedy corporation continues to trespass into wild and unroaded areas" (Rosebraugh, 2004, p. 60). Attacks at many U.S. locations have indeed continued since, including the August 2003 burning down of a 206-unit apartment complex that had been under construction in San Diego, causing roughly $50 million in damages (Ackerman, 2003, p. 143). Most recently, four attacks occurred in November and December of 2005, three in the United States and one in Greece, together causing an estimated $567,600 in damages (Ecological Resistance from Around the World, 2006). As a consequence of this frequent and escalating leaderless resistance, John Lewis, an FBI deputy assistant director and top official in charge of domestic terrorism, labeled "ecoterrorism," along with "animal liberation terrorism,"[3] as "the No. 1 domestic terrorism threat" in 2005 (Schuster, 2005).

Thus far, academic literature pertaining to leaderless resistance has focused on its use as an effective strategy for avoiding detection, infiltration, and prosecution by a powerful state (Garfinkel, 2003; Kaplan, 1997, pp. 90–93; Leader & Probst, 2003, p. 39). In this chapter, I argue that the strategy of leaderless resistance has another benefit—one most easily enjoyed by social movements that display a high degree of ideological diversity. The radical environmentalist movement, itself an incredibly diverse social movement, thus provides an ideal case study for examining this hitherto unexplored benefit of leaderless resistance.

My central argument is that leaderless resistance allows the ELF to avoid ideological cleavages by eliminating all ideology extraneous to the very specific cause of halting the degradation of nature. In effect, the ELF's use of leaderless resistance creates an "overlapping consensus" among those with vastly different ideological orientations, mobilizing a mass of adherents that would have never been able to find unanimity of purpose in an organization characterized by a traditional, hierarchical, authority structure. In short, in using leaderless resistance, the ELF allows its adherents to "believe what they will," while still mobilizing them to commit "direct actions" for a specific cause.

The Development of a Concept: Leaderless Resistance in America's Radical Right

Motivating Louis Beam's attempts to popularize leaderless resistance was his realization that the American radical right was reaching a low point in terms of its popularity and strength. He wrote Leaderless Resistance "in the hope that, somehow, America can still produce the brave sons and daughters necessary to fight off ever increasing persecution and oppression" (Beam, 1992, p. 12). Because the essay is still salient for understanding leaderless resistance today, I repeat a significant portion below. Beam writes:

> The concept of Leaderless Resistance is nothing less than a fundamental departure in theories of organization. The orthodox scheme of organization is diagrammatically represented by the pyramid, with the mass at the bottom and the leader at the top.... This scheme of organization, the pyramid, is however, not only useless, but extremely dangerous for the participants when it is utilized in a resistance movement against state tyranny. Especially is this so in technologically advanced societies where electronic surveillance can often penetrate the structure revealing its chain of command.... Anti-state, political organizations utilizing this method of command and control are easy prey for government infiltration, entrapment, and destruction of the personnel involved.... This understood, the question arises "What method is left for those resisting state tyranny?".... A system of organization that is based upon the cell organization, but does not have any central control or direction.... Utilizing the Leaderless Resistance concept, all individuals and groups operate independently of each other, and never report to a central headquarters or single leader for direction or instruction, as would those who belong to a typical pyramid organization. (1992, pp. 12–13)

Thus, according to Beam's original conception, leaderless resistance is only truly in effect when there is a complete absence of "top-down" authority structures. Simson L. Garfinkel later underscored this requirement by maintaining that "hub and spoke" organizations, in which partially independent cells receive commands from above, do not qualify as true leaderless resistance (2003).

Odinist David Lane also contributed to the development of the concept of leaderless resistance (Kaplan, 1997, pp. 89–90). In his article "Wotan is Coming," Lane describes his movement's need for an aboveground political arm—the function of which is to disseminate propaganda—as well as an underground militant arm that he called Wotan (for "will of the Aryan nation") (1993). Lane advised that Wotan should "draw recruits from those educated by the political arm," thus ensuring that adherents are in line ideologically with the rest of the movement (1993). He also stressed, however, that:

> When a Wotan "goes active" he severs all apparent or provable ties with the political arm. If he has been so foolish as to obtain "membership" in such an organization, all records of such association must be destroyed or resignation submitted. (Lane, 1993)

The benefits of this severance would be obvious to members of Lane's movement, who know well the dangers associated with the FBI's scrutiny.

Both Beam and Lane were ideologues with heavy personal commitments to particular streams of the racist far right, and it only makes sense that they would seek and endorse organizational strategies that would ensure the preservation and advancement of their respective ideologies in toto. Beam, for one, has no doubt that ideological purity is maintainable in non-hierarchical organizational structures, stating, "It is certainly true that in any movement, all persons involved have the same general outlook, are acquainted with the same philosophy, and generally react to given situations in similar ways" (1992, p. n.p). Such a generalization would raise the eyebrows of any serious student of social movements, and here the intellectually sophisticated Beam is uncharacteristically simplistic. Likewise, Lane's recommendation of a severance from Wotan "of all apparent or provable ties with the political arm" creates an organizational system that gives free reign to the centrifugal forces of ideological deviation that threaten all ideological groups, a fact that he either never realizes or chooses not to mention. As I will show below, this conduciveness of leaderless resistance to ideological diversity, which threatens to subvert the intentions of ideologues like Beam and Lane, has proven to be beneficial to the radical environmentalist movements like the ELF, whose sole aim is to mobilize many actions, the ideological justifications for which may be manifold.

Leaderless Resistance in the ELF

The ELF first began operating in the UK in 1992, started by a group of Earth First!ers who were frustrated by their organization's desire to abandon illegal tactics (Taylor, 2005, p. 521). By 1997, actions were occurring in the United States, and the perpetrators began delivering communiqués, claiming responsibility to environmental activists Leslie James Pickering and Craig Rosebraugh, first through their mailbox and telephone, and then through e-mail (Rosebraugh, 2004, p. 20). Rosebraugh and Pickering would then act as publicists for the perpetrators, conducting media interviews that would publicize the communiqués. Websites also play a major role in the ELF's exhortations of actions, by disseminating guidelines for action, by reporting the various direct actions that ELFers commit, and by providing instructions about how to commit direct actions successfully.

The ELF's deliberate employment of the leaderless resistance strategy is evident from statements made on its website:

> Because the ELF structure is non-hierarchical, there is no centralized organization or leadership. There is also no "membership" in the Earth Liberation Front. In the past...individuals have committed arson and other illegal acts under the ELF name. Individuals who choose to do actions under the banner of E.L.F. do so only driven by their personal conscience. These have been individual choices, and are not endorsed, encouraged, or approved of by the management and participants of this web site. (EarthLiberationFront.com, 2005)

There appears to be no intramovement communication between ELF cells, and demonstrations or events at which ELF adherents could congregate are markedly absent (Garfinkel, 2003).

Thus, the ELF does not recruit members to a preexisting organization, but rather encourages people to start their own micro-organizations to further ELF's ends. In an introductory video to the ELF, publicist Craig Rosebraugh advises, "There's no realistic chance of becoming active in an already existing cell.... Take initiative; form your own cell" (Barcott, 2002, pp. 56–59, 81). Similar to Beam, Rosebraugh advocates the leaderless resistance strategy because, unlike pyramidal or hub-and-spoke organizational structures, "if one cell is infiltrated or captured by authorities, the members cannot provide any information that might lead to the capture of other cells" (2004, p. 182). Earth First! leader Judi Bari's praise of the development of the ELF in the UK is also reminiscent of David Lane's recommendation of a separation between public and clandestine "arms" of his movement. Writes Bari:

> England Earth First! has been taking some necessary steps to separate above ground and clandestine activities. Earth First!, the public group, has a nonviolence code and does civil disobedience blockades. Monkeywrenching is done by [the] Earth Liberation Front (ELF). Although Earth First!ers may sympathize with the activities of elf, they do not engage in them. If we are serious about our movement in the U.S., we will do the same. Despite the romantic notions of some over-imaginative Ed Abbey fans, Earth First! is in reality an above ground group. We have above ground publications, public events, and a yearly national Rendezvous with open attendance. Civil disobedience and sabotage are both powerful tactics in our movement. For the survival of both, it's time to leave the night work to the elves in the woods. (1994a)

It is interesting that Bari does not advocate the abandonment of all sabotage per se. Rather, she advocates leaving it to the "elves" for strategic reasons. Thus, the ELF appears to exemplify the strategy of leaderless resistance outlined by far-right thinkers such as Louis Beam and David Lane, but under the auspices of an entirely different ideological framework.

The categories are ideal-typical, and any exemplars would therefore only be approximate. What is more, some groups clearly change their orientation toward leadership and thus may shift categories over time. A prime example of this would be al-Qaeda, which, at the time of September 11, 2001, was fairly pyramidal in its organizational structure. Since then, however, it has undergone a rhizomatic leveling such that it would now be best placed in either the hub-and-spoke (Sageman, 2004) or leaderless resistance categories.

Radical Environmentalism as a Call to Action

It is clear that the core motivation for radical environmental movements like the ELF is a call to action—"direct actions" specifically. Radical environmentalists gauge the success of their movement not in terms of the number of adherents it is able to attract, or whether it manages to develop a cogent philosophy or "worldview," or even whether it is able to successfully lobby governments to pass environmentally friendly laws. Rather, because the radical environmentalist goal is immediate change, its standard of success is gauged by the number of "direct actions" it can mobilize, and the efficacy of these actions in putting a halt to the ongoing degradation of the wilderness.

Historically, this call to action was a consequence of frustration with the ineffectiveness of the traditional forms of environmental protest that organizations such as the Wilderness Society and the Sierra Club were employing. By 1977, future Earth First! co-founder Dave Foreman had risen to become the Wilderness Society's chief congressional lobbyist, but his experiences in Washington soon served to disillusion him and he resigned his post (Taylor, 2005, p. 518). He had come to see many environmental groups as "becoming indistinguishable from the corporations they were supposedly fighting" (Bookchin & Foreman, 1991) and he regarded the lobbyists alongside whom he had been working as "less part of a cause than members of a profession" (Foreman, 1991, p. 17). Thus, in 1980, he and five friends went hiking in Mexico's Pinacate Desert where they formed Earth First!. The group's slogan, "No compromise in defense of mother earth!" meant to signal that within this organization there would be none of the "give and take" strategy of the Washington environmental lobby. The group Foreman envisioned would be committed to direct action—both in the form of civil disobedience and monkeywrenching—seeing it as the only viable option for staving off an ecological catastrophe.

Dave Foreman made clear his intention that Earth First! would give precedent to actions as opposed to ideas in his 1982 article "Earth First!," saying, "Action is key. Action is more important than philosophical hairsplitting or endless refining of dogma (for which radicals are so well known). Let our actions set the finer points of our philosophy" (Foreman, 1982, p. 349). To this day, Earth First! still holds to the ideal of allowing many divergent viewpoints as long as these different stances translate into direct actions:

> While there is broad diversity within Earth First! from animal rights vegans to wilderness hunting guides, from monkeywrenchers to careful followers of Gandhi, from whiskey-drinking backwoods riffraff to thoughtful philosophers, from misanthropes to humanists there is agreement on one thing, the need for action! (Foreman, 1982)

Thus, inclusion and action are two ideals to which Earth First! strives. The history of Earth First! demonstrates, however, that at times these two ideals can be less than complementary.

Factions Rather Than Actions

Keeping in mind the thesis of this chapter, namely that the radical environmentalist movement enjoys an increased ability to mobilize actions because of the ideological inclusiveness that leaderless resistance fosters, we would do well to recognize some of the difficulties that the movement suffered before certain parts of it evolved to shed its leaders. As Earth First! grew, ideological cleavages would indeed compromise its ability to keep actions—not ideas—in the forefront of the movement. A seemingly constant source of internal ideological discord within Earth First! was its eponymous journal. In its early years, Earth First!'s small format meant that there was room for the works of members of Earth First!'s governing body, "The Circle of Darkness," and little else. Thus, initially there was a certain level of ideological purity within the journal. The waters began to muddy, however, between December of 1981 and February of 1982, as the number of letters to the editor that the journal published went from "four to thirty-one per issue. In its new format, the paper disseminated not only the leadership's beliefs but also the often-divergent beliefs of the membership" (Lee, 1995, p. 59). This tolerance for the expression of divergent beliefs and values is a source of pride for Earth First!, but as the group grew in size, these newly influential members "exerted a centrifugal force on the group's structure" (Lee, 1995, p. 59). The Earth First! journal thus became the forum for many ideological debates very early in the organization's development.

Often these disputes would become strikingly apparent when representatives from various Earth First! chapters congregated at national conferences. These meetings had a tendency to devolve into hostile and unproductive debate among various factions. Attempts to make sure that each participant had a chance to voice his or her own opinion also took away from the meetings' constructiveness. Illustrative of this is Bari's recollection of a meeting at which Earth First!er Karen Wood proposed to change the structure of Earth First!'s editorial board. The meeting style was clearly far from productive. Bari recalled that after Karen Wood's proposal:

> The facilitator said, "Okay, that's one proposal, now let's have another." And she recognized another person with another proposal, then another, then another. If someone tried to just make a comment, the facilitator said, "Let's turn that into a proposal," until finally there were 23 proposals simultaneously on the floor, and the entire group was thoroughly confused. (Bari, 1994b)

Ethnographer Jonathan Purkis (2001) also has commented on Earth First! meetings he visited in Manchester, UK. He noticed that much of the meetings' inefficiency derived from the anti-authoritarianism that made potential leaders within the movement unwilling to step forward, give direction, and set rules. In his experiences, he noted that:

> The meeting would start rather haphazardly.... Someone, usually one of the core group, would spread the mail which the group had received out on the floor, and start the meeting with a remark such as: "these are the things we should discuss—do something about."... The lack of group minutes to refer to from one meeting to another certainly reduced the effectiveness of how meetings were carried out. The informality of these meetings was striking, sometimes including interruptions such as telephone calls to (or from) other "northern" groups and off-the-point remarks, which often went unchecked.... One of the core group—Owen (pseudonym)—had joked that group discussions were made on the basis of "a great deal of aimless discussion and banter." (Purkis, 2001)

It is clear that this egalitarian meeting style, combined with the ideological diversity of Earth First!'s adherents, at times severely compromised Earth First!'s ability to delineate its goals—let alone to work toward them.

Eventually, Earth First! split into two main factions. One faction, led by Judi Bari, Mike Roselle, and Darryl Cherney, focused on social justice issues and renounced treespiking and other forms of monkeywrenching, in part because the practices were potentially dangerous for loggers. The other faction, led by Foreman and Christopher Manes, remained focused on protecting biodiversity and supported the use of all forms of direct action. In Bron Taylor's analysis, the Foreman Manes faction are given the nickname "Wilders" because they believed "that tying environmental protection to other issues, such as social justice, anti-imperialism, or workers rights, alienates many potential wilderness sympathizers" (1994, p. 199). The other faction viewed Foreman's focus as being far too narrow ideologically and believed in a more holistic (Taylor terms them "the Holies") approach to environmentalism (Taylor, 1994, p. 199). A detailed account of this process of factionalization is beyond the scope of this chapter, but ultimately Taylor contended that the reason for the schism can be "traced to small but significant differences in beliefs about human nature and eschatology" (Taylor, 1994, p. 200). As this factionalization progressed, more energy was diverted toward debates about ideology and away from performing the direct actions that Foreman had envisioned as being Earth First!'s forte. He lamented, "Disagreements over matters of philosophy and style…threaten to compromise the basic tenets of Earth First!, or make [it] impotent" (Lee, 1995, pp. 106–107).

Foreman eventually left Earth First! altogether and started Wild Earth, a journal more in line with his specific ideological orientation.[4] The Earth First! journal continued, but still caused discord within the organization, airing a multitude of ideological disputes, which led to further instability in the movement and journal. One Earth First!er lamented,

> Now, Dave [Foreman] & crew are gone; and the new Earth First! marches on with its shining vision…. We have advanced so far that we have reached the point where Dave Foreman stood nearly ten years ago: We realize that not everything fits in one journal. (Matthew, 1996)

Thus, ideological cleavages were a constant problem for Earth First!, the first major radical environmental group in the United States. These cleavages diverted the movement's focus away from its initial goal of planning and instigating actions that would protect the wilderness from degradation. Despite this, Earth First! remains a potent—though less radical—force in the wider environmental movement milieu and continues to have its own successes and failures in relation to its current goals.

Benefits of Leaderless Resistance for the ELF

Bron Taylor gives the most authoritative account of the emergence of the ELF in his Encyclopedia of Religion and Nature, citing various Earth First! sources which claim that the ELF began as a radical offshoot of Earth First! in England (Taylor, 2005, p. 521). Taylor thus includes both Earth First! and the ELF under the same encyclopedic heading, signaling—what was in the beginning at least—a fundamental indistinctness between the movements. Clearly, today the ELF has outgrown this association with Earth First!, partly through its use of leaderless resistance, a strategy of recruitment that is well-suited to reaching beyond traditional ideological boundaries. The divergence of the two movements has meant that, while Earth First! has continued to moderate, looking less and less distinct from other formerly radical groups like Greenpeace, the ELF has produced ever-more extreme actions which have captured headlines around the world.

Both Ackerman and Taylor (Taylor, 1998, p. 14) argue that "prolific intra-movement debate" (Ackerman, 2003, p. 145) decreases the likelihood that members within a movement will begin to commit violent acts because debate tends to have a moderating effect on the extreme members and/or elements of organizations. Thus, for movements predicated on endorsing violent actions, the best strategy would be to limit opportunities for debate while being inclusive of a wide range of ideological positions. Below are some of the specific ways that leaderless resistance has enabled the ELF to be more ideologically inclusive.

First, the ELF moniker itself increases the range of ideological positions to which adherents can remain sympathetic, by enabling adherents to interpret the name in a way that suits their ideological orientation. For example, some radical environmentalists choose to conflate the animal liberation movement, represented by aboveground organizations such as People for the Ethical Treatment of Animals (PETA), with the radical environmentalist movement. For them, the Animal Liberation Front (ALF) and the ELF are merely different expressions of the same underlying ideology, and they see this unity represented by the similarity of the two movements' names. Other radical environmentalists, however, protest this union because they regard the actions of animal liberationists—who in the past have "liberated" exotic animals by releasing them into the wild—as being harmful to ecosystems. So, while some choose to see ELF and ALF as twin movements, others—for whom this pairing would be distasteful—can choose to see the ELF as entirely autonomous. Thus, when adherents of the ELF decide to engage in direct action, they can choose with whom they wish to associate ideologically.

The ELF moniker also lends itself to interpretations that are favorable to both sides of another prominent debate within the environmentalist movement, concerning the role that religion and/or myth ought to play in protest. Philosopher Kate Soper noted that there is a:

> Spectrum of positions in the green movement ranging from those who would dismiss any recourse to myth or magic as a capitulation to irrationalism that can only discredit its forms of protest, to those who would insist that these forms of thinking offer the most powerful and effective antidote to instrumental rationality. (1996)

While primarily political-rational-minded or secular adherents will read "ELF" as an acronym for "earth liberation front," those who have an affinity to the more mystical, pagan aspects of radical environmentalism will be more likely to read the ELF appellation in terms of its pagan symbolism, seeing themselves as mischievous "elves" who come to wreak havoc in the night (Taylor, 1998, p. 9). By being interpretable, the ELF moniker appeals to both ends of the sacred–secular spectrum, reducing the likelihood that someone will abandon his or her adherence to the movement because of disagreements about the role of religion and myth in environmental protest. Thus, the ELF name allows the movement to "cast its net wide" for adherents with very different ideological orientations.

Second, the ELF's ability to attract young men is enhanced by its limitation of ideological content on its website and in its publications. An overwhelming proportion of young men in an organization's constituency will provide a motivational predisposition for a general transition to more violent behavior (Ackerman, 2003, p. 148). This is a result of simple and measurable tendencies of young and male demographics. For example, a survey of U.S. district courts found that 92.9% of all defendants convicted for violent crimes in 2001 were male, while 78.4% of defendants convicted were between sixteen and forty years of age. Thus, given that violent actions are most likely to be perpetrated by those who are young or male, movements like the ELF which seek to instigate violent actions do best when their propaganda targets these demographics.

Since, however, young males do not tend to adhere to any particular ideology, and are distributed evenly throughout society, it would be difficult to provide an ideological basis for attracting young men specifically. Indeed, Chip Berlet, a senior analyst from the left-wing think tank, Political Research Associates, sees the ELF website as appealing more to young males' desire for glory rather than to any specific ideological beliefs they might hold. He sees the website as "a framework for recruiting young men to do this kind of stuff.... You come up with an exhortation of what a hero will do, and some person comes out and says 'I want to be a hero'" (Garfinkel, 2003).

The wording of ELF communiqués is often rebellious and playful, using themes such as Christmas in an irreverent way that would be appealing to young, disgruntled would-be heroes. Particularly striking in this regard was the communiqué sent to Rosebraugh after the burning of a U.S. Forest Industries office in Medford, Oregon, in 1998:

> To celebrate the holidays we decided on a bonfire. Unfortunately for US Forest Industries it was at their corporate headquarters office. On the foggy night after Christmas when everyone was digesting their turkey and pie, Santa's ELFs dropped two five-gallon buckets of diesel-unleaded mix and a gallon jug with cigarette delay; which proved to be more than enough to get this party started. This was in retribution for all the wild forests and animals lost to feed the wallets

> of greedy fucks like Jerry Bramwell, USFI president. This action is payback and it is a warning to all others responsible, we do not sleep and we won't quit. (Rosebraugh, 2004, p. 72)

What strikes one about this communiqué are not powerful ideological arguments—indeed, the ideological justifications are quite vague. Clearly of more impact for potential youthful recruits would be the almost comic-bookish style in which the communiqué was written. The arson is depicted as a mischievous "party" carried out by elfish subverters who act under the cover of darkness.

Ackerman points out that, of the few suspects who have been arrested or indicted for connections to ELF actions, "all but one have been male and most are teenagers or young adults"[5] (2003, p. 148). When one looks at these individuals, they are surprisingly bereft of long-standing and deep environmentalist commitments. For example, *New York Times* writer Al Baker had suspicions about how ideological were the motivations of Matthew Rammelkap (16), George Mashkow (17), and Jared McIntyre (17), all of whom plead guilty to arson conspiracy in 2001. He wondered if their ELF-claimed actions were "the work of a smart, devoted band of ecoterrorists or young vandals merely blowing off adolescent steam?" (Baker, 2001). Then there are Craig "Critter" Marshall (28) and Jeffrey "Free" Luers (22). Marshall, who is now serving a five-and-a-half-year sentence for fire-bombing a Chevrolet dealership in Eugene, Oregon, admitted to *New York Times* reporter Bruce Barcott (2002) that growing up, he "held political beliefs that weren't so much pro-environment as anti-authority" (p. 58). Similarly, Jeffrey Luers, now serving a twenty-two-year and eight-month sentence for his participation, remarked in an interview with Earth First! that "originally I was radicalized by anti-authoritarian, anarchist beliefs, as well as animal rights," and that his environmental radicalism came only in 1997 (SpritOfFreedom.org, 2006). Thus, one could question whether the ELF would have been able to mobilize these young males if it were more ideologically specific in its propaganda.

Another example of this strategy of limiting ideological content is the ELF's thirty-seven-page manual, *Setting Fires with Electrical Timers: An Earth Liberation Front Guide*. While it gives very detailed instructions on how to engage in acts of arson, this manual is nearly devoid of references to environmental issues or ideology. On the second page are instructions to copy and distribute the manual to "bookstores that specialize in animal rights, environmental and anarchist literature." After this very brief mention of the broad ideological orientation of its authors, the rest of the manual is devoted to technical issues such as creating a clean room to avoid leaving DNA evidence and soldering a digital timer for an incendiary device. By not explicitly stating ideological precepts, the manual lends itself to use by anyone, regardless of the person's ideological orientation. This open use is of little practical concern for the ELF, however, because, as Garfinkel (commenting on the Vail, Colorado arson) writes:

> Even if the ELF was not responsible for the Vail fire, ELF's claim of the fire gives it a powerful propaganda tool: a photograph of what appears to be the burning hotel appears on the front page of ELF's Web site. Even if people believing in ELF's ideology were not directly responsible for the fire, the existing of ELF and its ideology may have given the arsonists the additional motivation or cover to carry out the crime. (2003)

Today, actions from the ELF are very common, and fear of terrorism is rampant. In this climate, there may be no safer way to commit insurance fraud, or revengeful arson, or just go thrill-seeking, than to follow the ELF's guidelines, spray paint "the elves were here" at the site, and lead authorities up the garden path. Thus, the definition by the public and law enforcement of many of the ELF's acts as exclusively motivated by environmental concerns is itself part of the ELF's mobilization strategy. That the ELF gains notoriety and influence through the actions of those whose true motivations are far from certain underscores a foundational truism of sociological inquiry expressed poignantly by William Isaac Thomas: "If men define situations as real, they are real in their consequences" (Thomas & Thomas, 1929, p. 1).

Politics as a Contentious Issue among Radical Environmentalists

We have seen how leaderless resistance is beneficial to the ELF specifically, but there are many areas of debate that can be fractious for environmental organizations in general. Before closing this chapter, I consider just one of these areas—environmental politics—below.

Conventional wisdom is prone to seeing environmental concerns as existing primarily within the domain of left-of-center political interests. The presence of conservative anti-environmentalist organizations such as the Center for the Defense of Free Enterprise (CDFE), the "wise use" movement, along with the lack of concern for environmental issues by the Reagan and both Bush administrations reinforces this perception. John Gray summarized the conventional characterization of the relationship between conservatism and environmentalism:

> It is fair to say that, on the whole, conservative thought has been hostile to environmental concerns over the past decade or so in Britain, Europe and the United States. Especially in America, environmental concerns have been represented as anti-capitalist propaganda under another flag. (1993)

Today, the idea that environmentalism is an exclusively liberal cause continues to be popularly held despite some recent developments that would challenge such views. Thus, for many, the recent attempts by the Bush administration to open Alaska's Arctic National Wildlife Refuge to oil drilling represents merely another incident in continuance with a long legacy of environmental irresponsibility by conservatives in America.

Though it is true that those who hold positions of power within conservative movements have largely been unsympathetic to environmental causes, a conservative political orientation itself is not necessarily antagonistic to environmental concerns. Those not in power in the right wing (and thus of more interest for the study of leaderless resistance) are more likely to have interests and beliefs that are divergent from the mainstream of their movement. As Bruce Pilbeam showed, an environmental consciousness can be consistent with the general political philosophy to which conservatives subscribe. Furthermore, Pilbeam outlined how conservative thought may have an affinity even with many qualities of deep ecology—the philosophy that guides the thinking of many radical environmentalists (2003).

This potential affinity between conservatism and deep ecology makes the fact of Dave Foreman's Republican Party membership, his support of the Vietnam War, and his work as campaign manager for Barry Goldwater (Lee, 1995, p. 27) seem less surprising. Although the liberal Earth First!er Judi Bari saw "an inherent contradiction in Dave Foreman" (1994c), in fact, his example shows how conservative thought can be combined with radical environmentalist concerns to form a cogent worldview. Thus, Foreman's orientation is not merely an anomaly, a quirky exception to the general rules of where environmentalist concerns ought to fit within the political spectrum. Rather, he exemplifies how the politics of environmentalism often are incommensurable with the traditional left–right distinction that usually shapes political thought.

Recognition of this incommensurability also provides insight into the motivations of Canada's most prominent ecoteur, Wiebo Ludwig. On April 19, 2000, Ludwig was convicted of bombing a gas well and encasing another wellhead in concrete along with three other explosives-related charges in northwestern Alberta (Brooke, 2000), crimes for which he spent twenty-one months in jail. Two of these counts were for mischief by destroying property and possessing an explosive substance (Nikiforuk, 2001, p. 247). Interestingly, when committing direct actions, Ludwig used ideas that he gleaned from Dave Foreman's book, *Ecodefense: A Field Guide to Monkeywrenching*, such as covering his shoes with socks to avoid leaving tracks (Nikiforuk, 2001, p. 110).

A former Christian Reformed Church preacher, Ludwig was intensely conservative on social issues. While pastor of Goderich Christian Reformed Church, his strict views about male "headship" and the roles of women caused much dissention among his congregation. According to Nikiforuk, "He asked working women why they weren't home caring for children, and women with one or two offspring why they hadn't begotten 'a full quiver'" (2001, p. 2). For a time in 1999, rumors were circulating that Ludwig might run for leadership of the ultra-conservative Social Credit party in Alberta (Canadian Broadcasting Corporation, 1999). The late Green activist, Tooker Gomberg, who was a prominent liberal, spent some time camping with Ludwig, and summarized his feelings about the man as follows:

> I find myself staring into the fire for relief, trying to work through the paradox that, although this man is a patriarchal diehard, a fundamentalist, anti-gay—and arrogant—we have few differences on the ecological front. Dare I say I admire him? A few years back I stayed at his rambling farmhouse, where I marveled at the family's self-reliance. But he remains an imperfect hero. (2002)
>
> Thus, if one were to gather together a group of radical environmentalists, one can only assume that their discussions of politics would be lively, if not mutually vitriolic.

Only with a leaderless resistance strategy could people with political ideologies as divergent as Ludwig and Gomberg be mobilized to commit acts for a similar cause.

Conclusion

Social movements as different from one another as the American radical right and radical environmentalism are able to employ the strategy of leaderless resistance. The radical environmentalist movement's use of the strategy illustrates how it is conducive to intra-movement ideological diversity as well. Although the progenitors of leaderless resistance in these two social movements seek to assure potential followers (and perhaps themselves) that what coheres their respective movements is a shared ideology, the organizational structure (or lack thereof) of leaderless resistance means that there is, in fact, no way of determining if such a shared ideology actually exists. Once a social movement leader implements leaderless resistance, the movement becomes, in a sense, a "creature unto itself," and those who commit actions do so of their own ideological volition, completely separate from the wishes of those who are at times considered to be the movement's de facto leaders.

There is no doubt that, initially, the impetus for the ELF's adoption of the leaderless resistance strategy was the same as that of the American radical right: to avoid state detection, infiltration, and prosecution by powerful government agencies. Once implemented, however, it became clear that leaderless resistance also allows the ELF to avoid ideological cleavages by eliminating all ideology extraneous to the very specific cause of halting the degradation of nature, thereby eliminating opportunities for ideological debate. In effect, the ELF's use of leaderless resistance creates an overlapping consensus among those with vastly different ideological orientations, mobilizing a mass of adherents who would have never been able to work together in an organization like Earth First!, which is characterized by a more traditional organizational structure. In short, in using leaderless resistance, the ELF allows its adherents to "believe what they will" while still mobilizing them to commit many direct actions for a specific cause.

Since the initial writing of this chapter, there has been a rash of arrests and indictments against suspected ELF adherents. Based on the thesis presented here, one recommendation to investigators of terrorism is a caution against relying too heavily on ideological linkages among perpetrators of leaderless resistance actions. In leaderless resistance, the reasons for the formation of a new violent cell may have much more to do with group dynamics at the micro level and the psychological makeup/personal histories of violence-prone individuals rather than with the particular ideology to which perpetrators happen to subscribe or the subcultural milieu that they inhabit. An overreliance on ideological linkages in investigations of leaderless resistance is not only ineffective, but it can also elicit perceptions of harassment, contributing to persecutory ideation which in turn may serve to further radicalize fringe elements of movements that employ leaderless resistance.

Notes

1. Throughout this chapter I refer to "the ELF," but by this phrase, I do not intend to convey a sense that the ELF is characterized by significant levels of organizational unity or social cohesion. As this chapter will illustrate, rather than a "group" or an "organization," the ELF should only be seen as a collectivity in the most limited and virtual sense. Any conceptions of membership that are more robust than this would be misapplied in the case of the ELF.
2. Actions of radical environmentalists are less severe in that they aim not to kill human beings but rather to cause fear and to destroy property.

3. The FBI has consistently conflated the Animal Liberation Front (ALF) with the ELF. Although the ELF and ALF did release a communiqué claiming solidarity of action in 1993, it would be more precise to regard the two movements as separate for a number of reasons.
4. Wild Earth ceased publication in 2004.
5. It should be noted, however, that contrary to this trend, among those named in the January 19, 2006 indictment of eleven suspected ELF members were six women.

References

Ackerman, G. A. (2003). Beyond arson? A threat assessment of the Earth Liberation Front. *Terrorism and Political Violence, 15*(4), 143–170. Retrieved from http://dx.doi.org/10.1080/09546550390449935

Baker, A. (2001, February 18). A Federal case in Suffolk: Eco-terrorism or adolescence in bloom? *The New York Times.* Retrieved from https://www.nytimes.com/2001/02/18/nyregion/a-federal-case-in-suffolk-eco-terrorism-or-adolescence-in-bloom.html

Barcott, B. (2002, April 7). From tree-hugger to terrorist. *The New York Times.* Retrieved from https://www.nytimes.com/2002/04/07/magazine/from-tree-hugger-to-terrorist.html

Bari, J. (1994a). Monkeywrenching. In J. Bari (Ed.), *Timber Wars* (p. 285). Monroe, ME: Common Courage Press.

Bari, J. (1994b). Showdown at the Earth First! Corral. In J. Bari (Ed.), *Timber Wars* (pp. 207–208). Monroe, ME: Common Courage Press.

Bari, J. (1994c). Review: Dave Foreman's confessions of an eco-warrior. In J. Bari (Ed.), *Timber Wars* (p. 104). Monroe, ME: Common Courage Press.

Beam, L. (1983, May). Leaderless resistance. *Inter-Klan Newsletter Survival Alert.*

Beam, L. (1992, February). Leaderless resistance. *The Seditionist,* p. 12. Retrieved from http://www.solargeneral.com/library/leaderlessresistance.htm

Bookchin, M., & Foreman, D. (1991). *Defending the earth: A dialogue between Murray Bookchin and Dave Foreman, Steve Chase.* Boston, MA: South End Press.

Brooke, J. (2000, April 20). Radical environmentalist convicted of gas well blast in Canada. *The New York Times.* Retrieved from https://www.nytimes.com/2000/04/20/world/radical-environmentalist-convicted-of-gas-well-blast-in-canada.html

Burghardt, T. (1995). *Leaderless resistance and the Oklahoma City bombing.* San Francisco, CA: Bay Area Coalition for Our Reproductive Rights.

Canadian Broadcasting Corporation. (1999). *Program Summary for 13-10-99.* Retrieved from http://www.cbc.ca/insite/CANADA_AT_FIVE_TORONTO/1999/10/13.html

Ecological Resistance from Around the World. (2006, Spring). *Green Anarchy,* p. 22, 30.

Foreman, D. (1982). Earth First! In J. P. Sterba (Ed.), *Earth ethics: Introductory readings on animal rights and environmental ethics* (p. 349). Upper Saddle River, NJ: Prentice Hall.

Foreman, D. (1991). *Confessions of an eco-warrior*. New York, NY: Harmony.

Garfinkel, S. (2003). Leaderless resistance today. *First Monday*, *8*(3). Retrieved from https://doi.org/10.5210/fm.v8i3.1040

Gomberg, T. (2002, September 5). *Flawed messiah*. Retrieved from https://nowtoronto.com/news/story.cfm%3Fcontent%3D133452

Gray, J. (1993). *Beyond the new right: Markets, government and the common environment*. London: Routledge.

Kaplan, J. (1997). Leaderless resistance. *Terrorism and Political Violence*, *9*(3). Retrieved from https://doi.org/10.1080/09546559708427417

Lane, D. (1993). *Wotan is coming*. Retrieved from http://www1.ca.nizkor.org/ftp.cgi/people/l/lane.david/ftp.py?people/l/ lane.david//wotan-is-coming

Leader, S., & Probst, P. (2003). The Earth Liberation Front and environmental terrorism. *Terrorism and Political Violence*, *15*(4), 80–95. Retrieved from https://doi.org/10.1080/09546559708427417

Lee, M. F. (1995). *Earth First!: Environmental apocalypse*. Syracuse, NY: Syracuse University Press.

Matthew, B. (1996). Letter to the editor. *Earth First! 13*(6), 3.

Mitrovica, A. (2004). Droege is gone, but the hate lives on. *The Walrus*. Retrieved from http://www.walrusmagazine.com/article.pl?sid=05/04/22/1928218

Nikiforuk, A. (2001). *Saboteurs: Weibo Ludwig's war against big oil*. Toronto, ON: Macfarlane Walter & Ross.

Pilbeam, B. (2003). Natural allies? Mapping the relationship between conservatism and environmentalism. *Political Studies*, *51*, 490–508.

Pressman, J. (2003, December). Leaderless resistance: The next threat? *Current History*, *102*(668), 422–425.

Purkis, J. (2001). Leaderless cultures: The problem of authority in a radical environmental group. In C. Barker, A. Johnson, & M. Lavalette (Eds.), *Leadership and social movements* (pp. 160–177). Manchester & New York, NY: Manchester University Press.

Rosebraugh, C. (2004). *Burning rage of a dying planet: Speaking for the Earth Liberation Front*. New York, NY: Lantern Books.

Sageman, M. (2004). *Understanding terror networks*. Philadelphia, PA: University of Pennsylvania Press.

Schuster, H. (2005, August 24). *Domestic terror: Who's the most dangerous? Eco-terrorists are now above ultra-right extremists on the FBI charts*. Retrieved from http://www.cnn.com/2005/US/08/24/schuster.column/index.html

Soper, K. (1996, September). A green mythology. *CNS*, *7*(3), 120.

SpritOfFreedom.org. (2006). Retrieved from http://www.spiritoffreedom.org.uk/profiles/free/ef.html

Taylor, B. (1994). Earth First!'s religious radicalism. In C. K. Chapple (Ed.), *Ecological prospects: Scientific, religious, and aesthetic perspectives*, pp. 185-209. Albany, NY: State University of New York Press.

Taylor, B. (1998). Religion, violence and environmentalism: From Earth First! to the Unabomber, to Earth Liberation Front. *Terrorism and Political Violence*, *10*(4), 20.

Taylor, B. (2005). Earth First! and the Earth Liberation Front. In B. Taylor & J. Kaplan (Eds.), *Encyclopedia of religion and nature*, pp. 518-525. London & New York, NY: Thoemmes Continuum.

Thomas, W. I., & Thomas, D. (1929). *The child in America* (2nd ed.). New York, NY: Alfred Knopf.

Part III: Classic Writings on Political Repression

8. *Standing Up to Corporate Greed: The Earth Liberation Front as Domestic Terrorist Target Number One*

Anthony J. Nocella II and Matthew J. Walton

The current global political climate is steeped in fear of, and rhetoric about, political violence and terrorism. Scholars and practitioners must go beyond this fear and rhetoric, and instead seek a more nuanced understanding of political groups that utilize property destruction, kidnapping, assassination, armed struggle and other militant actions for political or ideological ends. Such understanding is important in order to slow down and reverse the current trend among legislative and policy-making bodies and leaders around the world, who increasingly marginalize, demonize and exclude such groups from arenas of debate, allegedly in the name of counterterrorism.

This chapter will examine the actions and philosophy of the Earth Liberation Front (ELF), particularly in relation to global capitalism. We seek to explain why the ELF does what it does, and why its actions have situated it atop the FBI Domestic Terrorist list, despite the fact that ELF guidelines specifically prohibit harming any human or nonhuman animals. Our argument is that ELF actions contain a compelling critique of capitalism, which is much more of a threat to "American values" and to the consumer-driven U.S. way of life in general, than other potential threats that seek to harm humans. In other words, maintaining the dominance of corporate power and the supremacy of market capitalism is more important to the U.S. government and intelligence agencies than protecting the lives of U.S. citizens.

Organization

We begin our chapter by examining the development of global capitalism in the 21st century. We then move on to give some historical background on the ELF as well as the philosophical underpinnings of the organization. We examine the demand for the ELF and show some examples of state repression of ELF activists and supporters. We then dissect the notion of "ecoterrorism," looking particularly at legislation that targets ELF-style actions. We will end by analyzing in more detail the ELF critique of capitalism, highlighting the threat that the ELF poses to this dominant economic and social paradigm.

Global Capitalism in the 21st Century

As we begin the 21st century it is useful to look back and examine the historical events and ideologies that have shaped the world we live in today. The 20th century, particularly the latter half, was characterized by industrialization, globalization, and technological development. All of these processes have been driven by one ideological agenda that has been sold to the global community as not just beneficial, but inevitable; this agenda is capitalism.

The phenomenal growth that came about as a result of the processes mentioned above could not have occurred without significant governmental influence and control over not only the economic sphere, but social relations as well. This level of control cannot be reached by any government without resorting to tactics that repress those elements of society, which seek to undermine the influence of market capitalism (Henderson, 1991). Some observers may feel that government control has become even more penetrating in light of the recent drastic shift in U.S. policy with regard to national security (Churchill, 2003). However, these tactics, specifically the depletion of civil liberties and limitations on social activism and freedom of speech, are only a continuation (albeit an intensification) of the repression that has necessarily existed within states from their formation (Goldstein, 2001; Schultz & Schultz, 1989).

The "War on Terror" is being fought by Western nations to protect everything that these former colonial states have acquired and achieved (mostly through destructive and intrusive means) over the past several hundred years. Through this lens, so-called global terrorism is the opposite of global capitalism. In response to aggressive capitalist globalization policies, intense forms of resistance are mounting against the great endorsers of corporate domination such as the United States and the UK. These resistance movements range from anti-Iraq war and social justice mobilization to Islamic fundamentalist forces such as al-Qaeda to the ELF.[1]

Background

The ELF was founded in Brighton, England, by members of Earth First! (an aboveground nonviolent radical environmental group) who refused to abandon criminal acts as a tactic when others wished to move the group into the mainstream (EarthLiberationFront.com, n.d.a). This forced a split in the Earth First! chapter in England and led to the creation of the ELF. The ELF approach was modeled on the Animal Liberation Front (ALF) and members utilized economic sabotage to directly attack their corporate enemies. The ELF's first action in the United States was on October 14, 1996 in Eugene, Oregon; a McDonald's had its locks glued and was spray painted, "Earth Liberation Front."

For the ELF and their supporters, the environmental crisis is real and severe; they believe that extraordinary times demand extraordinary actions. Arguing that the state and its approved legal channels of social and economic change are bureaucratic traps and dead-ends—and believing that the state is essentially the political arm of the ruling elite—ELF activists insist that the only way to stop exploitative industries is to attack their economic nerve center through costly acts of sabotage. Perhaps it is best to allow the ELF to explain itself:

> The Earth Liberation Front (ELF) is an international underground organization that uses direct action to stop the exploitation and destruction of the natural environment. The ELF realizes that all life on Earth is threatened by entities concerned with nothing more than pursuing economic gain at any cost. Therefore, the ELF uses clandestine guerrilla tactics in efforts to take the profit motive out of killing the Earth. (EarthLiberationFront.com, n.d.b)

> The ELF is organized into autonomous cells that operate independently and anonymously from one another and the general public. The group does not contain a hierarchy or any sort of leadership. Instead, the group operates under an ideology. If an individual believes in the ideology and follows the ELF guidelines, she or he can perform actions and become a part of the ELF. This means that anyone can be involved, even you.
>
> The ELF is structured in such a way as to maximize effectiveness. By operating in anonymous cells, the security of group members is maintained. This decentralized structure helps keep activists out of jail and free to continue conducting actions. (EarthLiberationFront.com, n.d.c)

The Philosophy of the ELF

ELF activists and supporters do not adhere to one particular ideology or theory. They are, however, most commonly associated with deep ecology, a way of living first articulated by the Norwegian scholar Arne Naess in the early 1970s. Deep ecology is often described in comparison to its counterpart, "shallow" ecology, or the conservation of resources based on utility for the human community (recycling, capping emissions, fuel standards, etc.). Deep ecologists are dedicated to the ideal of all living beings (plants, animals, even ecosystems as a whole) living together without being commodified as "resources" or used, oppressed, or destroyed for economic reasons. The theory also has strong critiques of capitalism, human overpopulation, materialism, and human overconsumption. Although some social movement theorists and environmental scholars write that radical environmentalists' motivation derives from a well-articulated philosophy of deep ecology, this is usually far from the truth. Rik Scarce explains that "most eco-warriors have no interest in a well-conceived philosophy or in any other explicit guideposts to tell them how to live their lives" (1990). While it is true that the tenets of deep ecology are compatible with many of the views of radical environmentalists, in the case of the ELF, it seems that actions are of primary importance, and a philosophical basis for these actions is only a secondary concern.

The organizational makeup of the ELF is rooted in anarchy, which results in a non-hierarchical, leaderless structure. ELF actions are governed only by the guidelines, which are posted in multiple locations on the Internet and distributed through various pieces of literature. Consequently, the motivational drive to protect the earth manifests differently within each cell and each member. An ELF member could be a southern Republican who does not want a highway in the back of her home or a parent who has been devastated to find out that his child is dying of a local pesticide that is being sprayed on a farm nearby.

The piece of literature that probably did the most to motivate individuals to adopt a radical environmental stance utilizing property destruction is Edward Abbey's (2000) *The Monkey Wrench Gang*. Abbey, an anarchist, writes about a group of pissed off environmentalists who all come to their radical convictions through diverse experiences and beliefs. They are fed up with what the "progress" of industrial capitalism is doing to the planet (and to their desert in particular) and set out to destroy billboards, bulldozers, bridges, and trains, and even dream of blowing up Glen Canyon Dam. The term "monkeywrench" was consequently taken up by Earth First! and is used to describe small acts of midnight vandalism, such as putting sand in the gas tank of an earth mover, flattening the tires of a bulldozer, or putting glue in locks of the U.S. Forest Service offices.

The Demand for the ELF

The ELF is an extremely topical and controversial group, yet to date there is very little analysis or understanding of them. In many ways, this is an advantage for activists, as it has hindered law enforcement efforts to infiltrate the organization, but it also makes it more difficult for other streams of radical and even mainstream environmentalism to come to an understanding of why ELF members do what they do. In the early part of 1997 a communiqué from the ELF was sent out; it read:

> Beltane, 1997
>
> Welcome to the struggle of all species to be free. We are the burning rage of this dying planet. The war of greed ravages the earth and species die out every day. E.L.F. works to speed up the collapse of industry, to scare the rich, and to undermine the foundations of the state. We embrace social and deep ecology as a practical resistance movement. We have to show the enemy that we are serious about defending what is sacred. Together we have teeth and claws to match our dreams. Our greatest weapons are imagination and the ability to strike when least expected.
>
> Since 1992 a series of earth nights and Halloween smashes has mushroomed around the world. 1000's of bulldozers, powerlines, computer systems, buildings and valuable equipment have been composted. Many E.L.F. actions have been censored to prevent our bravery from inciting others to take action.
>
> We take inspiration from Luddites, Levellers, Diggers, the Autonome squatter movement, the A.L.F., the Zapatistas, and the little people—those mischievous elves of lore. Authorities can't see us because they don't believe in elves. We are practically invisible. We have no command structure, no spokespersons, no office, just many small groups working separately, seeking vulnerable targets and practicing our craft.
>
> Many elves are moving to the Pacific Northwest and other sacred areas. Some elves will leave surprises as they go. Find your family! And let's dance as we make ruins of the corporate money system.

It is clear from this that the ELF does not only operate under the goal of defending Mother Earth, but also values building solidarity with other revolutionary groups that have common elements. They are facilitating an understanding of property destruction that moves from principle to practice, as a means to cause economic sabotage to corporations and to bring to their knees all those who profit from the destruction of the planet.

Repression of the ELF and Its Supporters

The U.S. government began strategic repression of the ELF and its supporters essentially since the organization's inception in this country in 1996. Individuals convicted of ELF-style actions have been given severe sentences and those people who publicly advocate for and defend the group have been harassed and have had their homes raided. The following communiqué is the first public notice by the ELF of solidarity with activists who are being repressed for their support of the underground organization. The communiqué claimed responsibility for torching four luxury homes on Long Island, New York, in resistance to gentrification. The estimated cost of damages was two million dollars.

> December 31, 2000
>
> Greetings Friends,
>
> As an early New Years gift to Long Island's environment destroyers, the Earth Liberation Front (E.L.F.) visited a construction site on December 29 and set fire to 4 unsold Luxury houses nearly completed at Island Estates in Mount Sinai, Long Island. Hopefully, this caused nearly $2 million in damage. This hopefully provided a firm message that we will not tolerate the destruction of our Island. Recently, hundreds of houses have been built over much of Mount Sinai's picturesque landscape and developers now plan to build a further 189 luxury houses over the farms and forests adjacent to Island Estates. This action was done in solidarity with Josh Harper, Craig Rosebraugh, Jeffrey "Free" Luers and Craig "Critter" Marshall, Andrew Stepanian, Jeremy Parkin, and the countless other known and unknown activists who suffer persecution, interrogation, police brutality, crappy jail conditions, yet stand strong.

This action is an example of the respect these underground environmental extremists have for the aboveground activists. Two of the activists in the list of above (Jeffrey "Free" Luers and Craig "Critter" Marshall) are now in prison for ELF-type actions.

Jeffrey "Free" Luers (who never claimed that he was involved with the ELF) is currently serving a twenty-two-year and eight-month sentence that began June 11, 2001. He was convicted of eleven felony charges for burning three SUVs at a dealership in Eugene, Oregon. His co-defendant, Craig "Critter" Marshall, is serving a five-and-a-half-year sentence. Luers is serving a greater sentence because police linked him to a past arson attempt at Tyree Oil Company.

Luers, a long-time nonviolent peace and environmental activist in his mid-twenties, has never harmed any individual, yet he received double what a rapist is typically sentenced to in the United States. Why is this? The symbolic arson Luers was convicted of draws attention to the fact that SUVs are contributors to air pollution and the destruction of the ozone layer. Actions like this threaten entire industries because they inject a necessary ethical consideration into consumerism. When potential buyers know the extended impact of their purchases they become more thoughtful consumers and are less susceptible to advertising and propaganda. These actions also seek to expose the dangers inherent in a capitalist system that hides the negative effects of its modes of production from its consumers.

While Luers does not affiliate with the ELF, he did take similar measures in making sure that his actions at the dealership would not harm anyone. At the trial, both the night watchman at the dealership and an arson specialist confirmed that the fire set to the three SUVs was not a threat to human beings. It is clear that what was on trial was not the burning of three SUVs, but rather Luers' environmental politics, which were brought up on a number of occasions by the police and prosecuting attorney. This trial was to set a precedent so that others would think twice about conducting such an extreme act (be it symbolic or not) against automobile, petroleum, and oil industries.[2]

Ecoterrorists?

According to the FBI, the ELF is the top "domestic terrorist" organization in the United States, considered more menacing to "American values" than violent neo-Nazi, militia, and anti-government groups. The ELF apparently merits such serious attention because in the last decade it has firebombed buildings, razed housing complexes under construction, burned Hummers and SUVs, and in various ways destroyed the property of industries that contribute to environmental problems such as habitat destruction and air pollution. Because their modus operandi involves illegal actions and property destruction (property being the most sacred icon of capitalist society), the ELF is an underground movement comprised of people who revile capitalism as a destructive social system, advocate radical environmentalism (or ecology), and form anonymous cells to carry out their strikes.

Many ELF activists and supporters consider themselves freedom fighters who defend Mother Earth against the increasingly damaging encroachments of capitalist industries—timber, automobile, housing, etc.—whose only concern is profit regardless of any social or ecological costs or consequence. Attacking the property of those who harm life and taking steps to avoid the harming of any form of life itself (humans or otherwise), the ELF rejects the stigma of being a "violent" or "terrorist" movement and turns the accusations against those in the corporate–state complex whose actions and policies kill animals, destroy ecosystems, and ultimately harm human beings too.

The "terrorist" label presents an interesting choice for the ELF and for those who undertake ELF-style actions. In the present political climate, usage and application of the term is used to terrorize the general population and eliminate any rational discussion regarding a group's motives or goals. For this reason, some people attempt to disassociate the ELF with the "terrorist" label, comparing them instead to other revolutionary groups such as the Zapatistas, the American Indian Movement, and the Irish Republican Army.[3]

Another strategy is to embrace the "terrorist" label in the name of solidarity. The ELF is fighting a war against global capitalism, one of the main tools of the U.S. Empire. In many ways, their struggle is the same as those militant groups (including al-Qaeda) that combat the exploitation of the land and possessions of people in developing countries. Actively adopting the "terrorist" label forces people involved in the more mainstream struggle for peace and justice to acknowledge that this war against empire is being fought on many different fronts and with a wide array of tactics.

The ELF and Capitalism

It cannot be denied that ELF actions have caused millions of dollars of damage in economic sabotage and, thus, the group represents a threat. However, the important question, in the context of labeling the group a domestic terrorist, is who exactly is the ELF threatening? The ELF represents no direct or overt threat to the U.S. government, like the many right-wing groups that have virtually disappeared from the DHS terror lists, despite targeting and threatening human lives. Rather, ELF actions hurt corporate abusers of the land, water, air, and animals. Again, the guidelines of the ELF specifically prohibit the harming of humans in the process of economic sabotage.

This line of argument becomes more convincing when we add in the fact that the ALF "…is a serious domestic terrorist threat" according to the FBI (Lewis, 2004). While ELF actions have caused massive amounts of capital loss for corporate interests, the monetary values associated with ALF actions are even less (although this figure continues to grow as both ALF and ELF members become more effective at what they do). Why then is this group also at the top of the domestic terror list? The answer is simple: The ALF and ELF are effective, decentralized, autonomous organizations that in their actions provide a clear and compelling critique of corporate capitalist society. Disregarding the level of actual damage they cause, every time they act, the lies and inequities contained within our current system of economic governance are laid bare. This critique is made even more insidious and effective because the very tools that serve to obscure the truth about the effects of capitalism to U.S. citizens (corporate- and government-controlled media) are being utilized to spread the gospel (through mainstream news reports and magazine articles on ELF actions).

"These are not your local Sierra Club folks," said Ms. Sandy Liddy Bourne, director of policy and legislation for the American Legislative Exchange Council (The Washington Times, 2004). The American Legislative Exchange Council reportedly helped draft approximately six "ecoterrorism" laws as recent as December 8, 2004. Laws that are currently being assisted by the American Legislative Exchange Council and U.S. Sportsman's Alliance include H.B. 433, commonly known as the "Animal and Ecological Terrorism Act" introduced into Texas legislation in February 2003 by Ray Allen (R-Grapevine). This bill defines a terrorist act as "…two or more persons organized for the purpose of supporting any politically motivated activity intended to obstruct or deter any person from participating in an activity involving animals or natural resources" (Texas H.B. 433). Actions that have been identified as "…intended to obstruct…" (Texas H.B. 433) include taking pictures of trees being cut down or trees being trucked off to a logging mill or cattle grazing on a ranch. New York, Pennsylvania, and Ohio have similar bills that are currently in the legislature. Ms. Bourne states that the reason for these bills is, "that ELF or ALF hit 20 states in 2003 with arsons, bombings, destruction of biotechnology labs, damage to genetically modified food crops and freeing of livestock" (The Washington Times, 2004).

This legislation expands the definition of a criminal act, comparable to what the USA PATRIOT Act does to the use of terrorism. The PATRIOT Act circumvents civil liberties and freedoms in order to investigate legal and illegal activities that fall under a broad, new (and intentionally vague) definition of terrorism. The "EcoTerrorism" bills go one step further in expanding the reach of the state and define as terrorism acts that are currently legal or that, until recently, were only misdemeanors (taking pictures, protesting logging companies, or sitting in front of a bulldozer). The message is clear: if you challenge or threaten the lumber, cattle, dairy, or vivisection industries and keep them from making a profit you are a terrorist.

Through their actions, the ELF has been able to expose an inevitable but commonly hidden result of capitalism. For this economic system to survive and continue to flourish, the government must convince its citizens that property (owned material) is deserving of rights and value equal to that of a human. This is essentially a reversal of the process of slavery, in which a human individual is degraded to the level of property. In order to defend the current world order, the government is forced to adopt legislation that endows corporations and property with certain legal status. Far from empowering the individual, capitalism eventually compels humans to grant almost equal status to their possessions and to intangible organizations in order to protect their acquisitions.

In testimony before the U.S. Congress in February 2002, former Northern American ELF spokesperson, Craig Rosebraugh (2004) ends his statement with the following:

> If the people of the United States, who the government is supposed to represent, are actually serious about creating a nation of peace, freedom, and justice, then there must be a serious effort made, by any means necessary, to abolish imperialism and U.S. governmental terrorism. The daily murder and destruction caused by this political organization is very real, and so the campaign by the people to stop it must be equally as potent. (Cox News Service, 2002)

It is clear that the ELF is not trying to reform the ways corporations interact with the environment. Their goal is the dismantling of multinational corporations that harm the natural environment and the complete collapse of the consumer-based, market capitalist economic order. Although this is a difficult message to bring to the masses, in targeting not only corporate criminals but also symbols of U.S. consumerism, the ELF is forcing people to confront their own complicity in the destruction of our natural world. Pete Spina, author of *Rethinking the Earth Liberation Front and the War on Terror* (www.infoshop.org, March 17, 2004), writes,

> "The FBI can't stop them, and their appetite for destruction is growing. Meet ELF, our biggest domestic terror threat." While über-glossies like Maxim usually don't delve much deeper than surfing chimpanzees or softcore hetero porn, March's issue contains an article detailing the exploits of the Earth Liberation Front, the decentralized group of militant environmental direct actionists who, together with their sister organization the Animal Liberation Front, have caused a total of $82,752,700 in property damage to SUV dealerships, ski resorts and other targets since 1996. Teresa Platt, executive director of fur industry lobby group Fur Commission USA, tells Maxim, "They've hit the common man—An SUV is the common man. They're hitting soccer moms." (EarthLiberationFront.com, 2004)

Conclusion

U.S. citizens should be deeply concerned with the implications this harassment of ELF activists and supporters has for civil liberties and the constitutional rights of those who oppose the obscene growth of corporate power.

While it is true that the ELF uses tactics that, when successful, cause millions of dollars in damage, their anti-corporate, anti-exploitation message contains many similar elements to other, less militant activist organizations. The response of the U.S. government to the ELF indicates a willingness to use any means necessary to protect and defend the current system that allows virtually indiscriminate corporate destruction of the natural world. Less militant activists may not personally agree with the tactics utilized by the ELF, but they must recognize that ELF-style actions have a place within a robust environmental/ecological movement. And furthermore, unchecked government repression of groups like the ELF strengthens the ability of corporations to continue to exploit the earth, lessening the effectiveness of more mainstream methods of protest (Cunningham, 2003; Lichbach, 1987; Moore, 2000). The anti-capitalist message of the ELF must be embraced by all of us who care for and wish to defend the earth.

Notes

1. We are not suggesting here that there is necessarily a continuum between moderate social justice mobilization (such as opposition to the war in Iraq) and the more militant tactics employed by Al Qaeda, simply that there is a spectrum of groups that utilize a variety of tactics and strategies to combat global capitalism.
2. For more information on Free's trial visit www.freefreenow.org.
3. While these groups may enjoy broader support on the left, they are by no means immune from the "terrorist" label themselves. Often the difference between a revolutionary and a terrorist is wholly dependent on both context and perspective.

References

Abbey, E. (2000). *The monkey wrench gang*. New York, NY: HarperCollins Publishers.

Churchill, W., II. (2003). Internationalists and anti-imperialists. *Social Justice, 30*(2).

Cox News Service. (2002, February 12). *Testimony before US House of Representatives Subcommittee on Forests and Forest Health*. Eunice Moscoso. Alleged EcoTerrorist Takes the Fifth Before Congress. Cox News Service.

Cunningham, D. (2003, September). The patterning of repression: FBI counterintelligence and the new left. *Social Forces, 82*(1).

DAM. (1997). *Collective Earth First! Direct action manual: Uncompromising nonviolent resistance in Defense of Mother Earth!* (1st ed.). Eugene, OR: DAM Collective.

EarthLiberationFront.com. (n.d.a). Retrieved from http://www.earthliberationfront.com

EarthLiberationFront.com. (n.d.b). Retrieved from http://www.earthliberationfront.com

EarthLiberationFront.com. (n.d.c). Retrieved from http://www.earthliberationfront.com

EarthLiberationFront.com. (2004, March 14). Retrieved from http://www.earthliberationfront.com/news/2004/031704a.shtml

Goldstein, R. J. (2001). *Political repression in modern America from 1870 to 1976.* Chicago, IL: University of Illinois.

Henderson, C. W. (1991, March). Conditions affecting the use of political repression. *The Journal of Conflict Resolution, 35*(1).

Institute for Deep Ecology. Retrieved from http://www.deep-ecology.org

Lewis, J. E. (2004). *Congressional testimony before the senate judiciary committee.* Federal Bureau of Investigation. Retrieved from http//www.fbi.gov/congress/congress04/lewis051804.htm

Lichbach, M. I. (1987, June). Deterrence or escalation? The puzzle of aggregate studies of repression. *The Journal of Conflict Resolution, 31*(2).

Moore, W. H. (2000, February). The repression of dissent: A substitution model of government coercion. *The Journal of Conflict Resolution, 44*(1).

Rosebraugh, C. (2004). *Burning rage of a dying planet: Speaking for the earth liberation front.* New York, NY: Lantern Books.

Scarce, R. (1990). *Eco warriors: Understanding the radical environmental movement.* Chicago, IL: Noble Press Inc.

Schultz, B., & Schultz, R. (Eds.). (1989). *It did happen here: Recollections of political repression in the America.* Berkeley, CA: University of California Press.

The Washington Times. (2004, December 7). Escalation of ecoterrorism seen in recent years. *The Washington Times.* Retrieved from http://www.washingtontimes.com/news/2004/dec/7/20041207-111414-8901r/

9. *Mapping Discursive and Punitive Shifts: Punishment as Proxy for Distinguishing State Priorities Against Radical Environmental Activists*

LAWRENCE J. CUSHNIE

Introduction

On New Year's Eve of 1999, Marie Mason burned down the Agriculture Hall on the campus of Michigan State University. The arson was in protest of the genetic engineering research carried out within. The research was part of a federally funded program to genetically modify foodstuffs for consumption in the United States. On February 5, 2009, Mason was sentenced to 22 years in prison. Prosecutors acknowledged that the fire was not set in an attempt to damage human life, yet Mason received the longest sentence ever for an act of environmental activism. Leading up to her sentencing, the Federal Bureau of Investigation (FBI) warned the press of the possibility of "terrorists" attending the court date to protest or otherwise interrupt the proceedings potentially through violent means (Potter, 2009). Similar intimidation tactics (not backed by any actual threats) were used by the federal government in the mid-1970s during the trials of various American Indian Movement (AIM) activists (Churchill & Vander Wall, 2001; Matthiessen, 1992). Federal prosecutors asked for a sentence of 20 years, the judge added another two for Mason's involvement with the Earth Liberation Front (ELF; Fox News, 2009). The judge reasoned that Mason's acts fit within the definition of terrorism constructed by the PATRIOT Act (*Public Law 107-53*, 2001) and the Animal Enterprise Terrorism Act (*Public Law 109-374*, 2006). Chief U.S. District Judge Paul Maloney also used a vague "terrorism enhancement" established by the Omnibus Counterterrorism Act of 1995 allowing broad discretion in sentencing (up to 20 years) for acts aimed at influencing the government and for endeavors found of a congressionally defined list of terrorist acts (*H.R. 896*, 1995). What political processes, climates, and strategies led to such a harsh penalty for Marie Mason? Why have courts in recent years issued several sentences to property-destroying environmental activists beyond those typically given for rape and murder? Why has the executive branch through federal law enforcement agencies been so aggressive in applying statutes (some already in existence for a decade, yet rarely used) that target property-destructive protest?

This chapter documents a variety of changes in political priorities and statutory weapons for prosecutors contributing to the rise in punitiveness against radical environmental activists. These circumstances include courts and judges carefully monitoring cues from the federal government as to how contentious political controversies are resolved in the legal realm. This link is most clear between publicized, concerted efforts on the part of federal law enforcement, demonstrated through the attorney general's Department of Justice's (DOJ) and Homeland Security's yearly strategic plans. In order to clearly identify the stakes (legal, philosophical, and existential), this chapter integrates discussions of the theoretical and normative place of property in American society. Specifically, one method of understanding the priorities of a community is to consider which crimes receive the most punitive sentences. While the severity of sentencing applied to environmental activity is a relatively new phenomenon, the trend represents punishments for the destruction of *things* on par with the destruction of *beings*.

In the United States, courts provide a multidirectional tool for competing environmental interests. Individuals may petition the court for grievances against private corporations and/or government interests, or they could find themselves as defendants for their activism. In the case of environmental activists, the interplay with American courts shifts over time. Federal law enforcement sets the agenda for the judiciary in their pursuit of various threats, and courts, over the past 10–15 years, responded with elevated sentences for similar crimes. Specifically, the government pursued higher penalties (in months of incarceration) for environmental activists over the past decade than in previous ones. The crimes are similar in tactics, scope, and severity, yet the sentencing of convicted environmentalists rose steeply. Understanding this trend through an evaluation of sentencing rates for similar crimes over the past two decades, focusing on instances of property destruction, arson, vandalism, etc., with activist motivations demonstrates a trend of increasing punitiveness.

Why has sentencing rates for similar acts of environmental activism increased? What factors explain this variation?

One possible reason for an increase in punitiveness involves activists shifting from "conventional" protest to activities destroying property and breaking laws. However, research identifies a dramatic shift in punitiveness even as tactics remain relatively stable. The shift occurred long after activists began using the tactic of property destruction. The political milieu and discourse surrounding "ecoterrorism" serves to increase attention to extralegal activism in defense of the environment. Domestic security forces (FBI) in conjunction with the federal legal apparatus (DOJ and Homeland Security) have made "ecoterrorism" a top priority since September 11, 2001 (Jarboe, 2002). Clear evidence of this emphasis exists in FBI press releases, congressional testimony, and revelations in the strategic agendas laid out by various federal entities discussing the threat of "ecoterrorism." This chapter argues that, while this new focus on activists leads to slightly more arrests and convictions, it disproportionately assigns more severe penalties to environmental radicals in comparison to pre-9/11 cases. Illuminating the agenda-setting power of federal law enforcement's response to radical environmental action demonstrates a realignment of federal priorities in the wake of 9/11 to reclassify destructive dissent as terrorism.

The directionality of this process is difficult to map. There are several possibilities for how the timing of massive international terrorist actions coincides with the rise in punishment for domestic political activists. One is that Congress passed a law targeting the specific threat of those affiliated with the perpetrators of the September 11 attacks, but utilized vague language and definitions, thus opening substantial legal space for pursuit of domestic agitators. Such a lack of specificity enabled federal overreach on the part of prosecutors utilizing outward looking congressional acts toward internal dissent. While the new legislation was publicly linked with the immediate tragedy, its existence and push for implementation preceded the events justifying its passage into law (Van Bergen, 2002). Regardless of timing and motivations, wide latitude is available for federal actors to pursue and prosecute a form of dissent as old as the country under the auspices of preventing terrorism. Considering American priorities toward the protection of property as a cause of more stringent penalties for activists in conjunction with a "War on Terror" is a necessary and logical next step toward a comprehensive explanation. This approach seeks to integrate understandings and antagonisms between property rights and the rights of protest and resistance. Utilizing a modern case study of how the American state confronts dissent through destruction undergirds the approach.

This chapter is divided into four sections. The first outlines the agenda-setting approach of the justice department's and federal law enforcement's mounting interest in and emphasis upon environmental activists utilizing direct action. The section provides an initial foundation and discussion about how concentrated federal efforts provide sentencing cues and priorities to courts. While some of these changes seem statute driven, in reality, a multitude of legislation salient to stricter sentencing was present for decades. Rather, the change is the result of increased political attention toward the War on Terror and the new priorities of the DOJ and the FBI. In other words, legislation such as the PATRIOT Act set a new agenda for federal actors, while also opening up the legislative past for previously underutilized statutes.

In the second section, a longitudinal data set documents the length of sentences in cases involving property destruction by environmental activists. The data reveals upward movement in the rhetoric of terror and fear from federal entities, which mirrors the increase in punitive sentences. Descriptions of environmental activists as terrorists and as significant threats to domestic security become the new standard. FBI, DOJ, and Homeland Security press releases, congressional testimony, and newspaper articles comprise the bulk of the data. From the case studies, a steady increase in terms of incarceration since early 2002 is immediately apparent.

The third section analyzes the data to hypothesize reasons for the observed changes. The analysis includes deeper interrogations of individual cases, analysis of discourse from the state, and considerations of the political landscape. These three areas enable the reconstruction of a political, a legal, and a law enforcement climate leading to longer rates of incarceration.

The final section considers the theoretical implications of increasing punishment for damage to property which explicitly rejects harm to individuals. The situation has not been one of a gradual rise in sentencing for environmentally motivated property crimes. Rather, sentencing vaults upward to a level reserved for rape, murder, and other violent crimes against sentient beings.

The chapter concludes with a discussion of the results and the implications the data provides for future interactions of activists and the courts. Effects are not a simple top-down description of increased state attention and condemnation, but a multidirectional interaction effect, in which courts are less responsive to the rights-claims of activists. The political climate allows for questionable prosecutorial tactics toward environmental activists, due to their participation in law-breaking activities against symbolic property targets. While a lack of sympathy is expected, the change in levels of punitiveness demonstrates a normative arena of contention. Property, as a sacrosanct symbol of the right to exclude in the liberal state, leads to emotional and reactive policies when property is destroyed in the course of protest. When an environment of terror complements these actions, we can expect a steep rise in the level of punitiveness for participants.

Federal Law Enforcement and Agenda Setting

Since the 1980s, the DOJ publishes yearly or semi-yearly reports on the status of foreign and domestic terrorist concerns. These reports offer a clear public agenda for FBI response to domestic incidents perceived as a terrorist threat. *Terrorism in the United States*, renamed *Terrorism* in 2001, significantly alters its labels and descriptions of environmental activists between 1996 and 2001. While groups such as the ELF are discussed as a significant threat going back to 1998, they are not anointed as "ecoterrorists" until the reports published in 2002. The 1998 DOJ *Terrorism in America* briefing uses an image from an ELF action in Colorado as the *cover* of the report, yet refers to ELF as "an extremist environmental movement" (DOJ, 1998). While their actions gain enough prominence to make the cover of the report, they are still described in terms of radical activists. Fast forward to 2002 and for the first time we see descriptions of "the challenge to respond to animal rights and ecoterrorism" (DOJ, 2002). Before 2001, "ecoterrorism" as a term was used sparingly in newspaper stories and other forms of popular media. In fact, the earliest use of "ecoterror" found using a popular internet search engine is by an environmental group in 1987 who named themselves the *Evan Mecham Eco-terrorist International Conspiracy*. This chapter also explores the strategic rhetorical use of "terror" attached to activists as one of several tools deployed by the federal government to realign destructive dissent with terror. The "War on Terror" provides a nebulous category to encompass many groups who contest federal power, especially when property is involved. This response fits within previously discussed historical cases of government attempts to combat controversial messages and actions of dissent.

The discursive shift beginning just after the 2001 attacks became much more significant after Congress responded to those attacks by giving federal law enforcement broad new powers to investigate and punish acts of "terrorism." The FBI began to use newly aggressive tactics similar to the ones used in an earlier generation with COINTELPRO (Churchill & Vander Wall, 2001). Significantly, however, federal law enforcement officials were much more open in announcing and taking credit for the tactics used in the post-2001 campaign against radical dissent. The FBI, in conjunction with the ATF to curb property destruction by environmental activists, launched Operation Backfire in 2004. The program targets environmental and animal rights activists participating in sabotage of industries harmful to the environment and animal welfare. Aligning the program with COINTELPRO is not due to its covert nature, but due to tactics law enforcement utilize to find, to arrest, and to prosecute activists. While Operation Backfire represents the clearest example of a policy shift from the federal government in the aftermath of 9/11 to combat domestic terrorism, various other crackdowns on public forums of protest demonstrate the extent that the control over discourse about dissent reaches (i.e., development of "free speech zones" at global economic conferences and vagrancy laws used against the Occupy movement).

Operation Backfire utilizes secret grand juries, FBI provocateurs, informants, unnamed sources, surveillance, preemptive arrests, and other tactics treading the border of legality. These are the same methods executed throughout the 1960s and 1970s against such groups as the Weather Underground, the Black Panther Party, the AIM, and the New Left more generally (Churchill & Vander Wall, 2001). While many of these actions fall in a gray area of legality, federal prosecutors legitimize them as necessary and relevant when the moniker of "terror" is attached to those being investigated. Similarly, these tactics appear more recently against various protest groups leading up to WTO, G8, and other international economic conferences including preemptory arrests and agent provocateurs.

After a shaming *60 Minutes* report ("Burning Rage" in 2005), it was clear that the FBI had failed to arrest anyone as part of Operation Backfire. While there was little public outcry at the time, the federal government was embarrassed by the combination of resources used and lack of tangible outcomes pointed out by CBS journalists (Bradley, 2005). Not long after, federal law enforcement dramatically ramped up both enforcement activity and publicity surrounding examples of "successes" targeting domestic dissenters as "terrorists." The most efficacious tactic in developing cases against activists involved threatening an informant with federal drug charges if he did not cooperate in secretly taping discussions with his conspirators and friends about events from previous years. Jacob Ferguson was flown around the country, while wearing a wire, in order to casually run into old acquaintances from his ELF days. Ferguson was a prolific arsonist and acknowledged his responsibility in most of the major actions perpetrated by an active ELF cell (Bernton, 2006). Indictments began raining down on members of a group dubbed "The Family" for various ecotage events going back to 1996. The eventual result was multiple convictions, helpful in reversing the image issue Operation Backfire suffered, as well as the imprisonment of 13 men and women (Bernton, 2006). These indictments, as well as the accompanying arrests and convictions, were widely publicized by the FBI through press releases, media interviews, and congressional testimony. These documents and statements conjure a picture of domestic terror cells conspiring to destroy the property of everyday American citizens as part of their radical environmental agenda. Understandably, the FBI does not mention or discuss motives for these illegal acts. Rather, the actions are lumped together within the larger "War on Terror." As Attorney General Alberto Gonzalez states, "Today's indictment proves that we will not tolerate any group that terrorizes the American people, no matter its intentions or objectives" (FBI, 2006). The method of pursuit, the tactics, and state descriptions of property destruction set a clear agenda for courts and judges to issue aggressive sentences to environmental activists.

Longitudinal Evaluation of Convictions

A total of 29 cases of property destruction, designated as acts of environmental activism associated with the ELF, from 1987 to 2012 constitute the case studies for analysis. The cases were found through a wide variety of sources. While the most dramatic cases were present across national news syndicates, federal government records provide the most salient examples. Since the focus of this chapter is on federal law enforcement's change in approach and veracity in sentencing, the cases promoted by the FBI (touted in press releases and press conferences) are most helpful. This demonstrates two important concepts: (1) federal agenda setting displayed in public dissemination of information including press releases and congressional testimony and (2) the shift in federal attention to these activists even as the research shows a continuing presence of these illegal actions stretching over decades. There are potential problems with this sampling method with an overreliance on federally controlled messaging and information. In other words, the entire universe of actions may not be present. Lower level offences taken care of at the city or county level might be excluded. However, since the argument is about federal attention to these acts, the sampling demonstrates shifts over time in the public attention granted to environmental activists. Another issue is the assigning of monetary damages that the events represent. These numbers are notoriously difficult to pin down with any real precision. As with large-scale drug busts, dollar amounts trend toward the dramatic. For this reason, sentencing rates rather than the monetary damages assigned for their actions provide a more accurate metric.

Seven of the incidents reach the sentencing stage before September 11, 2001, and 22 occur afterward. This date is chosen as the point of departure due to a concerted effort from federal law enforcement to crack down on environmental activists and any entities construed as terrorist elements.

Accompanying this higher level of attention is also a discursive shift. It is difficult to make a perfect comparison since many acts differ in levels of damage. This includes differences in cost (as ascribed by property value) and impact (as determined by symbolic importance). Therefore, the data shows a potential trend, rather than a clear outcome.

"Ecoterrorism" is the term used exclusively beginning in 2002 by the FBI and other federal institutions to label the destructive acts of environmental radicals (Jarboe, 2002). The 29 cases in the data set are found in press releases, newspaper articles, congressional testimony, environmental activist message boards, and civil rights newsletters. Simple comparisons of the mean and median of cases before and after September 11, 2001 illustrate a disparity and shift in the severity of sentencing. It is also important to describe the circumstances surrounding specific cases showing how the courts interpreted similar activism differently within a relatively short period of time. This chapter hypothesizes that the increase in rates of sentencing is attributed to the increased political attention from the federal government. Publicity surrounding federal law enforcement campaigns directs political attention to a specific issue increasing awareness and salience for the courts. In effect, the political climate contributes to actual legal outcomes and that these cases are demonstrative of such a trend. This is not a stunning or remarkable outcome in general terms concerning how political climate affects enforcement priorities; however, it is important in terms of the impact on the suppression of dissent more generally.

The analysis is divided into four main parts. First is a discussion of the results of the data gathered. The 29 cases demonstrate a steady rise in rates of sentencing. A variety of confounding factors present significant effects on the results: invocation of federal statutes, pleading guilty or not guilty, cooperation with law enforcement, testifying against other defendants, becoming an informant, previous convictions, and additional charges. Even with these considerations, an increase in severity of sentencing is present. Second, closer examination of a few individual cases builds a deeper account of the events surrounding specific verdicts. Disproportionately harsh penalties arise following 2001 compared to crimes before that year. Third, this chapter applies the logic of courts as political actors to understand how the political climate influences the supposedly insulated judicial realm. Finally, considering the implications of government labeling and FBI counterintelligence programs on the future of environmental activism and its prosecution helps to establish a framework for future analysis.

The data gathered represents a collection of the most prominent prosecutions of environmental activism, specifically described as "ecotage." Ecotage represents acts conducted to eliminate the profit motive of environmentally harmful actions. As Parson argues, "...ELF ecotage is also meant to question and confront the social, economic, and political realities of the world and to undermine them through their active problematization" (2008, p. 53). Ecotage can take many forms. Stereotypically, the word describes acts of arson and vandalism upon easily identifiable sources of environmental degradation. Debate remains within the activist community about whether these acts constitute a response to reduce the profit motive of individual issues or represent a larger revolutionary perspective. This distinction is unimportant to the federal government who reserves the legitimate authority to ascribe motive in their prosecutions. Whether the burning down of a planned community in environmentally sensitive wilderness represents an attempt to stop a specific instance of urban sprawl or its arson constitutes a larger struggle against commerce trumping protection of eroding ecosystems, federal law enforcement dictates the "proper" response.

Parson provides a helpful analysis of competing ideologies within the movement. Parson discusses the radical ecological traditions behind environmental activist groups such as Earth First! and ELF. He implicates three ideologies that help to encompass the reasoning and motivation behind these actions including deep ecology, social ecology, and green anarchism (Parson, 2008, pp. 54–58). Each provides a distinctive understanding of the place of the activist and motivation for their actions against corporations, research entities, urban sprawl, etc. This creates different priorities in target assessment for activists using property destruction as a tactic and complicates the portrait painted by the DOJ.

Thus, assorted actions fall within varying definitions of justified "ecotage" including animal release, vehicle sabotage, and tree spiking. Ideologies influencing activists lead to fluctuating understandings of legitimate resistance. Comparisons between ecotage and civil disobedience provide a persuasive evaluation of radical resistance, enabling a multifaceted understanding of actions and their potential justification (Vanderheiden, 2005, pp. 425–447). Vanderheiden develops spheres of defensible acts of ecotage which do not constitute terrorism, yet also fall outside of civil disobedience. His discussion is helpful in developing a spectrum of activism overcoming federally constructed binaries.

This data set suffers from many limitations. It is not an exhaustive list of all cases of ecotage and it is not necessarily representative of the entire population of cases. However, it does constitute the most salient cases due to the publicity surrounding them. These cases received the most attention in federal law enforcement press releases and testimony as well as availability from national news sources. As Table 9.1 demonstrates, acts of ecotage penalized by the courts before September 2001 have a mean sentence of 42.4 months and a median of 36 months. The shortest sentence out of the sample is 12 months while the longest is 84 months. These seven cases show a relatively homogeneous reaction by the courts for crimes involving property destruction.

Table 9.1. Sentence Lengths, in Months, of Environmental Activists—Property Crimes (See the Appendix for detail). (Source: Author)

Date Range	# of Convictions	Mean	Median	Shortest Sentence	Longest Sentence
1/1/91–9/11/01	7	42.4	36	12	84
9/12/01–3/31/12	22	92.9	81	6	262

Incidents after September 2001 experience a clear change. In Table 9.1, the mean has more than doubled to 92.9 months and the median is up to 81 months, similar to the largest penalty before September 2001. Across the cases, the shortest sentence is six months and the longest drastically increases to 262 months or 21.8 years. In addition, the sentences after September 2001 do not match with the more generally consistent convictions for vandalism and arson.

There are a variety of factors particular to the cases that might account for such changes, including value of those objects and/or structures vandalized or destroyed. It is difficult to rule out such factors completely with available information. Assigning value to damages is notoriously difficult, but publicized numbers tend toward the dramatic. Nevertheless, it seems that there is, at most, a small uptick in the amount of damage associated with the protest actions, and certainly not an increase proportional to the substantial increase in the level and length of incarceration. It is difficult (if not impossible) to make an accurate damage comparison, as figures are not reliable; however, there is little reason to believe that tactics intensified toward more substantial losses.

Few of the cases before September 11, 2001 involve the use of federal laws for sentencing; however, federal acts did exist before 2001 and were available to prosecute environmental activists. In 1995 and 1996, Congress passed the Omnibus Counterterrorism Act and the Federal Crime Bill and the Anti-Terrorism and Effective Death Penalty Act, respectively, in the wake of the Oklahoma City bombing (Singh, 2006, pp. 71–93). The acts articulate expanded definitions of domestic terrorist-related activities as well as federal sentencing guidelines. Most importantly, the birth of "terror enhancements" gave judges a tool allowing for an additional 20 years added to sentences at their discretion. Neither of these acts were mobilized against environmental activists prior to 2001. Thus, guidelines allowing for more punitive sentences were present, but remained quiescent. The RICO Act is also available to prosecute activists across causes, though it was originally written as a method to convict high-level mafia members well before 2001. The Animal Enterprise Protection Act passed in August of 1992 makes it a terrorist offense for commerce clause violations by anyone crossing state lines who "intentionally damages or causes the loss of any property (including animals or records) used by the animal enterprise, or conspires to do so" (*Public Law 102-346*, 1992). The law lay dormant for six years until it was used to convict Justin Samuel in 1998, which is one of the cases included in the sample. In 2006, Congress amended the law and renamed it the Animal Enterprise Terrorism Act (*Public Law 109-374;18 U.S.C. § 43*). Alterations to the statute went beyond simple naming to include further expansion of the definition of terrorism and enlarged powers for the courts to sentence wrongdoers. Examples of the discursive shift toward terrorism and the potential impacts of this key rhetorical change are elaborated upon later in the piece.

So what was the difference after 2001? Key changes include an increase in attention to acts construed as anti-capitalist, anti-American, violating copyright, and/or targeting property after 9/11; a reassertion of previously unused pre-9/11 statutes; and, most importantly, a shift in discourse and attention toward environmental activism from federal law enforcement. The move toward more aggressive pursuit of all types of "terrorism" made it much easier to facilitate and further a punitive agenda. Specifically, the discourse of "terror" justifies increased lengths of incarceration based upon more widely available sentencing guidelines at the federal level. Descriptions of terror also demonstrate a moral high ground for federal officials and allow the construction of activists as irrational or insane actors outside of political and/or ethical consideration.

In February of 2002, the Domestic Terrorism Section Chief testified before Congress naming the ELF and the Animal Liberation Front (ALF) as the two most dangerous domestic terror groups in the United States (Jarboe, 2002). In this instance, congressional testimony serves as the point of departure from reactionary policing and toward preemptive, concentrated, and organized prevention of actions by direct action environmentalists. Before this point, the crimes committed by members of ELF were prosecuted just as any other arson or act of property destruction, many times at the state rather than federal level. Following this address, rates and lengths of incarceration went up drastically. Environmental activists find themselves labeled as "terrorists" by the federal government in press releases, congressional testimony, and other public discourse. FBI monitoring of environmental activists became tactically similar to the COINTELPRO program of the 1960s and 1970s. In recent years, Operation Backfire was initiated to infiltrate and close down individual cells of the ELF. In coming pages, this chapter elaborates upon Operation Backfire and its varying outcomes.

ELF and ALF have never harmed or supported actions targeting sentient beings. In their own mission statement of sorts, they proclaim that their tenets require the step "to take all precautions against harming life" (Parson, 2008, p. 52). In other words, they repeatedly declare normative principles eschewing the targeting of sentient life and have so far lived up to that promise. Other, more violent groups neither profess to be non-violent nor demonstrate any commitment to similar ethical imperatives. ELF and ALF were elevated above the Ku Klux Klan, armed militias, violent anti-abortion activists, and the Aryan Brotherhood as the top domestic threat to the United States. According to congressional testimony (as of 2002) the ELF and ALF were responsible for over $40 million worth of property damage without harming a single individual or being (Jarboe, 2002). The author has been unable to find harm to sentient life in any of their actions since this addition to the congressional record. The reaction of the FBI and the federal government seems to protect economic interests, rather than address threats including hate-based rhetoric by violent organizations to incite fear and to destroy human life. Specifically, a study by the Combating Terrorism Center at West Point addressed the growth of acts perpetrated by domestic right-wing groups resulting in harm to human life.

In their study, members of far right organizations perpetrate a clear rise in violent acts against human beings. Each of these data points represents the attempt to physically attack a target. In sum, there were 4,420 violent incidents over the span of 22 years; 670 of the incidents resulted in fatalities and 3,053 resulted in physical injuries (Perliger, 2012, p. 87). During the same period of intense focus upon radical environmental and animal rights activists, actual harm was skyrocketing against human beings (typically of historically persecuted minority groups). Making a public statement that environmental activists constitute the number one domestic terror threat, while at the same time a steady rise of harm to life is perpetrated by another, sets a dangerous precedent. Thus, priorities of federal labeling and perceived threat level of "domestic terrorism" against inanimate objects versus sentient life come into question. It is not that federal law enforcement was not pursuing these violent, right-wing groups, but rather the public perception developed through press releases and congressional testimony emphasizes the danger of property destruction as a higher order threat. Setting the agenda in this way elevates the protection of property to a place that must be interrogated in the face of actual human violence.

However, this would be too base and stark a contrast. Rather, this example provides a set of priorities for domestic security forces in the United States that conjure interesting theoretical questions of law enforcement and sentencing priorities. A later section theorizes how excessive punishment of property crimes leads to demonstratively detrimental priorities for the state. Length of sentences change in relation to the level of cooperation from individuals in custody. Before 2001, individuals who assisted investigators would typically receive probation or short jail terms. After 2001, individuals who helped with an investigation were still given years in jail similar to non-political incidents of vandalism and arson. Federal prosecutors offer deals in which they promise not to pursue prosecution by federal terror statutes, yet still prosecute the individuals at rates that match or exceed pre-2001 levels. In other words, the standards shift toward increasing severity for the same crimes, even in the case of plea bargains.

A Tale of Two Actions

A discussion of two individual cases is helpful toward understanding the circumstances and the differing results of pre- versus post-9/11 convictions. Qualitative investigations assist in determining the context and the discourse surrounding each event. While it is not possible to draw firm conclusions about disproportionality by making comparisons across a small number of select cases, in the context of the data just presented, the additional details in this section lend additional plausibility to the claim that something changed after 2001. Earlier convictions of politically destructive acts lack implications of "terrorism" compared to later convictions. Terrorism connotes more than just a definitional characteristic of the actors and actions participating in political violence; it also gives wide leeway to those in pursuit. Defining an individual as "terrorist" removes rationality from them as a political- or conscience-driven actor. This allows for a wide variety of justifications in their surveillance, pursuit, and punishment. The moniker of "terrorist" is beyond existing laws because that individual is perceived as outside of societal norms to such an extent that they seek the overthrow or destruction of a political entity or innocent citizens. However, that description is rarely controlled by the one labeled as terrorist. The state decides who counts as an enemy and thus who is worthy of aggressive pursuit and prosecution.

Besides the discursive power of the term "terrorism," there are also legal ramifications for defendants. Most prominent are "terror enhancements" available with wide judicial discretion in their application. Accompanying legal statutes are a wealth of government resources, at the ready, with the directive to capture and punish. Thus, the character of state actions varies drastically from typical policing. The stakes are seemingly higher in the case of fighting terrorism rather than the preservation of law and order. The difference in convictions is a result of a variety of factors, but the most salient factor seems to be the divisive political climate surrounding each incident and state-directed implications as to what these actions represent; that is, the difference between controlling activists and punishing terrorists.

In a 1997 indictment, Douglas Ellerman received 16 federal counts including purchasing, constructing, and transporting five pipe bombs as well as setting fire to a fur breeding facility in Utah (Jarboe, 2002). Ellerman's sentence was seven years in prison. Ellerman admits to being part of a radical environmental organization, yet he was not prosecuted under enhanced federal statutes. All of the information necessary to use federal guidelines toward increased sentencing (as well as the federal statutes themselves; i.e., the 1995 and 1996 congressional acts) were present in this case. They chose not to. Why would prosecutors decide to not throw the book at an admitted member of a radical organization who participated in every step of the process eventually leading to almost $1 million in destruction? The answer lies in the political climate. In 1997, the word "ecoterrorism" was not part of the federal government's lexicon even as new domestic terror statutes were in effect. The term itself originates from the Center for the Defense of Free Enterprise in an attempt to set the agenda in the face of the growing environmental protest movement of the 1980s and early 1990s (Potter, 2011, p. 55). Ron Arnold (who takes credit for coining the phrase) used the term to describe any "crime committed to save nature" (Potter, 2011, p. 55). Activists participating in property destruction were convicted based upon existing statutes dealing with arson, incendiary devices, and vandalism. Statutes did exist at the federal level which could be applied in these cases, yet none were invoked.

The shift after 2001 emerges when comparing Ellerman's case with that of Eric McDavid. On March 6, 2008, Eric McDavid was convicted on charges of conspiracy to destroy property by fire or explosion. He was sentenced to 20 years in prison. The charges stem from the planning and preparation to destroy four targets symbolic of supporting environmental degradation (Scott, 2008). McDavid was arrested before any damage occurred due to an undercover, independent contractor working for the FBI. "Anna" was a paid informant who asked for the position with federal law enforcement after years of work as a volunteer infiltrator of left-leaning movements (Todd, 2008). The sentence McDavid received is longer than the average sentence for murder (19 years) in the United States. What differences in the two cases led to such divergent outcomes? Both men conspired with other individuals to destroy property as a means of protest. Both men purchased the materials necessary to make incendiary devices. Both men planned (or assisted in planning) attacks to guarantee maximum damage. One of the perpetrators, Ellerman, was successful in his plot and destroyed almost $1 million worth of property. The other was arrested before he was able to carry his plan to fruition. The resulting prison sentences differ by 13 years, with the longer sentence being given for an action that did not even take place.

These differences are attributable to the discursive shift since September 11, 2001, which puts direct action environmentalism and property damage in the same category as terrorism. A quote from the prosecuting attorney in the McDavid case is revealing:

> Today's severe punishment of nearly 20 years in federal prison should serve as a cautionary tale to those who would conspire to commit life-threatening acts in the name of their extremist views. (Scott, 2008)

This statement demonstrates the federal government's concern with making an example of McDavid, rather than simply prosecuting a planned arson. Groups, such as the ELF, condemn practices that could potentially harm innocent life. The FBI has acknowledged that fact (Jarboe, 2002). One can imagine the difference in outcomes if the Ellerman case shifted ten years into the future. Ellerman participates in a conspiracy to destroy property, purchases and assembles the materials necessary for destruction, and carries out the act successfully. He receives seven years for his crimes. An examination of the academic literature concerning the integration of the legal and political realms is helpful toward understanding discrepancies between these case studies.

A key factor to consider is judicial decision making determining the length of sentence and whether or not to use additional federal guidelines. While prosecutors make recommendations for length of sentence, judges retain discretion after a jury assigns a conviction or a guilty plea is entered. In these specific cases, a wide range of options are available to a judge not available in cases involving harm to individuals. The fact that judges become the ultimate arbiters of which type of sentence, sentence length, and application of federal statutes is important to identifying the various actors who react to volatile political climates. As a supposedly insulated figure within the legal realm, one would expect sentencing rates to remain static unless the specific laws pertaining to arson change. Since these laws remain the same, the change in sentencing results from other factors.

The strategic approach in judicial behavior literature acknowledges that judges make decisions based upon their perceptions of whether or not a decision will be viewed as legitimate by the government and the public (Baum, 2006). While this literature tends to focus on the Supreme Court, its application to federal justices is also enlightening. Judges are aware of the standards and expectations criminal cases can set. Even though the criminal court system does not specifically function upon a system of precedent, other decisions in similar cases are still pertinent. If a contemporary issue is salient due to attention in the media, acknowledgment in official government channels, and attempts to influence public opinion, judges will also be aware. For instance, if "ecoterrorism" is publicly discussed by the federal government as a problem requiring sustained attention and renewed focus, judges may feel pressured to issue decisions consistent with contemporary understandings of environmental activists as terrorists. Courts mediate issues that fluctuate in saliency. In 1997, Douglas Ellerman was considered part of a fringe group of activists who destroyed property in an attempt to make a political point. He was dealt with as other vandals or arsonists regardless of his affiliation. In 2008, Eric McDavid was arrested in an atmosphere of heightened political and legal awareness of the threat posed by "ecoterrorism." Courts react to the discursive shifts of the federal government. Political situations can sometimes find their resolution in the courts, and the courts can take their cues from the political realm.

Why were federal prosecutors successful in increasing rates of sentencing for activists? What strategies and tactics led to a clear rise in punitiveness of sentences? Three main factors accounted for the change. First, the discursive shift from "activist" to "terrorist" assisted federal law enforcement and prosecutors in gaining a favorable position in political and legal opinion. This tactic restructures law enforcement's position beyond legal authority to a place of moral authority. Descriptions of "countering terrorism" dismiss the environmental concerns in question as secondary or simply not pertinent. It also removes rationality from actors described as "terrorist." Second, the FBI undertook a counterintelligence program—Operation Backfire. Operation Backfire originally directed its attention at one specific cell, but expanded its operations after successfully disbanding their original target. Operation Backfire symbolizes the archetype of the federal government's interaction with and against "ecoterrorists." It demonstrates a marked change from simple prosecution to active infiltration. Third, time itself is an actor. The salience of these groups increases as they register as a more substantial threat to the federal government.

Operation Backfire is the physical manifestation of time and the discursive shift mentioned above. The FBI spearheaded the plan assisted by ATF and other law enforcement organizations, in order to target and infiltrate activist cells. The task force was originally conceived to target a specific cell of activists responsible for some of the most highly publicized attacks on private property. These included the $12 million arson of a Vail ski resort expansion threatening lynx habitat, the disabling of a high-tension power line near Bend, Oregon, as well as acts spanning across Wyoming, California, and Washington. After completing their objective, the FBI continued Operation Backfire as a semi-clandestine mission to pursue similar radical entities such as the individuals responsible for arson at the University of Washington Center for Urban Horticulture (Bartley & Carter, 2008). Activists, independent media outlets, and the National Lawyers Guild denounced the tactics used by the FBI during this campaign (Flynn, 2006; National Lawyers Guild, 2006).

Months after the patriotic fervor sparked by 9/11 (allowing for overarching support of the PATRIOT Act), attacks from civil liberty groups grew in response to the expansive powers granted to the federal government and, more specifically, the executive branch. The pursuits of "ecoterrorists" fell under its expansive language and provided a legal basis for engaging in questionable levels and methods of surveillance as well as the opportunity for newly appropriated federal funds for law enforcement. The PATRIOT Act also sets a point of emphasis for federal attention to any movements or actors threatening the United States after 9/11. Its passage marks a sense of legitimation for ramped up federal attention and pursuit of dissidents.

Discursive Shifts and Theoretical Implications

Describing someone as a "terrorist" serves an explicitly rhetorical purpose in contemporary discourse, though the very language and imagery the term conjures obscure its rational analysis: (figure it implies a moral claim for their aggressive pursuit and prosecution unconstrained by the conventional limits set upon military or law enforcement action (Vanderheiden, 2005, p. 425).

The discursive use of "ecoterrorist" helps to justify surveillance and aggressive prosecution of environmental activists. By utilizing the term "terrorist," the government signals its retention of "the legal powers to pursue activists free from the constraints of conventional civil liberties" (Vanderheiden, 2005, p. 427). Vanderheiden's reference to "legal powers" involves various federal statutes constituted before and after September 11, 2001, giving wider leeway to federal prosecutors and increased funds for law enforcement. Thus, defining an organization as supporting terrorism or participating in terrorism serves a variety of functions. The term signals to the public, political, and legal realms that direct action environmentalists do not deserve the same rights as others; it provides the government with a moral claim to back their actions; and it introduces individuals into the legal system and exposes them to punishment beyond regular criminal prosecution. As discussed in previous sections, "terror enhancement" sentencing grants discretion for added punishment in terms of decades, rather than months. Expanded definitions of terrorism also appear in the PATRIOT Act justifying detention without trial and expanded search and seizure provisions, all of which grant the federal government expanded instruments in pursuit of environmental radicals.

Terrorism has a wide variety of definitions, but an understanding of it as "the calculated use of violence or threat of violence to attain goals that are political, religious, or ideological in nature…through intimidation, coercion, or instilling fear" provides a resonate starting point (Chomsky, 2003, p. 69). It is important to begin from a more generalized definition of terrorism in order to articulate how federal understandings shift in the 21st century. Typically, "terrorist" refers to individuals who do not recognize noncombatant immunity (Walzer, 1977). Inciting fear and intimidation among innocents is a clear goal. Applying this definition to radical environmental activists requires amplification in a variety of directions. First, violence is perpetrated upon property rather than people. This removes the *purposeful* threat to human life. Second, the goals are ideological in nature and toward specific actors. Their specific attacks are linked to instances of environmental degradation typically with corporations as targets. While messaging is meant to reach the general public, they do not represent a threat to "noncombatants"—that is, the average citizen. In testimony before the Senate Judiciary Committee, John Lewis, the deputy assistant director of the FBI, defines domestic terrorism as:

> acts of violence that are a violation of the criminal laws of the United States or any state, committed by individuals or groups without any foreign direction, and appear to be intended to intimidate or coerce a civilian population, or influence the policy of a government by intimidation or coercion, and occur primarily within the territorial jurisdiction of the United States. (Lewis, 2004)

The state's definition may successfully encompass many of the actions of activists already mentioned; however, the rhetoric itself is suspect. A better application of the definition postulated by the FBI throws a vast net of inclusion that resonates with violent groups of the far right more so than the property destroyers of the far left.

This is not to say that environmental and animal liberation activists are perfectly legitimate political players while participating in law-breaking, but it does ask important questions about legitimate levels of punitive sentences for acts bestowed with moral dimensions by the federal government—especially considering the real consequences of prison. Many of the activists convicted after 2001 are being held in Communication Management Units (CMU). CMUs were set up in 2006 to control the communication of convicted individuals with relationships to terrorist organizations or who committed terrorist acts (Johnson & Williams, 2011). The majority of prisoners held in these facilities are aligned with modern Islamic radical groups; however, various environmental activists have found themselves confined in these highly restricted areas (Center for Constitutional Rights, 2013). The facilities are notorious for their intensely controlled, solitary environments. The philosophical implications of punishment for destruction of property going beyond punishment for the destruction of beings are critical elements when studying the suppression of dissent.

Targeting Property: Implications of Destruction

One of the more interesting and controversial implications of property destruction as a political tactic involves the deep roots of liberalism and capitalism in the United States. A Lockean understanding of property as a fundamental right bestowed upon man from God is present in the founding philosophical tenets of American liberal democracy (Locke, 1980). Property is the primary unit of the economic system, the symbol of accomplishment, and the mark of status for individuals in the United States (Veblen, 1994). When property suffers public defacement and destruction, the reactions of citizens as well as the state are clearly disapproval. Property destruction moves beyond a simple act of rebellion or a violation of the legal code; it has the potential to be perceived as an attack upon a normative paradigm of Americanism.

Modern examples of this alternative form of political participation receive concentrated attention from the federal government against the backdrop of the "war on terror." In a post-9/11 legal environment, actions traditionally dealt with through preexisting statutes (i.e., vandalism, criminal mischief, arson, etc.) are now within the purview of federal prosecution and increasing levels of punishment (i.e., PATRIOT Act). For instance, the United States labels property destruction by environmental activists (such as ELF) as acts of terrorism meant to incite fear among the general populace (Yang, 2005). Prosecutions and sentencing reflect PATRIOT Act statutes expanding the criteria for what constitutes a terrorist act (Yang, 2005). The federal government perceives property as an entity, which when destroyed, represents a more general attempt to incite fear or attack the foundations of modern society through expanding definitions of terrorism. These assumptions relate to fundamental understandings of property and the place it holds in capitalist economies. In essence, the preservation of property is so sacrosanct that larger-scale attempts to destroy it result in national fear and terror. In other words, violence against property constitutes an attack upon the normative tenets of the United States rather than as an act of conscience. Press releases from the U.S. government discussing radical environmentalism describe destructive acts in a similar nature (Yang, 2005). Any discussion attempting to reorient property destruction, as a method to enter the political arena, must confront issues associated with the status of property in the United States.

ELF concentrates upon symbolic and functional targets for destruction. They call attention to specific instances of environmental degradation as well as reveal topics of larger environmental concern. Ironically, their actions serve to protect property owned or controlled by third parties (i.e., air, water, forests, etc.). Under a Lockean ideal these acts constitute irrational meddling where one's interest is in what one owns, and nothing else. Their many actions include the destructions of a ski resort in Colorado, a massive construction site in San Diego, a Hummer dealership in Southern California, and a rural cluster development in Washington State. These four events committed by loosely organized cells of activists, in the case of the Aspen ski resort, the San Diego event, and the "green" rural cluster development, attempt to draw attention to three specific cases previously challenged in formal legal channels. The destruction of the car dealership in West Covina, California, was an attempt to spread a further reaching, symbolic message against disproportionate consumption of fossil fuels by luxury automobiles and the tax breaks available to owners due to federal loopholes (Plungis, 2002). Whether or not that message resonated with attentive members of public is questionable. While some may ask "why?" when perpetrators carry out such a large-scale destructive act, many were likely to question the rationality of the actors behind the vandalism.

ELF actions reveal the complicity of the state in environmental degradation. Therefore, the use of legal and political channels to contest their messages reinforces a government monopoly on defining legal and rational acts of participation. The federal contestation and response was not in an argumentative form, but rather through three key methods—labeling, surveillance, and punishment. The performative element of any given act is an attempt to seize the public's attention in regard to an issue deemed too important to overlook (Parson, 2008; Vanderheiden, 2005). ELF tactics seek to generate aesthetic awe in the experience of individuals witnessing such dramatic acts of protest. However, with the federal government launching campaigns like Operation Backfire under heavy publicity, aesthetic awe can quickly turn to witnessing irrationality, unchecked militantism, or terror within the discursive choices of state actors. This brief discussion illustrates the role performance plays in acts of dissent through destruction. It is this theatrical element which lends itself to the current level of attention from federal security forces. These actions challenge a fundamental American perspective as to the sanctity of private property. By attacking a seemingly definitional component of American culture, federal response will rise to meet it—especially in an era of terror.

Conclusions

Since September 11, 2001, the federal government's campaign against radical environmental activists (who participate in ecotage) has drastically increased sentencing rates. Lengths of sentences usually reserved for murderers and rapists now appear in the convictions of arsonists and of vandals. The culmination of several factors accounts for the new levels of punitiveness.

The specific causes include shifts in governmental discourse, concentrated law enforcement activity, and large-scale changes in the political climate. Research showed that reference to "ecoterrorists" was not consistently apparent until after the events of 9/11. A new frame emerges during the "war on terror" to justify inordinate amounts of resources and attention to domestic threats upon the status quo. Environmental activists using property destruction as political protest are symbolically important targets for punishment and control. The federal government's concern with quelling dissent is especially pertinent when such actions are accomplished through anti-capitalist means. Operation Backfire is the clear implementation of discourse, policing, and punishment toward controlling dissenting elements of the population. The FBI's campaign is successfully infiltrating and discrediting the fringes of the environmental movement.

The result of these new federal efforts is a significant rise in the level of punishment for property crimes with environmental associations. Agenda setting and judicialization of politics literature discuss how the political climate has direct affects upon the actors within the legal realm as well as the legal institutions themselves. Increased sentences over time for similar actions are directly related to the discursive shift from law enforcement at the federal level and has a substantial chilling effect upon political dissent.

Appendix

Date	Name	Plea	Sentence
9/1/91	Marc Leslie Davis	Guilty	72m, $19,821
9/1/91	Margaret Katherine Millet	Guilty	36m, $19,821
9/1/91	Marc Andre Baker	Guilty	12m, $5,000
9/1/91	Ilse Washington Asplund	Guilty	12m, $2,000
7/3/95	Rodney Adam Coronado	Guilty	57m, $2mil
6/19/97	Douglas Joshua Ellerman	Guilty	84m, $750,000
9/16/98	Justin Samuel	Guilty	24m, $364,106
2/11/02	Mark Warren Sands	Guilty	216m, 2.82m
7/23/04	George Mashkow	Guilty	12m
7/26/04	Jared McIntyre	Guilty	42m, $300k
8/30/05	Peter Young	Guilty	36m,
9/1/06	Matthew Rammelkamp	Guilty	6m
6/13/07	Chelsea Dawn Gerlach	Guilty	108m
6/13/07	Joyanna L. Zachler	Guilty	92m
6/13/07	Nathan Fraser Block	Guilty	92m
6/13/07	Sarah Kendall Harvey	Guilty	46m
6/13/07	Daniel Gerard McGowan	Guilty	84m
6/13/07	Stanislas Gregory Meyerhoff	Guilty	156m
6/13/07	Jonathan Mark Christopher Paul	Guilty	51m
6/13/07	Suzanne Savoie	Guilty	51m
6/13/07	Darren Todd Thurston	Guilty	37m
8/3/07	Kevin M. Tubbs	Guilty	151m
5/8/08	Eric McDavid	Not Guilty	235m
6/19/08	Brianna Waters	Not Guilty	78m
7/18/08	Jennifer Kolar	Guilty	60m, 7.1m
9/19/08	Lacey Phillabaum	Guilty	36m
10/20/08	Frank Ambrose	Guilty	108m, 3.7m
2/5/09	Marie Mason	Guilty	262m
3/20/12	Justin Solondz	Guilty	84m

References

Anti-Defamation League. (2007). *Ten ecoterror arsonists sentenced in Oregon.* Retrieved from http://archive.adl.org/nr/exeres/97d40331-06fc-43e7-a8b8-40657fc5085e,18aa02a5-13a1-44f3-b4d4-6176e1ecc36c,frameless.html

Bartley, N., & Carter, M. (2008, June 20). UW arsonist, Brianna Waters, sentenced to 6 years. *The Seattle Times.* Retrieved from http://seattletimes.nwsource.com/html/localnews/2008008001_uwarson20m.html

Baum, L. (2006). *Judges and their audiences: A perspective on judicial behavior.* Princeton, NJ: Princeton University Press.

Bernton, H. (2006, May 9). An activist-turned-informant. *The Seattle Times.* Retrieved from http://seattletimes.com/html/localnews/2002977591_ecoarrests07m.html

Best, S., & Nocella, A. J., II. (Eds.). (2006). *Igniting a revolution: Voices in defense of the earth.* Oakland, CA: AK Press.

Bradely, E. (2005, November 13). *Burning rage: Ed Bradley reports on extremists now deemed biggest domestic terrorist threat.* Retrieved from http://www.cbsnews.com/news/burningrage/

Center for Constitutional Rights. (2013). *CMUs: The federal prison system's experiment in social isolation.* Retrieved from http://ccrjustice.org/files/CCR_CMU_Factsheet_March2013.pdf

Chomsky, N. (2003). Terror and just response. In J. P. Sterba (Ed.), *Terrorism and international justice.* New York, NY: Oxford University Press.

Churchill, W., & Vander Wall, J. (2001). *Agents of repression: The FBI's secret wars against the Black Panther Party and the American Indian Movement.* Cambridge: South End Press.

CNN. (2008, March 11). Officials: Earth Liberation Front members indicted in fire. *CNN Online.* Retrieved from http://www.cnn.com/2008/CRIME/03/11/elf.indictments/index.html?iref=allsearch

Del Gandio, J., & Nocella, A. J., II. (Eds.). (2014). *The terrorization of dissent: Corporate repression, legal corruption, and the animal enterprise terrorism act.* Brooklyn, NY: Lantern Books.

Department of Justice. (1998). Terrorism in the United States. *Counterterrorism Threat Assessment and Warning Unit, National Security Division* (1–29).

Department of Justice. (2001). Terrorism 2001/2002. *Homeland Security/Federal Bureau of Investigation* (1–49).

ELF Press Office. (2009). *ELF press office calls activist's 22 year sentence cruel and unusual.* Retrieved from http://mostlywater.org/media_release_elf_press_office_calls_activists_22_year_sentence_cruel_and_unusual

Federal Bureau of Investigation. (2006). *Eco-terror indictments: "Operation backfire" nets 11.* Retrieved from http://www.fbi.gov/news/stories/2006/january/elf012006

Federal Bureau of Investigation. (2009). *Michigan State University eco-terrorist sentenced in arson case.* Retrieved from http://www.fbi.gov/detroit/press-releases/2009/de020509.htm

Federal Bureau of Investigation. (2013). *Crime in the United States 2013.* Retrieved from http://www.fbi.gov/about-us/cjis/ucr/crime-in-the-u.s/2013/preliminary-semiannual-uniform-crime-report-january-june-2013/tables/table-4-cuts/table_4_offenses_reported_to_law_enforcement_by_state_oklahoma_through_wisconsin_2013.xls

Flynn, J. (2006, May–June). Operation backfire backfires on FBI. *Earth First! Journal.* Retrieved from http://news.infoshop.org/article.php?story=20060510125753518

Fox News. (2009). Activist gets nearly 22 years in prison for Michigan State University arson. *Fox News Online*. Retrieved from http://www.foxnews.com/story/2009/02/05/activist-gets-nearly-22-years-in-prison-for-michigan-state-university-arson/

Jarboe, J. (2002). Congressional testimony "the threat of eco-terrorism". *Before the House Resources Committee, Subcommittee on Forests and Forest Health*. Retrieved from http://www.fbi.gov/news/testimony/the-threat-of-eco-terrorism

Johnson, C., & Williams, M. (2011, March 3). "Guantanamo north": Inside secretive U.S. prisons. *National Public Radio News*. Retrieved from http://www.npr.org/2011/03/03/134168714/guantanamo-north-inside-u-s-secretive-prisons

Kessler, R. E. (2004, July 27). Sentence draws gasps. *Long Island News*. Retrieved from http://www.newsday.com/news/sentence-draws-gasps-1.730566

Lewis, J. (2004). Animal rights extremism and ecoterrorism. *Congressional Testimony*. Retrieved from http://www.fbi.gov/news/testimony/animal-rights-extremism-and-ecoterrorism

Locke, J. (1980). *Second treatise of government* (C. B. Macpherson, Ed.). Indianapolis, IN: Hackett.

Matthiessen, P. (1992). *In the spirit of crazy horse*. London: Penguin Books.

Moynihan, C. (2015). Man convicted of environmental terrorism is freed. *New York Times*. Retrieved from http://www.nytimes.com/2015/01/09/us/man-convicted-of-environmental-terrorism-wins-early-release.html?_r=0

National Lawyers Guild. (2006). National lawyers guild condemns operation backfire as unconstitutional. *NLG Press Release* published on Common Dreams NewsCenter. Retrieved from http://www.commondreams.org/news2006/0612-10.htm

Nocella, A. J., II, & Walton, M. (2005). Standing up to corporate greed: The Earth Liberation Front as domestic terrorist target number one. *Green Theory & Praxis*, *1*(1), 2–19.

Parson, S. (2008). Understanding the ideology of the Earth Liberation Front. *Green Theory & Praxis*, *4*(2), 50–66.

PATRIOT Act. (2001). *Public Law 107-53*. Retrieved from http://www.gpo.gov/fdsys/pkg/BILLS-107hr3162enr/pdf/BILLS-107hr3162enr.pdf

Perliger, A. (2012). Challengers from the sidelines: Understanding America's violent far-right. *The Combating Terrorism Center at West Point*.

Peterson, E. (2009). Press release—"Defendants sentenced to prison in ELF action from 2000." *Federal Bureau of Investigation*. Retrieved from http://milwaukee.fbi.gov/dojpressrel/pressrel09/mw022609a.htm

Pickering, L. J. (2007). *The Earth Liberation Front 1997–2002*. Binghamton, NY: Arissa Media Group.

Plungis, J. (2002, December 18). SUV, truck owners get a big tax break. *USA Today*. Retrieved from http://usatoday30.usatoday.com/money/autos/2002-12-18-suv-tax-break_x.htm

Potter, W. (2006). Animal enterprise protection act. *Green is the New Red*. Retrieved from http://www.greenisthenewred.com/blog/aepa/

Potter, W. (2009). Environmentalist sentenced to 21 years as a "terrorist". *Green is the New Red.* Retrieved from http://www.greenisthenewred.com/blog/marie-mason-sentenced/1023/
Potter, W. (2011). *Green is the new red.* San Francisco, CA: City Lights Books.
Regan, L. (2008, March 6). Briana waters found guilty on two counts. *Olympia Civil Liberties Resource.* Retrieved from http://olycivlib.blogspot.com/2008/03/briana-waters-has-been-found-guilty-on.html
Reuters News Service. (2002, February 12). Arizona "eco-arsonist" gets 18 years in prison. *Los Angeles Times.* Retrieved from http://articles.latimes.com/2002/feb/12/news/mn-27647
Richmond, T. (2005, August 31). Animal activist accused of freeing thousands of mink to plead guilty and serve two years. *Associated Press.* Retrieved from http://legacy.signonsandiego.com/news/nation/20050831-1449-
Rosebraugh, C. (2004). *Burning rage of a dying planet: Speaking for the Earth Liberation Front.* Brooklyn, NY: Lantern Books.
Scott, M. (2008). Eco-terrorist given nearly twenty years in prison. *Department of Justice.* Retrieved from http://sacramento.fbi.gov/dojpressrel/pressrel08/sc050808.pdf
Singh, N. (2006). The afterlife of fascism. *Southern Atlantic Quarterly, 105,* 71–93.
Smith, C. (2003, November 26). Grooming an ELF. *Willamette Week Online.* Retrieved from http://www.wweek.com/portland/article-2692-grooming_an_elf.html
Todd, A. (2008, May). The Believers. *Elle Magazine.* Retrieved from http://veganxjen.livejournal.com/4471.html
United States Congress. (1992). Animal Enterprise Protection Act. *Public Law 102-346.* Retrieved from http://www.nal.usda.gov/awic/legislat/pl102346.htm
United States Congress. (1995). Omnibus Counterterrorism Act of 1995. *H.R. 896.* Retrieved from http://www.govtrack.us/congress/bills/104/hr896
United States Congress. (2006). Animal Enterprise Terrorism Act. *Public Law 109–374;18 U.S.C. § 43.* Retrieved from https://www.govtrack.us/congress/bills/109/s3880/text
Van Bergen, J. (2002). The USA PATRIOT act was planned before 9/11. *Truthout.* Retrieved from https://ratical.org/ratville/CAH/PAplndbefore.pdf
Vanderheiden, S. (2005). Eco-terrorism or justified resistance? Radical environmentalism and the "war on terror." *Politics and Society, 33*(3), 425–447.
Veblen, T. (1994). *The theory of the leisure class.* New York, NY: Dover Publications.
Walzer, M. (1977). *Just and unjust wars.* New York, NY: Basic Books.
Winzelberg, D. (2004, August 1). Earth liberation member gets 3 ½ years for arsons. *New York Times.* Retrieved from http://www.nytimes.com/2004/08/01/nyregion/in-brief-earth-liberation-member-gets-3-1-2-years-for-arsons.html?src=pm
Yang, D. (2005). Pasadena man who participated in firebomb attacks on SUVs sentenced to over eight years in federal prison. *U.S. Department of Justice.* Retrieved from http://www.justice.gov/usao/cac/Pressroom/pr2005/059.html

10. *Speaking About "Ecoterrorists": Terrorism Discourse and the Prosecution of Eric McDavid*

JOSHUA M. VARNELL

The number one domestic terrorist threat currently facing the United States, according to the FBI, are radical animal rights/environmentalist (107th Congress, 2002; 108th Congress, 2004; 109th Congress, 2005a; Best & Nocella, 2004; Del Gandio & Nocella, 2014; Loadenthal, 2013; Smith, 2008). This has been an often-recited refrain in Congressional hearings and FBI press releases and memos, a refrain echoed by many inside and outside the halls of the federal government. For example, shortly following the attacks of September 11, Alaska Representative Don Young stated that he believed the attacks may have been carried out by radical ecoterrorists linked to the WTO protests in Seattle in 1999 (Ruskin, 2001). In a 2012 speech, then presidential hopeful Rick Santorum claimed that the radical environmental movement had created a "reign of environmental terror," creating a boogie man out of the hydro-fracturing process, a process Santorum claimed to be completely safe (Guillen & Summers, 2012).

Terrorism has come to be understood as the major threat facing the United States and the Western world in the 21st century. It is seen as an existential threat to civilization. Today, it is even claimed that a dangerous "terrorist ideology" has come to influence public education in the United States. In Oklahoma, conservatives attacked high school AP history as "un-American," "dangerous," and the ideological indoctrination of "terrorism." Dr. Ben Carson commented on the Oklahoma AP History course, stating: "I think most people, when they finish that course, they'd be ready to go sign up for ISIS" (Gambino, 2015, para. 22).

As Edward Said (2001) has noted, "[t]errorism is anything that stands in the face of what we want to do...people's movements of resistance against deprivation, against unemployment, against the loss of natural resources, all of that is termed terrorism" (para. 8–9). It is because these causes would and do directly challenge the foundations of the modern liberal–democratic state that they are understood as terrorism. Terrorism is most often applied to groups and individuals who criticize or attack the status quo. All too often, the terrorism discourse has come to be employed when capitalism, or the near religious faith in the free market, is directly challenged. This pronouncement was seen in George W. Bush's proclamations after 9/11 that the best way to fight against terrorism is to go out shopping, to continue consuming. Capitalism is understood as the foundation of Western civilization and the battle against terrorism is often represented as a "clash of civilizations," to borrow Huntington's famous phrase. It has come to represent a clash of good v. evil.

This chapter sets out to explore the effects of the terrorism discourse in the investigation of and prosecution of Eric McDavid through a critical discourse analysis (CDA). McDavid was arrested in early 2006 for conspiracy to destroy the Nimbus dam. In May of 2008, McDavid was sentenced to nearly 20 years of prison after receiving a terrorism enhancement. The terrorism discourse has important effects for who we as a society consider a terrorist and who is authorized to speak about terrorism. As an ideological tool, the terrorism discourse allows elites (social status, economic, and political elites) to effectively secure and protect the status quo by providing what Noam Chomsky (1998) termed a "grave enemy" to channel the active fears and discontent of the population. Today, the radical environmental and animal rights movements are portrayed as the "grave enemy" of domestic terrorism, with the ALF/ELF being the FBI's number one priority for 15 years. This designation is being pushed by economic and political elites who believe that the position advocated by these movements is a direct threat to their positions (105th Congress, 1998; 107th Congress, 2002; 108th Congress, 2004; 109th Congress, 2005a, 2005b, 2006; Arnold, 1983, 1997; Cong. Rec. Oct. 14, 1988).

Critical Discourse Analysis

In this section I want to briefly set out and summarize the main tenets of CDA as a theoretical and methodological tool designed to investigate the social effects of discourse. Drawing from a number of major figures in the field of CDA, we can identify six main tenets: (1) CDA concerns itself with social problems; (2) discourse is a social practice, understanding discourse as a social practice implies a wider investigation of social context; (3) CDA concerns itself with power relations in discourse and how discourse (re)produces social inequalities and/or social injustice; (4) discursive events are situated within a dialectical relationship to situation(s), institution(s), and social and political structures; (5) discourse may have ideological effects. To uncover such effects, it is necessary to explore, investigate, and reveal the interpretations of discourse and the social effects of a particular discourse; (6) CDA is both practice and theory, engaged in actively challenging social and political domination (Blommaert, 2005; Fairclough & Fairclough, 2012; Hammersley, 1997; Keller, 2013; Kress, 1990). CDA limits itself to interpretation, understanding, and explanation and not to a nomothetically oriented goal; it is not, as Fairclough and Wodak (1997) state, a "dispassionate and objective social science, but [CDA sees itself] as engaged and committed. It is a form of intervention in social practice and social relationships" (p. 258).

CDA is a theoretical and methodological approach which holds that there exists a fundamental relationship between discourse and society, that discourse is a social practice (Blommaert, 2005; Fairclough & Fairclough, 2012; Fairclough & Wodak, 1997; Kress, 1990). In turn, because CDA understands discourse to be a social practice, the researcher is not divorced from this practice, so that, there is a fundamental relationship between analysis, and the practices and events analyzed (Fairclough & Wodak, 1997; Kress, 1990). In this respect, researchers play an active role in discourse (re)production. This is because CDA understands the researcher to be an agent embedded in social structures and institutions, which influence their choice of and understanding of social problems, and that their particular situation requires them to be committed to emancipatory social and political change. This means that from the CDA perspective, researchers cannot position themselves outside of the practices and events which they study, that there exists no truly "objective" position from which one may observe and describe the world (Fairclough & Fairclough, 2012; Fairclough & Wodak, 1997; Hammersley, 1997; Keller, 2013; Kress, 1990; Van Dijk, 2001).

Because CDA understands discourse to be a "form of social practice" (Fairclough & Wodak, 1997), discourse is seen as being shaped by and shaping society, so that social and political structures are both outcome and medium of discourse (Fairclough & Fairclough, 2012). This means that when analyzing any particular discursive event or practice, the researcher must be aware that discourses are relevant only with respect to context. Discourses are historically rooted, and culturally and ideologically embedded as well as being "connected intertextually to other discourses" (Keller, 2013, pp. 25–26). Discourses are powerful social practices which produce ideological effects because they are representative of reality, that is, they create meaning by representing the world in particular and specific ways. Discourses organize the world around us by creating understandings for events, processes, individuals, identities, common sense and by putting subjects into "imagined" relationships, to borrow Althusser's (2001 [1970]) formulation. Discourses form the basis for how agents understand the world and act as social agents. Hegemonic discourses (re)produce social knowledge, embedded within them are ideological perspectives which maintain the status quo.

CDA's goal is to uncover the social and ideological effects of discourse by demonstrating the way in which hegemonic discourses obscure alternatives. Hegemonic discourses often portray their ideological assumptions as "rational," "normal," "benign," "neutral," "natural," and/or simply as "common sense." Such representations are essential for legitimating discourses because alternatives are then seen as "irrational," "unnatural," and/or "unrealistic" (Van Dijk, 1993; Wodak & Meyer, 2009). For example, such subtle forms of domination like racism, sexism, and speciesism are opaque and taken for granted, supported, and (re)produced through specific discourses. Such forms of domination were simply accepted as common sense or natural until they were challenged (Van Dijk, 1993). As both practice and theory, CDA actively engages in exposing the ideological function of discourses which reproduce such forms of domination in social and political practices. CDA is also a productive discourse designed to alter and change social, economic, and political relationships so that they are more equitable and just.

In this chapter, I seek to employ a CDA approach to uncover how the terrorism discourse was ideologically employed against Eric McDavid, with its core ideological assumptions reproduced within the ecoterrorism discourse. Such a discourse was used to legitimate both FBI tactics and federally prosecute Eric McDavid as a domestic terrorist. In a larger respect, I hope that such an analysis will help to destabilize the ecoterrorist discourse which is currently used to delegitimize radical environmental and animal rights organizations and activists by painting them as irrational and violent existential threats to Western society. Such representations of reality are inherent to the terrorism discourse, having social and political stock as common-sense understandings of reality. Yet, as critical research has demonstrated, the terrorism discourse itself is highly vulnerable to destabilization.

Data

Eric was freed from prison in January of 2015, after FOIA requests revealed that the FBI, and likely federal prosecutors, intentionally withheld evidence in his case. Using the terrorism discourse, federal prosecutors, relying on a confidential informant as their primary source of information, portrayed Eric as a domestic terrorist mastermind bent on the destruction of the United States (Habeas Hearing, 2015; Holpuch, 2015; Pilkington, 2015; Potter, 2015). Data for this chapter is drawn from the trial transcripts of Eric's trial which ran from September of 2007 through May 2008 when he was sentenced and the January 2015 Habeas Hearing in which evidence from FOIA requests was presented to the court. Additional data is drawn from trial documents, including law enforcement declarations, law enforcement reports, petitions, juror declarations, habeas petitions, appeal briefs, and news reports.

Terrorism Discourse

Discourse has a profound effect on the way in which we understand the world, because of its power to construct reality. Discourse is a productive activity, meaning that discourse acts to produce "meaning-structures of our reality" (Keller, 2013, pp. 71–72). This means that discourse is constructive of reality. The way in which we understand reality is informed by how we speak about, understand, and think about the world around us. Today the terrorism discourse has an outsized role in social and political discussion, occupying a role of importance equal to discussions of democracy or climate change. It is because of the power of discourse that the terrorism narrative "function[s] to construct and maintain a specific understanding of, and approach to, terrorism and counterterrorism and the 'knowledge' generated in the field has certain academic, political, and social effects" (Jackson, 2009, p. 69).

Critical studies on terrorism have revealed and uncovered the core assumptions of the contemporary terrorism discourse, which informs our understanding (Della Porta, 2013; Gunning, 2007b; Jackson, 2007a; Jackson, Breen Smyth, & Gunning, 2009; Schmid & Jongman, 2005; Silke, 1998, 2009; Stampnitzky, 2013). This research demonstrates that the concept of "terrorism" is highly malleable, politically biased, and often ideologically driven. This is the result of a field of investigation that "rather than looking like a discipline or a closed 'cultural field,' terrorism expertise is constructed and negotiated in an interstitial space between academia, the state, and the media. The boundaries of legitimate knowledge and expertise are particularly open to challenges from self-proclaimed experts from the media and political fields, and this has had significant consequences for the sorts of expert discourses that tend to be produced and disseminated" (Stampnitzky, 2013, p. 47).

Discussions of terrorism since the 1970s have increasingly come to focus on describing acts and incidents as irrational, illegitimate, and evil, and those described as terrorists have come to be understood as pathological, irrational, and evil (Della Porta, 1995, 2013; Della Porta & Klandermans, 1992; Gunning, 2007b, 2009; Jackson, 2007a, 2009; Loadenthal, 2013; Ranstorp, 2009; Silke, 1998, 2009; Stampnitzky, 2013). This is because much of the discussion about terrorism has become tied to moral judgments (Stampnitzky, 2013, p. 8). In turn, conventional definitions of terrorism go to great lengths to exclude the state, most often read Western states, from being included within the definition of terrorism. Terrorism has become an identity marker, "where the identity of the actor rather than the act itself defines the designation of 'terrorism'" (Miller & Mills, 2009, p. 417). This understanding, however, is simply the recognition that we cannot understand the actions or individuals because they are irrational, evil, nihilistic, abnormal, and strictly not like us (Crenshaw, 2014; Miller & Mills, 2009; Silke, 1998, 2009).

The discourse on terrorism is essentially a refusal "to grant terrorism and terrorists the consideration of whether or not such actions may be justifiable—for, if they are justifiable, they are no longer 'terrorism'" (Stampnitzky, 2013, p. 4). Critical studies of terrorism and the field of terrorism expertise have revealed that the conventional understanding of psychological abnormality, immorality, and irrationality is simply not borne out by evidence. In fact, many studies point to the way in which many acts labeled as terrorism are provided justifications, with many justifications being rational and in many cases sounding like justifications used by states to explain state acts of violence (Gunning, 2007a, 2007b, 2009).

Furthermore, if the definition of terrorism was consistently applied, we would have to acknowledge that "there have been a number of historical cases where terrorism has been used on behalf of causes most Western liberals would regard as just" (Wilkinson & Steweart, 1987, p. xiv). Or, as Herman (1982) has argued, that the "sub-rosa" violence carried out with U.S. acquiescence, and in many cases outright support, pales in comparison with what is contemporarily labeled as "terrorism." Critical studies have revealed that while the terrorism discourse is highly unstable and contradictory, it continues to persist driven by an overblown threat that is represented as unpredictable, imminent, and one capable of mass destruction that seeks to destroy the Western world (Jackson, 2007a, 2009; Mueller, 2009; Stampnitzky, 2013). This discourse finds resonance in the mass-media because the media overwhelmingly promotes a "discourse of fear" (Altheide, 2003), and media outlets overwhelmingly rely on experts who are "ideologically conservative" and have deep connections to the state or think tanks linked to government agencies (Miller & Mills, 2009). The discourse itself serves important purposes for state and corporate elites.

Far from identifying a unique form of political violence, the terrorism discourse acts to demonize actors and silence oppositional voices who criticize Western states' claims to enlightened progress and claims of freedom, justice, and fairness. The discourse on terrorism has produced a discourse that, while not simply constructed to support the state's demonizing of political opponents, "is at the same time a highly complex and intertwined set of narratives and rhetorical strategies that aims to reinforce the authority of the state and reify its disciplinary practices" (Jackson, 2005, p. 178).

Before turning to an analysis of how the terrorism discourse is used against activists to justify questionable law enforcement tactics and how the discourse was used to prosecute Eric McDavid, I turn to a detailed discussion of the Eric McDavid's case as this case serves as an example of the social effects of the ecoterrorism discourse. Understanding the contours and context of the case will help us make sense of the terrorism discourse's application as well as provide context for the case under investigation.

The Case of Eric McDavid

In August of 2004, Eric McDavid, then a young college student and budding anarchist, traveled from his home in northern California to Des Moines, Iowa, for the annual CrimethInc. Convergence (*U.S. v. McDavid*, 2007, pp. 207–208). This yearly convergence of anarchists attracted anarchists from across the United States engaging in several days of discussions about the major tenets of anarchism from the foundations of anarchist philosophy to the role of violence in the movement to more practical guides for living an anarchist lifestyle. It is here that Eric first met a young, and radical, anarchist known as "Anna." Wearing a camouflage skirt, this young lady with bright pink hair instantly impressed Eric. Anna sees in Eric a young man deeply committed to anarchism, but inexperienced. Eric and Anna spend days together getting to know one another, and at the end of the convergence the two travel to New York to protest the Republican National Convention. Anna, however, is no political activist, she is a confidential informant working in coordination with the FBI. Both Anna and the FBI initially misidentify Eric as a leader in the anarchist movement, but ultimately a benign individual they conclude (Declaration of Walker, 2012; Memo in Support of *Brady* Claims, 2014).

Anna was first approached by the FBI in the fall of 2003. She was then a 17-year-old Miami community college student whom the FBI asked for help in infiltrating left-leaning protest movements in order to report on illegal activity. Anna was the main source of evidence and the primary witness in the government's case against Eric McDavid (Todd, 2008; *U.S. v. McDavid*,

2007, p. 195). Anna first came to the attention of the FBI following a class report she presented on the Free Trade Agreement of the Americas (FTAA) protests for a political science course (*U.S. v. McDavid*, 2007, p. 199). In that class, a former Florida State Highway Patrol Officer, impressed by her report, showed a copy of it to his superiors, who in turn shared it with the FBI. The FBI asked Anna to work as a confidential informant, attending protests and reporting back on any illegal activity taking place during the protests. In the case against Eric McDavid, Anna was able to provide evidence of an ongoing conspiracy that involved plans to build explosives and bomb federal institutions—a threat framed as a national bombing campaign.

On January 13, 2006, following several months of investigations, wiretapping, and electronic surveillance, Eric McDavid, Lauren Weiner, and Zachary Jenson were arrested in a K-Mart parking lot in Auburn, CA. The case presented by federal prosecutors painted a picture of Eric as a violent anarchist terrorist intent on attacking the federal government by whatever means necessary in pursuit of his extremist political views. The case against Eric rested on the testimony of Anna and wiretaps that seemed to present Eric as the organizer of a bombing conspiracy that targeted the Nimbus Dam, the United States Forest Service Institute of Forest Genetics in Placerville, CA, and cell phone towers.

The FBI was able to produce much of the evidence in the case through electronic surveillance of a cabin procured by the FBI for the group. Anna made the cabin available to the group to plan through their winter bombing campaign, providing an opportunity to bring all the suspects together at one place and record their movements. The cabin, located in Dutch Flats, CA, allowed the group to work and plan over six days from January 6th through January 12th of 2006, with the FBI diligently monitoring the progress of the conspiracy just down the road in their command post. While the FBI portrayed the investigation as the dismantling of a major domestic terrorism cell that justified the FBI's investigative techniques, the facts of the case reveal a far more nuanced discussion and considerable questions about the actual threat posed. Anna's role as a confidential informant highlights the highly suspect nature of using confidential informants in domestic terrorism investigations, as well as raising questions about the actual efficacy of the FBI's counterterrorism operations, specifically if the FBI engaged in the investigation of a legitimate security concern, or simply acted to suppress political opponents.

Confidential Informants

Since September 11, 2001, the federal government has increased law enforcement budgets, expanded the criminal code, created new agencies, and pursued domestic terrorists with an increased vigor, all justified under preventing another terrorist attack on domestic soil. In turn, the FBI's mission has been updated from one of criminal investigation to one focusing primarily on counterintelligence as the Bureau takes the lead on many domestic terrorist investigations. The updated mission of counterintelligence focuses on foiling threats before they can come to fruition (Ashcroft, 2002).

Cunningham (2004) has noted that this updated mission is one in which "the Bureau [...] stresses agents' ability to anticipate future threats, often indiscriminately targeting suspects for their ostensible hidden activities" (p. 8). Extensive FBI investigations have focused on disrupting terrorist networks through intelligence gathering strategies employing counterterrorism tactics. The transformed mission of the FBI has meant that directors and Special Agents in Charge (SAC) dedicate significant resources to identifying and disrupting terrorist networks by employing counterintelligence tactics, similar to those in the previous COINTELPRO operations of the 1960s and 1970s (Cunningham, 2004). In pursuit of its updated mission as a counterintelligence agency, the FBI has come to rely heavily on confidential informants, who are individuals paid by the FBI to infiltrate suspect communities and report back on "terrorist" activity. However, what is growing increasingly clear is that these investigations rest on suspect police work and political bias. Suspects are targeted because of ethnic identity, religion, or political ideology (Center for Human Rights and Global Justice, 2011; Greenwald, 2010; Human Rights Watch, 2014; Kamat & Soohen, 2010).

Law enforcement and the FBI justify the use of confidential informants in terrorism cases based on the terrorism discourse portrayal of terrorism as a shadowy and unpredictable event. The terrorism discourse has influenced the way in which the FBI understands the threat of terrorism and how, in turn, it responds to that threat. As former federal prosecutor, David Raskin, states in a *New York Times* interview: "There isn't a business of terrorism in the United States [...] You're not going to be able to go to a street corner and find somebody who's already blown something up [...] Therefore, the usual goal is not to find somebody who's already engaged in terrorism but find somebody who would jump at the opportunity if a real terrorist showed up in town" (Shipler, 2012, para. 7–9). As the Raskin quote makes clear, there exists no terrorist infrastructure from which security agencies can monitor. Because terrorism is understood to be a "special" kind of violence, one that is unpredictable, hidden, and strikes without warning, traditional law enforcement tactics are inadequate in combating the threat of terrorism. This threat narrative presents terrorism as only being able to be overcome through intensive information gathering (Ackerman & Yuhas, 2015).

Focusing on a preventative model of policing has meant that the FBI must focus on the processes that lead to violent terrorism, which has meant looking for sources that produce terrorists. The terrorism discourse holds that ideology plays an important role in motivating or influencing individuals to engage in terrorist behavior. Smith (2008) points out that in "2002, an FBI memo indicated that potential terrorist groups included 'anarchists,' 'animal rights extremist[s],' and 'environmental extremist[s]'" (p. 16). In addition, Smith found that prosecutors and law enforcement agencies have been advised that

> [a]n effective way to begin tracking potential ELF members is to track active members of other environmental organizations with similar ideologies [...] Earth First! is one group which might be tracked, in part because it support[s] an environmental preservation philosophy. A hint as to what other ideologies—besides "environmental preservation"—might provide grounds for terrorist investigations surfaced in a report published by the Heritage Foundation. The report suggests that it is likely that people will be killed by environmentalists if the philosophy of Deep Ecology is not challenged at the philosophical level. (p. 18)

In essence what this discourse does is present ideology as an important marker of violent behavior. Infiltrating groups that represent subversive and terrorist ideologies becomes an important aspect of the preventative model. Using confidential informants is an attractive tactic for the FBI for several reasons. Informants provide easy access to suspect communities because they are often drawn directly from the communities they are charged with infiltrating. They can sweep up all manner of information without regard to criminal activity, because they are not restricted by the same guidelines that control undercover operations. Informants are a low-risk, high-reward tactic for investigations. Not only does the FBI not have to employ a large intelligence gathering apparatus, but the high conviction rate of cases involving informants makes it an attractive tactic.

The guidelines that direct the use of confidential informants are devised by the U.S. Attorney General's office and implemented in the Domestic Investigative Operational Guidelines (DIOG); yet attorney general guidelines have been significantly scaled back since 2002 (USDOJ, 2008). In conjunction with the USA PATRIOT Act and the Animal Enterprise Terrorism Act (AETA), domestic law enforcement agencies have been granted unprecedented powers of surveillance along with a wide latitude in investigative operations (Black & Black, 2004). The result has been an increased focus by the FBI on suppressing critical political dissent of subversive groups, with an overwhelming focus by the federal government on animal rights/environmental activists coming to be known as the "Green Scare" (Best & Nocella, 2004, 2006; Del Gandio & Nocella, 2014; Kuipers, 2009; Loadenthal, 2013; Lovitz, 2010; Potter, 2011).

Attorney General John Ashcroft first articulated the justification for revising of FBI guidelines in a May 2002 speech. In that speech, Ashcroft asserted that the FBI was burdened by unduly harsh restrictions on its activities, restrictions that provided cover to terrorists. Essentially Ashcroft argued in this speech that the FBI needed to be allowed to engage in any activity that terrorists could engage in so as to allow the Bureau to adequately gather intelligence of ongoing terrorist plots (2002). The threat of terrorism is represented as one that can only be overcome by intelligence gathering tactics; limiting those tactics means that the FBI would be hindered in their ability to thwart terrorist plots. As Ashcroft notes, "[t]hese restrictions are a competitive advantage for terrorists who skillfully utilize sophisticated techniques and modern computer systems to compile information for targeting and attacking innocent Americans" (2002). The FBI makes clear that the use of confidential informants plays an essential role in counterterrorism operations as a valuable and much needed source of information. An FBI spokesperson stated in a 2005 *Washington Post* article that "[c]onfidential informants and other confidential human sources are critical to the FBI's ability to carry out our counterterrorism, national security and criminal law enforcement missions.... A source can have a singular piece of information we could not otherwise obtain, enabling us to prevent a terrorist act or crime, or apprehend a fugitive" (Eggen, 2005).

Questioning the Efficacy of Informants as a Tactic

A 2005 report from the Office of the Inspector General (OIG) reviewed the FBI's compliance with the attorney general's 2002 guidelines and indicated serious failures (Eggen, 2005; USDOJ OIG, 2005). This review, covering 120 cases, found that the "most significant problems were failures to comply with the Confidential Informant Guidelines. For example, we identified one or more Guidelines violations in 87 percent of the confidential informant files we examined" (USDOJ OIG, 2005, p. 2). While many of the violations were minor in nature, the high proportion of cases that exhibit some sort of violation should give us pause. A 2011 report by the NYU School of Law's Center for Human Rights and Global Justice found that the use of confidential informants has resulted in a 97% conviction rate for cases that employ informants; however, the cases that rely on confidential informants are also marked by excessive concerns over the FBI's role in facilitating the very crimes they investigate (Center for Human Rights and Global Justice, 2011).

Many cases represent constructed threats that relied on FBI know-how, funding, and resources. There are also considerable concerns over the choice of targets, with the FBI focusing on marginalized individuals facing personal hardships. The conclusion of the report states that many of these cases appear to simply be cases of entrapment. A July 2014 report by Human Rights Watch echoed much of what was in the Center for Human Rights and Global Justice report, stating that many domestic terrorism cases indicate that confidential informants play key leadership roles and it's likely, with the assistance of the FBI, constructed entire plots (Human Rights Watch, 2014). However, proving entrapment in court requires overcoming an excessively high standard, in which the defense must prove no predisposition to commit the crime (Center for Human Rights and Global Justice, 2011; Kamat & Soohen, 2010).

This prospect is often complicated by the fact that the FBI, law enforcement agencies, and prosecutors rely on evidence that cannot be "fairly contested" (Human Rights Watch, 2014). This procedural hurdle is raised when prosecutors or law enforcement agencies withhold valuable information (*New York Times* Editorial Board, 2015). In turn, much of the information produced by informants is classified by the FBI. This means that for those charged with terrorism-related crimes, they are likely to be convicted even in the face of serious investigative and procedural flaws, because they do not have access to evidence that might otherwise be exculpatory or evidence that might demonstrate investigative violations. While many critical reports into terrorism cases focus on the American Muslim community, anyone who finds themselves under investigation as a terrorist faces the same problems (Center for Human Rights and Global Justice, 2011; Human Rights Watch, 2014; Kamat & Soohen, 2010). With the FBI insisting that the greatest domestic terrorist threat facing the nation comes from radical animal rights and environmental activists, it comes as no surprise that these tactics have been employed against these activists as well. The threat posed by animal rights and environmental activists is apparently so pressing that the FBI has attempted to insert informants into vegan potlucks, claiming these as hotbeds of extremist and terrorist activity (Potter, 2008).

Anna, the FBI, and the Construction of a Threat

During Eric's trial, Anna was presented as an unimpeachable witness. The FBI and federal prosecutors painted a picture of Anna as a heroic young woman who waded into danger for the love of country. Without her bravery and assistance, prosecutors claimed, the United States would have faced a devastating ecoterrorist attack. However, many in the anarchist and environmental communities saw Anna as entrapping Eric in a romantic affair that ultimately led him into a conspiracy plot. While the truth probably lies somewhere in the middle of these two representations, it does appear that Anna played a much larger role in the conspiracy than originally admitted by the federal government, given the evidence released through FOIA requests (*U.S. v. McDavid*, Brady Memo, 2014; *U.S. v. McDavid*, Habeas Hearing, 2015; *U.S. v. McDavid*, Habeas Petition, 2012).

Anna was a young woman clearly affected by growing up in the aftermath of September 11, 2001, a world hyper-sensitive to the "terrorism" threat. Anna became an informant for the FBI at the age of 17, just two years after 9/11, and after earning her GED and beginning her first semester of college. In a May 2008 *Elle* magazine interview, Anna describes how she left high school at 17, earning her GED amidst her parents' "acrimonious divorce" (Todd, 2008, p. 267). She describes growing up a middle child of three, from a middle-class family. Describing her parents as Vietnam-era protesters, she is quick to note, though, that this was a long time ago, that she is a self-described "hawk," the result she says of growing up in the aftermath of 9/11 (p. 267). At 15, Anna dedicated herself to joining military counterintelligence after witnessing the tragic events of 2001. She notes in the *Elle* interview that this was the result of her being a unique teenager, politically aware and savvy, and ready to do her patriotic duty, stating: "My friends and I saw that plane fly into the World Trade Center, and we thought right away that it was (some Palestinian) terrorist group [...] Keep in mind, we were teenagers reading *The Economist*" (p. 267).

Anna jumped into her new role with the FBI without hesitation, certain that the focus on animal rights and environmentalists was justified because they posed a serious terrorist threat; "to believe that these people aren't capable of harm or serious attack is not giving them enough credit" (p. 270). She so fully dedicated herself to her new role that she went far enough to get a tattoo on her shoulder of a skull and black flag (p. 270). Anna's first investigative successes came in June and July of 2004 while attending the G8 Summit and then the Democratic National Convention (DNC) protests. It is at G8 that Anna first met Zachary Jensen, and according to Anna, Zachary helped "score" her entry into the 2004 CrimethInc. Convergence along with others she met at the 2004 DNC protest (Todd, 2008, p. 270; *U.S. v. McDavid*, 2007, p. 207).

During the trial, Anna describes, and misrepresents, entry to the CrimethInc. Convergence as a complex process of shadowy meetings and coded messages that eventually ended in a formal invitation for those who were thoroughly vetted (*U.S. v. McDavid*, 2007, p. 227). Anna represents the anarchist movement in her testimony as a highly organized and centralized entity, with a leadership that enforced strict protocols and extensive background checks. CrimethInc. Convergences, however, were widely publicized and open to attendance. The only restriction was that law enforcement agents were not welcome.

While Anna was infiltrating the anarchist movement, she also came to have a profound respect for the movement and individuals she later described often as "disgusting" and "dirty" (Todd, 2008; *U.S. v. McDavid*, 2007, p. 245). In particular, Anna was impressed by the movement's egalitarian nature, stating that "[o]ne of the best things about this movement is the way women are treated and viewed [...] They reject typical standards of beauty [...] They focus on a woman's independence, her passion, her conviction. And she is treated as an equal" (Todd, 2008, p. 272). Anna found in the movement the very quality of respect and equality that was lacking within the confines of the FBI (p. 323). Anna notes that on several occasions she felt as if the FBI was dismissive of her because of her gender. None of this came to light in the trial and was only relayed later by Anna in her *Elle* interview. While the FBI's male-centered culture may have played a role in agents being dismissive of Anna's ability, FOIA revelations reveal that many FBI agents were skeptical of the truthfulness of her reports. A FOIA request by Eric's lawyers, as well as a declaration from Special Agent Nassan Walker, agent in charge of the case, reveals that there had been internal FBI requests for Anna to take a polygraph test to confirm her reports. It seems several agents were skeptical about the validity of her claims; however, the polygraph request was refused by Anna's handler, Special Agent Ricardo Torres (*U.S. v. McDavid*, Brady Memo, 2014; *U.S. v. McDavid*, Declaration of Walker, 2012).

Anna was first assigned to work under Agent Torres' direction in early 2005, and the two grew close almost immediately. Torres spoke highly of Anna in the *Elle* magazine article, saying, "She was so young, and she wasn't an agent [...] but everything she said would happen, happened. I was able to verify every bit of information she passed on to us" (Todd, 2008, p. 323). Agent Torres and Anna became so close that Anna confided in Torres concerning very personal and traumatic events in her life. Feeling safe with Torres,

Anna revealed to Torres that she had been the victim of a sexual assault in college (Todd, 2008, p. 323). While we have no knowledge of the actual sexual assault, it does appear that this event was significant enough to cause Anna distress during the investigation. Anna claims that the sexual assault had a profound impact on her behavior in the Dutch Flats cabin; she felt the stress of working undercover was too much, stating: "I was experiencing some kind of flashback, to being in a situation with a man who wouldn't leave me alone," she said, reminding Agent Torres of her sexual assault (p. 324). These revelations in themselves raise concerns about Anna's internal state, her position as a vulnerable subject, and the responsibility of the FBI in such a situation.

Evidence from the trial transcripts additionally raises serious questions about the competency of Agent Torres as Anna's handler in the case. Under cross-examination, Torres revealed that he had no training in undercover operations or the use of confidential informants. More concerning, he was unaware of the U.S. Attorney General guidelines that outline confidential informant use, or recent reviews by the OIG that raised concerns about the FBI's use of confidential informants and entrapment (*U.S. v. McDavid*, 2007, pp. 643–650; USDOJ OIG, 2005).

It now seems very likely that Anna's actions during the investigation were highly suspect and indicate that she and the FBI worked very hard at constructing a terrorist threat and entrapping three individuals (*U.S. v. McDavid*, Brady Memo, 2014; U.*S. v. McDavid*, Habeas Hearing, 2015). Anna, with FBI funding, bankrolled the entire enterprise, paying for the food, supplies, and travel expenses for the group, as well as supplying FBI laptops and a chemistry set (*U.S. v. McDavid*, 2007, pp. 840–841). No one in the group other than Anna had any stable source of income. Eric and Zach often traveled by hitch-hiking or train hopping and without the Dutch Flats cabin would have been homeless (*U.S. v. McDavid*, 2007, pp. 907, 996–997, 1070). Zach lived on food stamps at the time and he and Eric practiced a freegan lifestyle, a trait the prosecution raised many times to demonstrate their radical natures in resisting modern norms. Lauren lived on a small stipend provided by her parents, who also paid for Lauren's living expenses while she was in art school in Philadelphia (*U.S. v. McDavid*, 2007, pp. 775–778, 794).

In addition, Anna had to drive both Lauren and Zach to California in early January of 2006, or the two would have had no other way of traveling west, and they would have been stranded in California without Anna (*U.S. v. McDavid*, 2007, pp. 849–850). During the drive from Washington D.C. to Dutch Flats, California, in January of 2006, both Lauren and Zach would testify that they felt Anna was in charge of the group, leading them (*U.S. v.*

McDavid, 2007, p. 1028). In fact, Zach Jensen, during the trip from Washington DC to California, states in audio recordings that he felt Anna was leading the group into a trap. He said he felt Anna was doing this because of something "bad" that had happened to her in the past (*U.S. v. McDavid*, 2007, p. 1028). At the cabin, Anna urges all the members to take part in the explosives development (*U.S. v. McDavid*, 2007, pp. 845–846). Lauren testifies to the fact that she and Zach were terrified at the prospect and were berated by Anna until they agreed to take a more active role in the construction of the explosives (*U.S. v. McDavid*, 2007, pp. 845–846). Anna even states in her testimony that had she not pushed the group to act or move forward, they would have "dillydallied" and got nothing done (*U.S. v. McDavid*, 2007, p. 494).

In addition, following the trial, numerous jurors stated that they believed Anna played a much larger role than was admitted by federal prosecutors and that the FBI overstepped in their investigation (Kuipers, 2012; *U.S. v. McDavid*, Carol Runge, Juror Deceleration, 2008; *U.S. v. McDavid*, Diane Bennett, Juror Declaration, 2008; Todd, 2008). Jurors were also presented with two contradictory statements during their deliberation concerning Anna's role as an informant, with one set of instructions stating that Anna was not an FBI informant and one statement saying Anna was an agent under the direction of the FBI. The confusing nature of the instructions put the jurors in a position that they felt left them no alternative but to find Eric guilty. Appeals courts refused to consider juror declarations or the errors in instruction as grounds for retrial.

But what now seems most damning in the case are the FOIA revelations that uncovered numerous letters from Anna to Eric, in which Anna seems to be pushing and cajoling Eric and in which Anna seems to be promising a romantic relationship if Eric progresses with the conspiracy (Democracy Now, 2015; Pilkington, 2015; Potter, 2015; *U.S. v. McDavid*, Brady Memo, 2014; *U.S. v. McDavid*, Habeas Hearing, 2015). Federal prosecutors claim that the withholding of evidence was unintentional and they were unaware of the evidence being held by the FBI (*U.S. v. McDavid*, Habeas Hearing, 2015). The FBI claims the evidence was non-exculpatory and did not warrant release to the defense. During Eric's Habeas hearing, Judge England expressed a cautious skepticism about both claims and pushed several times for federal prosecutors to answer why such a mistake would or could take place (*U.S. v. McDavid*, Habeas Hearing, 2015).

Reproducing the Terrorist Discourse in Trials

While the terrorism discourse justifies the implementation of questionable security tactics to uncover terrorist activities, it also plays an important role in the representation of individuals designated as terrorists in trials. From the very beginning the McDavid case was framed by the federal government as a successful counterterrorism operation. The government portrayed Eric McDavid as a violent domestic terrorist, convinced of both his ability to carry out a terrorist attack and in his commitment to a "terrorist philosophy." McGregor Scott, U.S. Attorney, stated after the trial that if the defendants would have "succeeded in blowing up Nimbus Dam [...] It would make New Orleans look like a Sunday pancake breakfast" (The Eric McDavid Story, 2008; Todd, 2008, p. 323).

Actually, destruction of the dam would have resulted in nothing more than a "trickle," claims Jeff McCracken, spokesperson for the dam (Todd, 2008, p. 323). How did the federal government use the terrorism discourse to prosecute Eric McDavid in a case that resulted in no actual destruction of property or the death of citizens? To answer this question, it's important to analyze the terrorism discourse that has grown around the environmental movement; often accepted uncritically, it is taken for granted that the ALF/ELF are "terrorists" writ large.

While the hegemonic discourse on ecoterrorism is highly unstable and contradictory, it retains its power as useful and remains meaningful partly through its employment in trials. This gives courts a particular role in pronouncing on the inherent moral judgments within the discourse, acting not only as a site of moral reinforcement but also as sites of political control and political neutralization. Court cases provide evidence of the continuing danger and threat from terrorism, which, in turn, provides the justification for the increased domestic security measures. Trials of "ecoterrorists" reinforce and reproduce the hegemonic discourse by demonstrating that defendants are inherently violent, acting irrational, and are simply evil. Motivation and explanation become irrelevant because the discourse of terrorism provides a self-explanatory and circular logic; terrorism is the result of terrorists.

Over and over studies have consistently disputed the conception of radical animal rights and environmentalists as engaging in direct violence. Most actions are minor violations of law and at the most they are cases of property damage. Vanderheiden (2005) points out that the moral transgression inherent to discussions of terrorism is the use of violence against a civilian population who is not the direct target of the violence. Such violence, Vanderheiden notes, is meant to serve as a threat to a secondary target of individuals, that is, if they do not adequately respond they will be met with future violence. Studies of the actions carried out by the ALF/ELF have consistently rejected the narrative of violence so often employed by opponents of these groups because they do not seek to injure or kill (Amster, 2006; Carson et al., 2012; Hirsch-Hoefler & Mudde, 2014; Johnson, 2007; Vanderheiden, 2005).

Furthermore, the criminal direct actions of the ALF/ELF are not directed indiscriminately, the target of such actions is the intended recipient, and the destruction of property in such instances is not intended to signify future violence aimed at harming individuals. Actual violence in the "ecoterrorist" discourse is replaced with arguments of potential violence by those opposed to the movements. Such potential violence is often demonstrated through reference to ideology or philosophical position. With respect to the ALF/ELF, these actors often display an anarchist perspective, one that is anti-capitalist and anti-corporate. Joosse (2012) and Mcleod and Detember (1999) have both demonstrated that within news framing, anarchists are often trivialized by focusing on their "abnormal" appearance and behaviors, and represented as an inherently violent threat to the state and corporations. These misconceptions have also been reproduced in research.

Borum and Tilby's (2005) research into anarchist violence reproduces the conception of anarchists and anarchism as inherently violent and abnormal; they state that "people with unusual attitudes, behaviors, and views of the world frequently (and disproportionately) are drawn to counterculture movements and extremist groups […] These individuals would likely be engaging in criminal or violent behavior, regardless of their circumstances. Affiliating with a movement or ideal, however, gives them a reason and adds some sense of legitimacy" (pp. 205–206). Borum and Tilby's discussion demonstrates how ideology acts as a signifier of inherent violent behavior. Anarchists cannot be understood as being drawn by social justice and political or moral considerations; rather, they are simply engaging in movement activity as a way to legitimate or justify their own pathological violent behavior; in short, terrorists simply behave as they do because they are evil.

Finally, an interesting aspect of the "ecoterrorism" discourse in trials is the use of a moral equivalency argument in comparing defendants to clearly violent but ideologically dissimilar cases. The result is odd portrayals of violent actions, rhetoric, or ideologies as equivalent to the crimes committed by environmental activists. Within the hegemonic discourse, differences in groups or ideologies is overlooked or strained attempts are made to demonstrate how the ideologies held by terrorists are simply "terrorist" ideologies. This type of comparison eliminates from the discussion the foundation of actions, the non-violent guidelines of the ALF/ELF, and the fact that no individual has been harmed in direct actions carried out by the ALF/ELF. Further, these portrayals attempt to portray the state as the progressive defender of social justice, ignoring the states' actual position or role in constructing and reinforcing social injustice.

Portraying Eric as a "Terrorist"

The portrayal of Eric as a domestic terrorist was successful because since the 1980s, radical environmentalists and animal rights activists have been portrayed as dangerous and violent. During the trial of Eric McDavid, the most overt portrayals of this discourse came in the state's sentencing memo and in Judge England's comments during the sentencing hearing. Federal prosecutors stated in their sentencing memo: "McDavid's home-grown brand of eco-terrorism is just as dangerous and insidious as international terrorism. A 20-year term of imprisonment demonstrates that the public does not tolerate those who would generate fear and inflict massive property damage in order to oppose government policy" (*U.S. v. McDavid*, Government Sentencing Memo, 2008, p. 6).

Such a portrayal reproduces the terrorism discourse's assertion that "terrorism" is a serious and shadowy threat to the Western world. In many instances, we see assertions and references to international terrorism as an existential threat to Western civilization, with 9/11 serving as the ultimate reference point. The second half of the federal prosecutor's statement introduces the idea that the primary goal of terrorism is to produce an emotional response of fear in order to produce a policy outcome. The assumptions underlying this is that terrorism is a symbolic act directed at an audience beyond the main target. Terrorist targets then serve as referents. While this might help explain some actions, many actions have multiple goals and are directed at multiple audiences. The ALF/ELF, far from simply directing their action symbolically at a larger audience, are acting directly on the audiences they target for their message. The idea here is that Eric's actions would have been directed at producing a general fear among the larger population, misrepresenting the activist community's goals and motivations. The ALF/ELF have taken great pains to avoid physical harm to individuals, believing that such actions would most likely undermine their goal and message. The aim is certainly not to simply incite fear in a population. The goals are often twofold: to raise awareness of a particular issue by exposing obscured corporate and state behavior and to increase the cost of doing business.

Judge England's remarks during sentencing also reproduced conventional terrorist discourse:

> The Court has considered the kinds of sentences available, and the need for the type of sentence involved. There have not been many cases that have involved domestic terrorism. This is one of the newer cases. As indicated, this is a *new* world after September 11, 2001. And, again, I cannot help but recall the audio transcript or audio recording of Mr. McDavid indicating that there will have to be collateral damage at some point in time. And that's referring to human lives, and IEDs, which is the talk that we listen to, we hear of when referring to actions that are taking place 6,000 miles away in Iraq, and what people are undergoing at that point in time. (U.S. v. McDavid, Sentencing Hearing, 2008, pp. 55–56, emphasis added)

Judge England reproduces the idea that 9/11, a "new" kind of unprecedented violence, has ushered in a new world. Much terrorism scholarship has made claims to a "new" terrorism ushering in a "new" world, a terrorism of profound violence unexperienced in previous eras. However, those events described as terrorism today are strikingly similar to past events and past descriptions of terrorism. The claim to "newness" has come to represent terrorism since the 1990s, and certainly after 2001, as something altogether different from previous forms of political violence. Judge England also introduces into the discussion references to the Iraq war and improvised explosive devices (IEDs).

These references reinforce the war narrative present in many terrorism discussions. Terrorism is essentially the resistance to the Western civilizing project, reproducing the language of clashing civilizations or a war pitting good against evil, the "War on Terror." The use of military language like IEDs additionally helps to reinforce the image of terrorism as unpredictable violence. The use of IEDs was a key referent in discussions of terrorism emerging in Iraq as a form of indiscriminate, illegitimate, and unpredictable killing. Once again we have the conflation of attacks specifically designed to destroy property and kill to actions that simply target property. Direct actions are, according to the ALF/ELF, responses to violence perpetrated by the state and corporations against all living creatures and the environment. They are motivated by the belief that capitalism is inherently immoral and that actions justified simply with reference to capitalism are inherently wrong. The focus of the ALF/ELF on attacking capitalism, its symbols, institutions, and its foundations, however, does have the effect of being used to justifying the state and corporate claim that these organizations and individuals are an existential threat to Western civilization and are inherently violent.

A second important feature of the terrorism discourse reproduced in the McDavid trial was the continued use of language that demonstrated an irrational and abnormal character inherent to all terrorists. The dominant image that has emerged of terrorists is one of an irrational, psychologically disturbed, evil, misanthrope. Dominant portrayals of domestic terrorist's abnormality are indicated by reference to ideological persuasion. Ideology plays an important role in the terrorism discourse as it acts both as evidence of terrorism and individual abnormality. For Eric this meant that descriptions of anarchy implied an irrational and abnormal character. The result is a description of individuals who demonstrate unusual behaviors or attitudes, the goal being to show how terrorists are not like "us." The criminal complaint filed against Eric and his co-defendants refers to anarchy or its derivatives 26 times in 15 pages. It then goes on to describe the dangerous nature of anarchism and linking this to the ELF, and according to federal agents a known terrorist organization: "ELF adherents share a strong philosophical connection to the anarchist movement. The anarchist movement seeks to end the current system of government, economy and replace them with systems characterized by a lack of authoritarian/hierarchical relationships" (U.S. v. McDavid, Weiner, & Jenson, Criminal Complaint, 2006, p. 3). During the trial, anarchy played an important role as a signifier of violence and abnormality. The first witness for the prosecution was former police officer Bruce Naliboff whose testimony covered a description of "anarchism" and the ALF/ELF. Naliboff described anarchism to the jury as a "lifestyle choice," but did recognize that many anarchists advocate for political and social change (*U.S. v. McDavid*, 2007, p. 182).

The description of anarchism as a lifestyle choice has several consequences. Primarily, by equating anarchism as a lifestyle choice, it disarms anarchism as a critical political discourse. It trivializes anarchism, it becomes nothing more than a personal choice akin to tastes or preferences, reducing its meaning to the level of a personal characteristic. The goal of the terrorism discourse is to demonize and delegitimize opposition voices. This seemingly incompatible representation is the same process identified by Joosse (2012), who found a "transgression of binary categor[ies]" led to a "semiotic excess" (p. 84). Thus, during the trial, anarchism was portrayed as both morally perverse and dangerous, as well as a trivial lifestyle choice. If anarchism is a "lifestyle choice" it has no claim to legitimacy as a position from which individuals may act for social and political change. The result is to remove the foundations from which individual activists act. Trivializing anarchism removes from the discussion grievances. It becomes irrational for individuals to claim general political and social grievances as arising from "personal choices." Motivation and explanation are explicitly organized outside the conversation as irrelevant.

Anarchism during the trial came to be an indicator of Eric's abnormality and violence. Demonstrating this abnormality, prosecutors repeatedly made references to how Eric lived. During opening statements, Stephen Lapham, assistant U.S. Attorney, spent a considerable amount of time describing the lifestyle habits of Eric McDavid and, by extension, his anarchism as abnormal, making sure that the jury understood that Eric lived abnormally: "Food he got from dumpster diving, or he would get from begging or getting it free from some source" (*U.S. v. McDavid*, 2007, p. 116). The oddity of Eric's lifestyle was often raised to demonstrate that he chose to live a life that was outside the norm.

In making clear that his lifestyle was not the result of circumstance, but of choice, prosecutors stated: "It's not as if they were homeless and paupers because of their circumstances. They chose to travel and live the way they did. It was a choice" (*U.S. v. McDavid*, 2007, p. 1277). Anna as well participated in this process of constructing an image of abnormality describing how she had to construct a "dirty" and "disgusting" image to fit into the activist community (*U.S. v. McDavid*, 2007, p. 245; Todd, 2008). It is, of course, not enough to demonstrate oddity or abnormality of individual habits and choices. This abnormality has to also be demonstrative of a larger more insidious and violent nature.

The terrorism discourse represents individual "terrorists" as inherently violent and drawn to subversive or extremist ideologies that provide them motive, legitimacy, and cover for their violent natures. Responding to the assertion by Eric's family and friends that he was a "kind" and "gentle" individual, the prosecutors stated: "Clearly, the defendant became a different person than his friends and family recall from his youth. He began attending CrimethInc meetings and anarchist gatherings" (*U.S. v. McDavid*, Government Sentencing Memo, 2008, p. 16). The underlying assertion is that being "kind" or "gentle" cannot co-exist with subversive ideologies. To be an anarchist is to be neither kind nor gentle, but is to be suspected of violence, to be suspected of terrorism. Terrorists cannot be seen as kind, gentle, or compassionate, as this might inject into the conversation the similarity between terrorists and "us." To do so would in turn result in questioning how individuals like "us" might become engaged in these activities. If terrorists can be kind and gentle, then they may be justified in their actions.

Prosecutors provided plenty of evidence during the trial to demonstrate that Eric was a violent and dangerous individual. Two events during the trial became particularly important for demonstrating Eric's violent nature, yet both incidents were unverifiable. The first was a road trip to Chicago in which Anna drove Eric to Chicago following the 2005 CrimethInc. Convergence, and Anna claimed that Eric threatened to kill her with a knife. The second incident took place in the Dutch Flats cabin the night prior to Eric's arrest. Both Anna and the FBI claim that Eric waved a knife in front of Anna's face as she slept. The first incident could never be verified or confirmed because the only witness was Anna, and she was not wearing a body wire at the time. The second incident, however, took place in the Dutch Flats cabin, which had been fully wired with surveillance equipment, yet no audio, video recordings, or notes exist from law enforcement monitoring in the HQ. The federal government, the FBI, and Anna all claim that these incidents took place, but no evidence was presented in court to support these claims.

In addition to these two events, prosecutors demonstrated Eric's violent nature by returning once again to the group's discussion of "collateral damage." During that discussion Eric raised a nuanced view that accounted for the possibility of unintended casualties; ultimately, Eric concludes that this should be avoided at all costs to the best of the group's ability (Kuipers, 2012). Federal prosecutors, however, represented this discussion as evidence of violence, stating: "No emotion. It's just a fact. And, as you hear in that recording, it's murder, and the Government will call it murder. He is aware of that" (*U.S. v. McDavid*, 2007, p. 1276). A theoretical discussion, then, became direct evidence of violence.

Collateral damage was an important and ongoing discussion for the prosecution during the trial. The goal for prosecutors was to decouple the legitimating effects of "collateral damage" when used by states to explain their actions from McDavid's discussion. Collateral damage is the unintentional killing of civilians. The effect is to obscure the fact that an operation resulted in the death of civilians. The use of the term often implies the necessity of a particular military operation that did not intend to kill civilians. Intent becomes the reference point from which to judge an action. Federal prosecutors went a long way in making sure that collateral damage did not obscure the fact that this meant the death of civilians or that discussing the possibility of collateral damage was tantamount to advocating for the killing of individuals. This discussion helped to reinforce the idea of terrorism as illegitimate violence. It also helps to reinforce the idea that the state cannot engage in terrorism and that terrorism is only carried out by non-state actors. Again, terrorism is defined in actor-based terms.

Finally, the trial of Eric McDavid employed an odd comparison between defendants and cases that clearly sought the harm of individuals. Eric McDavid's crime of conspiracy was compared during sentencing and judgment to crimes committed by members of the white supremacist movement and the militia movement. The State sought to portray these crimes motivated by a right-wing ideology and specifically designed to kill civilians to those of Eric, who conspired to destroy property in support of the environmental movement.

Three cases in particular were raised by federal prosecutors as analogous to McDavid's crime of conspiracy: the case of Kevin Patterson and Charles Kiles, the case of Matt Hale, and the case of Jack Dowell (*U.S. v. McDavid*, Government Sentencing Memo, 2008). In response to the Defense Sentencing memo, federal prosecutors claimed that Eric's crime was not comparable to other "ecoterrorists" as his crimes were of a different nature. Kevin Ray Patterson and Charles Kiles were convicted of conspiring to destroy gas storage tanks. Patterson and Kiles were members of a right-wing millennial militia. Their goal was to hasten the collapse of the corporate U.S. government in hopes of restoring Constitutional order. The two planned to destroy gas storage tanks on Y2K in the belief that the new millennium would usher in a wave of chaos and destruction. Their hope was to cause mass civilian casualties in what they believed would be nationwide coordinated attacks by right-wing militias seeking to restore Constitutional order.

Matt Hale, founder of the World Church of the Creator, conspired to murder a federal judge in his tax evasion case. Hale, an avowed white supremacist, advocates for the murder of marginalized groups and left-wing activists. One of Hale's followers went on a multistate shooting spree targeting minority citizens after the Illinois Bar Association denied Hale his law license, and Hale has been described as the "face of hate" in the United States. Jack Dowell was convicted of burning down a Colorado IRS building. Dowell was at the time a member of the Constitutional Law Group and the Army of the American Republic.

These types of comparisons in the Eric McDavid case are no anomaly. During sentencing for Daniel McGowan, federal prosecutors compared the arson committed by McGowan and his fellow defendants under the moniker of the ELF/ALF to the burning of Southern churches by the Ku Klux Klan (*U.S. v. McGowan*, Terrorism Enhancement Hearing, 2007). Comparing activists in the environmental and animal rights movements to avowed violent right-wing groups and organizations has several important effects. First, comparisons of right-wing and racist crimes and rhetoric which directly advocates for the killing of individuals connects violence to an avowed non-violent movement. Another effect of this comparison is the tying of what many accept as the irrationality of right-wing militia ideology and supremacist ideology to animal rights and environmentalists. The inherent racism in these right-wing movements is now widely accepted as an irrational foundation for social and political organization. By tying these movements together, federal prosecutors present both movements as irrational and violent. Finally, it constructs an image of the state as a defender and advocate for civil rights. This obscures the fact that the animal rights and environmental movement have drawn both tactically and philosophically from the civil rights movement and liberation movements. It also ignores the many historical examples of state intransigence and outright resistance to civil rights. This comparison of Eric's crime of conspiracy to right-wing groups is also odd given the state's insistence that Eric's crimes were not comparable to other ecoterrorism cases after a trial that sought to present the conspiracy as a clear-cut case of ecoterrorism. There are two important explanations for this portrayal. First, if prosecutors would have compared Eric's crimes to that of other ecoterrorists, they would not have had a connection to violence. Second, comparing Eric to other ecoterror cases would have presented examples of a sentencing range far lower than what the state advocated. Both of these aspects would have jeopardized the terrorist portrayal and in turn the terrorism enhancement applied to Eric during sentencing.

Conclusion

I hope that what the case of Eric McDavid demonstrates is the way in which questionable assumptions in the terrorism discourse were simply recycled to present Eric as a dangerous threat. The terrorism discourse itself is based on flawed data and assumptions that have no basis in empirical fact. Rather, the terrorism discourse has been used by political and economic elites with ties to agri-business and biomedical research to delegitimize activists and silence them. September 11, 2001 was widely seen as an intelligence failure, a failure that has reinforced the belief that domestic security requires an extensive intelligence gathering apparatus. Confidential informants, long a useful tool for law enforcement, have become important and powerful tools for meeting the new demands of intelligence gathering in the era of the War on Terror. Intelligence becomes the primary arena in which terrorism is fought because the terrorist discourse represents the threat as a shadowy and insidious threat. Because of this, terrorism must be confronted prior to its actual manifestation, which means predicting who will become a terrorist. Confidential informants can easily access suspect communities with few resources and little risk to the FBI. From the FBI's point of view, the overall success of confidential informants in terrorism investigations is demonstrated in the high conviction rate of cases that rely on confidential informants as the primary source of information.

The success of these prosecutions, however, is most likely the result of several interrelated factors. Federal prosecutors are statistically more likely to win convictions. The evidence produced by confidential informants is often difficult to verify, even for agents in charge of the investigation. Additionally, evidence produced in investigations employing confidential informants cannot be fairly contested. Given the few restrictions and limited oversight of confidential informants, this makes it difficult to verify the information passed by confidential informants in the early assessment stages of an investigation. Finally, cases that employ an informant make it difficult for defendants to prove entrapment. An entrapment defense places a high burden on defendants to prove they had no predisposition to commit the crime for which they are charged. The difficulty of the entrapment defense is compounded because defendants may not question government conduct until they have proven no predisposition (Target and Entrapped; Human Rights Watch).

These concerns arose in the trial of Eric McDavid and demonstrated the suspect nature of evidence procured through the use of a confidential informant. Confidential informants also play an important role in the reproduction of the terrorism discourse by providing confirming evidence for law enforcements' focus on specific groups. Confidential informants do not simply serve an informational gathering role; they play an active role in the crimes. In many instances confidential informants are suspected of moving crimes forward by ensuring that suspects are progressing through the conspiracy. In the case of Eric, there exists many instances of Anna being the prime mover in the conspiracy by pushing and cajoling the other members to move forward with the conspiracy, providing resources, and even actively bringing the members together from across the country.

Federal courts are hardly neutral sites of determining facts and ascertaining truth. Federal courts are embedded within the political and social structure. As such, institutional mechanisms operate to protect the institution and the larger system. Because the ALF/ELF are understood as threats to the system, they threaten powerful elite groups with interests within the system, they have become targets for repression. Because the terrorism discourse is hegemonic, federal prosecutors need only to link the defendant's characteristics with already known and understood terrorist characteristics. The pervasiveness of the terrorism discourse means that label itself brings forward the image of irrational, pathological violence. Through prosecutions like Eric McDavid's, courts serve to reinforce the social understanding of terrorism and its application to the number one domestic terrorist threat, the ALF/ELF. Such characteristics and representations are readily reproduced in the mass media and within government agencies, law enforcement, and legislators at both the federal and state level. The terrorism discourse presents a simplified pathway from radicalism to violence, with ideology simply serving as cover for pathologically violent individuals. Much of the terrorism discourse reproduces reductionist theories of violence that are rooted in a predisposition to violence as a function of psychological deviancy. Such deviancy is an important function of the overall discourse as it "others" those targeted.

While the terrorism discourse linking environmental and animal rights movements is hegemonic in its portrayal of activists as terrorists, it is by no means uncontested. All discourse is open to challenge as discourse is a process continually in flux and open to continuous articulation and rearticulation. The terrorism discourse itself is a mixture of contradictory characteristics based on flawed data and unverifiable assumptions. It acts to construct an overblown and misrepresented threat to the state. As Jackson (2009) has articulated, the terrorism discourse is less about understanding and responding to a real threat and more about "controlling wider social and political dissent, restricting human rights, and setting the parameters for acceptable public debate; and altering the legal system" (p. 79). But it is also at these points that the discourse can be challenged, where fissures in the discourse can be exposed.

The terrorism discourse when applied to radical environmentalists and animal rights activists who hold a non-violent stance risks conflating acts of civil disobedience engaged in out of compassion with acts of heinous violence and aggression. In turn, such a discourse operates to obscure real violence committed by agri-business and biomedical corporations when they use animals and natural resources as commodities by naturalizing their acts as common sense. When we challenge such conceptions and ask what is meant by terrorism, how is it employed, what its effects are, and who is silenced by the discourse, we engage in the process of counterhegemonic discourse, as I hope I have accomplished here.

References

105th Congress. (1998). *Acts of ecoterrorism by radical environmental organizations.* Hearing before the Subcommittee on Crime of the House Committee on the Judiciary. Retrieved from http://commdocs.house.gov/committees/judiciary/hju59927.000/hju59927_0.HTM

107th Congress. (2002). *Eco-terrorism and lawlessness in the national forests.* Subcommittee on Forests and Forest Health, Committee on Resources, U.S. House of Representatives. Retrieved from https://www.gpo.gov/fdsys/pkg/CHRG-107hhrg77615/pdf/CHRG-107hhrg77615.pdf

108th Congress. (2004). *Animal rights: Activism vs. criminality.* U.S. Senate Committee on the Judiciary Retrieved from https://www.gpo.gov/fdsys/pkg/CHRG-108shrg98179/pdf/CHRG-108shrg98179.pdf

109th Congress. (2005a). *Oversight on eco-terrorism specifically examining the Earth Liberation Front (ELF) and the Animal Liberation Front (ALF).* United States Senate Committee on Environmental and Public Works. Retrieved from https://www.gpo.gov/fdsys/pkg/CHRG-109shrg32209/pdf/CHRG-109shrg32209.pdf

109th Congress. (2005b). *Eco-terrorism specifically examining stop Huntington animal cruelty ("SHAC").* United States Senate Committee on Environmental and Public Works. Retrieved from https://www.gpo.gov/fdsys/pkg/CHRG-109shrg39521/pdf/CHRG-109shrg39521.pdf

109th Congress. (2006). *Animal Enterprise Terrorism Act.* Subcommittee on Crime, Terrorism, and Homeland Security, Committee on the Judiciary, U.S. House of Representatives. Retrieved from https://www.gpo.gov/fdsys/pkg/CHRG-109hhrg27742/pdf/CHRG-109hhrg27742.pdf

Ackerman, S., & Yuhas, A. (2015). FBI told its cyber surveillance programs have actually not gone far enough. *The Guardian.* Retrieved from http://www.theguardian.com/us-news/2015/mar/25/fbi-review-commission-technological-abilities-human-intelligence

Altheide, D. L. (2003, Fall). Mass media, crime, and the discourse of fear. *The Hedgehog Review*, 9–25.

Althusser, L. (2001 [1971]). Ideology and ideological state apparatuses. In L. Althusser (Ed.), *Lenin and philosophy and other essays.* New York, NY: Monthly Review Press.

Amster, R. (2006). Perspectives on ecoterrorism: Catalysts, conflations, and casualties. *Contemporary Justice Review*, *9*(3), 287–301.

Arnold, R. (1983, February). Eco-terrorism: Environmental extremists have declared guerrilla war on developers...and the environmental mainstream stands by silently. *Reason*, 31–36.

Arnold, R. (1997). *Ecoterror: The violent agenda to save nature—the world of the unabomber.* Bellevue, WA: Free Enterprise Press.

Ashcroft, J. (2002). *Remarks of Attorney General John Ashcroft.* Retrieved from http://www.justice.gov/archive/ag/speeches/2002/53002agpreparedremarks.htm

Becker, M. (2006). Rhizomatic resistance: The Zapatistas and the Earth Liberation Front. *Green Theory and Praxis*, *2*(2), 22–63.

Best, S., & Nocella, A. J., II. (2004). Introduction. In S. Best & A. J. Nocella II (Eds.), *Terrorists or freedom fighters?: Reflections on the liberation of animals.* New York, NY: Lantern Books.

Best, S., & Nocella, A. J., II. (2006). *Igniting a revolution: Voices in defense of the earth.* Oakland, CA: AK Press.

Black, J., & Black, J. (2004). The rhetorical "terrorist": Implications of the USA Patriot Act for animal liberation. In S. Best & A. J. Nocella II (Eds.), *Terrorists or freedom fighters?: Reflections on the liberation of animals.* New York, NY: Lantern Books.

Blommaert, J. (2005). *Discourse: A critical introduction.* New York, NY: Cambridge University Press.

Borum, R., & Tilby, C. (2005). Anarchist direct actions: A challenge for law enforcement. *Studies in Conflict & Terrorism*, *28*, 201–223.

Center for Human Rights and Global Justice. (2011). *Targeted and entrapped: Manufacturing the "homegrown threat" in the United States*. New York, NY: NYU School of Law. Retrieved from http://chrgj.org/documents/targeted-and-entrapped-manufacturing-the-homegrown-threat-in-the-united-states/

Chomsky, N. (1998). *The common good* [Interviews with David Barsamian]. Berkeley, CA: Odonian.

Cong. Rec. 14 Oct. 1988: S16036-01. Retrieved from https://1-next-westlaw-com.libproxy.library.wmich.edu/Document/I41150BC0631D11D9A612E62709B7A6DF/View/FullText.html?originationContext=document&contextData=%28sc.Keycite%29&cacheScope=null&transitionType=DocumentItem&searchWithinQuery=McClure&chunkSize=S&docSource=909b1037b663476394d01dbb547d516e&needToInjectTerms=False&searchWithinHandle=i0ad82352000001521860c9f72faf62f1

Crenshaw, E. (2014). American and foreign terrorists: An analysis of divergent portrayals in U.S. newspaper coverage. *Critical Studies on Terrorism*, *7*(3), 363–378.

Crenshaw, M. (2000). The psychology of terrorism: An Agenda for the 21st century. *Political Psychology*, *21*(2), 405–420.

Cunningham, D. (2004). *There's something happening here: The new left, the Klan, & FBI counterintelligence*. Berkeley, CA: University of California Press.

Del Gandio, J., & Nocella, A. J., II. (2014). Introduction: Situating the repression and corruption of the AETA. *The terrorization of dissent: corporate repression, legal corruption, and the Animal Enterprise Terrorism Act*. Brooklyn, NY: Lantern Books.

Della Porta, D. (1995). *Social movements, political violence, and the state*. Cambridge: Cambridge University Press.

Della Porta, D. (2013). *Clandestine political violence*. Cambridge: Cambridge University Press.

Della Porta, D., & Klandermans, B. (Eds.). (1992). *International social movement research vol. 4. Social movements and violence: Participation in underground organizations*. Greenwich, CT: Jai Press Inc.

Democracy Now. (2015). "Eco-terrorist" freed 10 years early after feds withheld evidence on informant's role [Interview with McDavid]. *Democracy Now*. Retrieved from http://www.democracynow.org/2015/1/14/exclusive_eco_terrorist_freed_10_years

Eggen, D. (2005). FBI agents often break informant rules. *The Washington Post*. Retrieved from http://www.washingtonpost.com/wp-dyn/content/article/2005/09/12/AR2005091201825.html

The Eric McDavid Story. (2008). *Indybay: Central Valley*. Retrieved from http://www.indybay.org/newsitems/2008/01/18/18473322.php

Fairclough, N., & Fairclough, I. (2012). *Political discourse analysis*. New York, NY: Routledge.

Fairclough, N., & Wodak, R. (1997). *Discourse studies: A multidisciplinary introduction*, London: Sage Publications.

Federal Bureau of Investigation (FBI). (n.d.). *Definitions of terrorism in the U.S. Code*. Retrieved from https://www.fbi.gov/about-us/investigate/terrorism/terrorism-definition

Finsen, L., & Finsen, S. (1994). *The animal rights movement in America: From compassion to respect*. New York, NY: Maxwell Macmillan International.

Gambino, L. (2015). Oklahoma educators fear high school bill will have "devastating" impact. *The Guardian*. Retrieved from http://www.theguardian.com/us-news/2015/feb/20/oklahoma-ap-history-bill-devastating-dan-fisher

Greenwald, G. (2010). The FBI successfully thwarts its own terrorist plot. *Salon*. Retrieved from http://www.salon.com/2010/11/28/fbi_8/

Guillen, A., & Summers, J. (2012). Rick Santorum slams "reign of environmental terror." *Politico*. Retrieved from http://www.politico.com/news/stories/0212/72681.html

Gunning, J. (2007a). A case for critical terrorism studies? *Government and Opposition*, *42*(3), 363–393.

Gunning, J. (2007b). *Hamas in politics: Democracy, religion, violence*. London: Hurst.

Gunning, J. (2009). Social movement theory and the study of terrorism. In R. Jackson, M. Breen Smyth, & J. Gunning (Eds.), *Critical terrorism studies: A new research agenda* (pp. 156–177). London: Routledge.

Hammersley, M. (1997). Foundations of critical discourse analysis. *Language & Communication*, *17*(3), 237–248.

Herman, E. S. (1982). *The real terror network: Terrorism in fact and propaganda*. Boston, MA: South End Press.

Holpuch, A. (2015). Man convicted of "eco-terrorism" freed amid claims FBI hid evidence. *The Guardian*. Retrieved from http://www.theguardian.com/us-news/2015/jan/09/eric-mcdavid-eco-terrorism-fbi-evidence

Horne, T. (2006). The radical environmental movement: Incorporating empire and the politics of nature. *Green Theory and Praxis*, *2*(1), 31–51.

Human Rights Watch. (2014). Terrorism prosecutions often an illusion. *Human Rights Watch*. Retrieved from http://www.hrw.org/node/127456

Jackson, R. (2007a). Constructing enemies: "Islamic terrorism" in political and academic discourse. *Government and Opposition*, *42*(3), 394–426.

Jackson, R. (2005). *Writing the War on Terrorism: Language, Politics and Counter-terrorism*. Manchester, UK: Manchester University Press.

Jackson, R. (2007b). The core commitments of critical terrorism studies. *European Political Science*, *6*(3), 244–251.

Jackson, R. (2009). Knowledge, power, and politics in the study of political terrorism. In R. Jackson, M. Breen Smyth, & J. Gunning (Eds.), *Critical terrorism studies: A new research agenda* (pp. 66–83). London: Routledge.

Jackson, R., Breen Smyth, M., & Gunning, J. (2009). Introduction: The case for critical terrorism studies. In R. Jackson, M. Breen Smyth, & J. Gunning (Eds.), *Critical terrorism studies: A new research agenda* (pp. 1–10). London: Routledge.

Joosse, P. (2012). Elves, environmentalism, and "eco-terror": Leaderless resistance and media coverage of the ELF. *Crime Media Culture*, *8*(1), 75–93.

Kamat, A. (Producer), & Soohen, J. (2010). Entrapment or foiling terror?: FBI's reliance on paid informants raises questions about validity of terrorism cases. *Democracy Now*. Retrieved from http://www.democracynow.org/2010/10/6/entrapment_or_foiling_terror_fbis_reliance

Keller, R. (2013). *Doing discourse research: An introduction for social scientists*. London: Sage.

Kress, G. (1990). Critical discourse analysis. *Annual Review of Applied Linguistics*, *11*, 84–99.

Kuipers, D. (2009). *Operation bite back: Rod Coronado's war to save the American wilderness*. New York, NY: Bloomsbury.

Kuipers, D. (2012). Honey stinger. *Outside*. Retrieved from http://www.outsideonline.com/1911086/honey-stinger

Lichtblau, E. (2014). In Cold War, U.S. spy agencies used 1,000 Nazis. *New York Times*. Retrieved from http://www.nytimes.com/2014/10/27/us/in-cold-war-us-spy-agencies-used-1000-nazis.html?_r=1

Loadenthal, M. (2013). Deconstructing "eco-terrorism": Rhetoric, framing, and statecraft as seen through the insight approach. *Critical Studies on Terrorism*, *6*(1), 92–117.

Lovitz, D. (2010). *Muzzling a movement: The effects of anti-terrorism law, money, and politics on animal activism*. New York, NY: Lantern Books.

Mcleod, D. M., & Detember, B. H. (1999). Framing effects of television news coverage of social protest. *Journal of Communication*, *49*(3), 3–23.

Memo in Support of *Brady* Claims, *U.S. v. McDavid*, No. 2:06-CR-00035-MCE (E.D. Cal. 2014).

Miller, D., & Mills, T. (2009). The terror experts and the mainstream media: The expert nexus and its dominance in the news media. *Critical Studies on Terrorism*, *2*(3), 414–437.

Mueller, J. (2009). *Overblown: How politicians and the terrorism industry inflate national security threats and why we believe them*. New York, NY: Free Press.

New York Times Editorial Board. (2015). How to force prosecutors to play fair. *New York Times*. Retrieved from http://www.nytimes.com/2015/02/16/opinion/how-to-force-prosecutors-to-play-fair.html?_r=0

Nocella, A. J., II, & Walton, M. J. (2005). Standing up to corporate greed: The Earth Liberation Front as domestic terrorist target number one. *Green Theory and Praxis*, *1*(1), 2–19.

Pilkington, E. (2015). Role of FBI informant in eco-terrorism case probed after documents hint at entrapment. *The Guardian*. Retrieved from http://www.theguardian.com/us-news/2015/jan/13/fbi-informant-anna-eric-mcdavid-eco-terrorism

Potter, W. (2008). FBI looking for informants to infiltrate vegan potlucks [Web log post]. *Green is the New Red*. Retrieved from http://www.greenisthenewred.com/blog/fbi-informant-vegan-potluck/437/

Potter, W. (2011). *Green is the new red: An insider's account of a social movement under siege.* San Francisco, CA: City Lights Books.

Potter, W. (2015). Eric McDavid released from prison, feds withheld evidence [Web log post]. *Green is the New Red.* Retrieved from http://www.greenisthenewred.com/blog/eric-mcdavid-released-prison/8129/

Ranstorp, M. (2009). Mapping terrorism studies after 9/11: An academic field of old problems and new prospects. In R. Jackson, M. Breen Smyth, & J. Gunning (Eds.), *Critical terrorism studies: A new research agenda* (pp. 13–33). London: Routledge.

Rosebraugh, C. (2004). *The burning rage of a dying planet: Speaking for the Earth Liberation Front.* New York, NY: Lantern Books.

Ruskin, L. (2001). Rep. Dong Young (R-AK) links terrorist attacks to WTO protest. *Anchorage Daily News.* Retrieved from http://www.iatp.org/news/rep-don-young-r-ak-links-terrorist-attacks-to-wto-protest

Said, E. (2001, July–August 19). They call resistance "terrorism." *International Socialist Review.* Retrieved from http://isreview.org/issues/19/Said_part2.shtml

Schmid, A., & Jongman, A. (2005). *Political terrorism.* Piscataway, NJ: Transaction.

Shipler, D. K. (2012). Terrorist plots, hatched by the FBI. *New York Times.* Retrieved from http://www.nytimes.com/2012/04/29/opinion/sunday/terrorist-plots-helped-along-by-the-fbi.html?pagewanted=all&_r=3

Silke, A. (1998). Cheshire-cat logic: The recurring theme of terrorist abnormality in psychological research. *Crime & Law, 4*(1), 51–69.

Silke, A. (2009). Contemporary terrorism studies: Issues in research. In R. Jackson, M. Breen Smyth, & J. Gunning (Eds.), *Critical terrorism studies: A new research agenda* (pp. 34–48). London: Routledge.

Smith, R. K. (2008). Eco-terrorism?: A critical analysis of the vilification of radical environmental activists as terrorists. *Environmental Law, 38*(2), 537–573.

Stampnitzky, L. (2013). *Disciplining terror: How experts invented "terrorism."* Cambridge: Cambridge University Press.

Todd, A. (2008, May). The believers. *Elle Magazine*, p. 266–325.

USDOJ (U.S. Department of Justice). (2008). *Attorney general guidelines for FBI domestic operations.* Retrieved from http://www.justice.gov/sites/default/files/ag/legacy/2008/10/03/guidelines.pdf

USDOJ OIG (U.S. Department of Justice, Office of the Inspector General). (2005). *The Federal Bureau of Investigation's compliance with the Attorney General's investigative guidelines.* Retrieved from https://oig.justice.gov/special/0509/final.pdf

U.S. v. McDavid, No. 2:06-CR-00035-MCE. (E.D. Cal. 2007).

U.S. v. McDavid, Carol Runge, Juror Declaration, No. 2:06-CR-00035-MCE. (E.D. Cal. 2008).

U.S. v. McDavid, Declaration of Ellen Endrizzi, No. 2:06-CR-00035-MCE. (E.D. Cal. 2012).

U.S. v. McDavid, Declaration of Nasson Walker, No. 2:06-CR-00035-MCE. (E.D. Cal. 2012).

U.S. v. McDavid, Diane Bennett, Juror Declaration, No. 2:06-CR-00035-MCE. (E.D. Cal. 2008).

U.S. v. McDavid, Government Sentencing Memo, No. 2:06-CR-00035-MCE. (E.D. Cal. 2008).

U.S. v. McDavid, Habeas Hearing, No. 2:06-CR-00035-MCE. (E.D. Cal. 2015).

U.S. v. McDavid, Sentencing Hearing, No. 2:06-CR-00035-MCE. (E.D. Cal. 2008).

U.S. v. McDavid, Weiner, & Jenson, Criminal Complaint, No. 2:06-MJ-0021. (E.D. Cal. 2006).

U.S. v. McGowan, Terrorism Enhancement Hearing, No. 6:06-CR-60124-AA. (D. Or. 2007).

Van Dijk, T. A. (1993). Principles of critical discourse analysis. *Discourse and Society*, *4*(2), 249–283.

Van Dijk, T. A. (2001). Critical discourse analysis. In D. Schriffirn, D. Tannen, & H. E. Hamilton (Eds.), *The handbook of discourse analysis* (pp. 352–371). Malden, MA: Blackwell Publishers.

Vanderheiden, S. (2005). Eco-terrorism or justified resistance?: Radical environmentalism and the "War on Terror." *Politics and Society*, *33*(3), 425–447.

Wagner, T. (2008). Reframing ecotage as ecoterrorism: News and the discourse of fear. *Environmental Communication*, *2*(1), 25–39.

Wilkinson, P., & Stewart, A. M. (1987). *Contemporary research on terrorism*. Aberdeen: Aberdeen University Press.

Wodak, R., & Meyer, M. (2009). *Methods of critical discourse analysis*. London: Sage.

Part IV: Current Perspectives

11. Radical Environmentalism as Teacher: A Pedagogy of Activism

Meneka Repka

Introduction

As both a teacher and an animal/Earth liberation activist, I occupy a conflicting space that unintentionally supports a Western industrial model of education, but also seeks to undermine that system through radical tactics against capitalism and imperialism. Although I teach in a traditional system of schooling, I feel just as engaged in the politics of teaching as I do in street activism. My teaching practice draws heavily on critical pedagogy (Freire, 1997), ecopedagogy (Kahn, 2010), anarchist models of learning (Drew & Socha, 2015), and the emerging field of critical animal pedagogy, which builds on critical animal studies (CAS) principles (Nocella, Sorenson, Socha, & Matsuoka, 2014). My interest in considering the pedagogical applications of the Earth Liberation Front (ELF) tactics is an outgrowth of my involvement in local protests against industries that profit from causing harm to humans, nonhumans, and the Earth. By examining the ELF through the lens of educational praxis, I hope to decouple the necessity for education to be synonymous with only actions within a framework of legality. Because public school systems generally work toward moulding students into law-abiding citizens, it is rare for students to question whether what is legal is necessarily just. In this conceptual chapter, I examine how the organizational structure, group dynamics, and collective values of ELF are inherently pedagogical. Education is a fundamental thread that runs through the overarching goals of ELF; the second ELF guideline states that activists must strive to "reveal and educate the public on the atrocities committed against the Earth and all species that populate it" (Rosebraugh, 2004, p. 18). This statement has certainly been interpreted literally, as many activists have devoted their time and energy into creating accessible public lectures, workshops, information brochures, under- and aboveground journals, and internet resources in addition to the verbal education that takes place during protests. While all of these actions are unequivocally an integral component of ELF, there is also much to be learned from simply observing how ELF members work. In addition to suggesting that the structural framework of ELF is applicable to educational spaces, I draw parallels between how ELF members are treated by the public and by law enforcement and how students are treated in formal schools. In each of the four sections of this chapter, I address the current state of schooling, contrast this model with how ELF functions, and finally recommend what elements can be drawn from ELF and into places of learning.

Because my experiences of teaching and learning have occurred in a North American context, my discussion is focused within the paradigm of this North American system (though it is not far-fetched to speculate that many arguments would apply to other Western models as well). As structures that function alongside other institutions to uphold capitalist values, schools seem like fundamentally infertile places for radical tactics. As Crozier, Huntington, and Watanuki (1975) observe, "those institutions which have played the major role in the indoctrination of the young in their rights and obligations as members of society have been the family, the church, the school, and the army" (p. 162). Systems of schooling are reproductions of cultural norms (Bourdieu, 1973, 1974, 1979, 1991; Dewey, 1902; Pedersen, 2010) and have important social functions. The current industrial model that typifies most traditional public schools is the residual effect of the Prussian military style of education, which was conceived to promote efficiency, competition, and obedience (Meshchaninov, 2012). Over time, this assembly line of systemic information transfer (from the teacher authority to the student receptacle) sought to prepare students for passive roles in society that would ultimately cycle back to reinforce the same system. In contrast, ELF actions serve to disrupt, resist, and obscure rampant capitalism. In this chapter, I propose that there are four significant ways that a teaching and learning praxis can take up ELF values: collaboration, non-hierarchical leadership, rejection of punitive justice systems, and intergenerational and community learning.

Collaboration

Traditional systems of Western industrial schooling depend upon the fragmentation and separation of ideas, voices, and bodies. Subjects are taught individually and are further segmented into schedules. Information is compartmentalized and groups of students are expected to collectively switch between the ideas of one subject area to another. Students implicitly learn that the knowledge for particular subjects must occur within the constraints of a specific room and time period. Furthermore, the assumption of this industrial model is that the most salient commonality between students is their age, justifying another level of categorization. Even within classrooms, students are separated by being forced into competition through standardized exams and reward systems for obedient behaviour.

ELF, on the other hand, has consistently demonstrated that building alliances and collaborative opportunities and acting in solidarity with other groups are much more effective means of reaching common goals. While the ELF may initially seem to be a divisive group, unable to cooperate with the more conventional strategies of its predecessor, Earth First!, the actions of both ELF and EF! represent true collaboration. As Molland (2006) notes, decentralized environmentalism experienced immense growth in Britain in the early 1990s, particularly through EF! demonstrations. With the intent to maintain popularity and public support, EF! chose to focus primarily on consistent street protests, sit-ins, and recruitment and training of new activists (Molland, 2006, p. 49). It was informally decided that those who wanted to engage in illegal ecotage activities would do so under the newer ELF group. Despite this split, both EF! and ELF members kept the common goal of preventing further environmental damage at the forefront. Rather than attempting to compete with or delegitimize one another, EF! and ELF found ways to synergistically disrupt and expose corporate and government groups that were guilty of unrelenting environmental harm. Though their tactics were different, “Elves would mingle in with the EF! activists and whilst the EF! activists dropped their banners and blockaded the premises, the Elves would be busy gluing the locks of the buildings that the EF! activists were occupying” (Molland, 2006, p. 51). As well, during an EF! protest against the unnecessary building of a supermarket chain store, ELF members participated by surreptitiously leaving a cart full of frozen meat to thaw and later instigating an arson attack on the store (Molland, 2006, p. 51). It is also clear that the original British ELF members supported rather than competed with similar actions in North America. Both the Canadian Earth Liberation Army (ELA) who targeted trophy hunters and environmentally destructive industries, and later the American Elves who started sabotaging gas stations and McDonald’s restaurants were openly accepted and encouraged in ELF publications (Molland, 2006, p. 55).

ELF draws inspiration for both its name and actions from the Animal Liberation Front (ALF). The ALF developed in the late 1970s in England with the purpose of causing non-violent economic harm to people and industries known to exploit nonhuman animals (Molland, 2006, p. 49). ELF founders strategically chose to emulate the ALF to make their goals and actions apparent. In a capitalist society, it is conceivable that this could have caused tensions between the two groups; animal rights activists are typically associated with individual sentient beings, while environmentalism considers entire species and ecosystems as a whole (Laws, 2006, p. 144). However, ELF and ALF members had numerous collective successes. In the late 1990s, American activists used both ELF and ALF banners together by spiking trees (inserting a nail to prevent logging), firebombing nonhuman animal research labs, and releasing wild horses destined for slaughter (Molland, 2006, p. 56). These actions not only revealed the interconnected relationships between Earth and animal liberation, but also demonstrated the strength of collective direct action.

The spirit of collaboration that ELF activists embrace is very much needed in today's K–12 schools. In my experience as a teacher, "collaborative learning" still occurs within a tense capitalistic environment that praises competition and hierarchy. Students collaborate on "assignments" because they are told to by a teacher authority; these assignments are later assigned a numeric value to further stratify students. In a truly collaborative environment, students would work in the same way that ELF does by choosing who to collaborate with based on common interests and goals. By placing students under conditions where they are required to be motivated only by grades or other external rewards, educators are displacing the possibility for students to learn because they are actually curious or interested in resolving injustices.

Non-hierarchical Leadership

Understanding how ELF operates in contrast to traditional structures of public education can also be beneficial in establishing more holistic and anti-capitalist educational models. North American state-sponsored schooling is rooted in the Prussian-industrial model, which was unequivocally developed for the purpose of establishing a docile and submissive populace (Meshchaninov, 2012). The application of industrial values such as efficiency and conformity to education rested upon an insistent dependency on hierarchy. "The students," writes Meshchaninov (2012) "feared the teacher, who in turn feared the principal, who in turn feared the superintendent, who in turn feared his supervisor, up until the King" (p. 4). Each stage of this bureaucracy would be consistently surveilled, managed, and evaluated by people or groups with increasing shades of power. Evidence of the Prussian-industrial origins of public schooling is still clear in the structure of today's classrooms. Students are generally expected to line up to enter and leave specific rooms, they are seated in rows, and they look to the teacher as an absolute authority from whom knowledge is to be obtained. This intellectual dependence on the teacher authority as the ultimate source of truth effectively prevents free thought and protects the system of industrial education.

Because the goal of traditional institutions of schooling is to uphold the status quo, participants in the system have been conditioned to remain fundamentally static. Those who resist the system are either removed or socialized to accept a state of apathy, as evidenced by elaborate reward and punishment systems enforced by teachers and administrators (i.e. honour roll, detention, phone calls home).

Similarly, the goal of mainstream environmental groups is incremental change, so members advocate reform but the system as a whole remains largely unchallenged. The goal of ELF, however, is social transformation, which means that members must be revolutionary and radical. It is clear that ELF has been largely successful in achieving its goals, and these accomplishments suggest that much can be gleaned from analysing ELF's organizational structure. Rosebraugh (2004) highlights the competence of ELF to realize its aims as he observes, "the vast majority of ELF actions-including the most spectacular and financially devastating [...] have yet to be solved" (p. 176). Although ELF is not completely immune from perhaps unintentionally adopting frameworks that are isolating to those who do not fit into a Western cis-heteropatriarchal model (see Starr, 2006), the intent of the ELF is to establish a world without these hierarchies. ELF represents a departure from some mainstream environmental organizations because it is not tied to corporate funding and is therefore in a position to "reject bureaucratic models of change" (Somma, 2006, p. 37). Starr (2006) remarks that ELF's structure is a jarring departure from much of what has become familiar in labour and community organizing since the 1960s, particularly in its embrace of what many would call "undignified" low- or no-budget physical spaces, unwillingness to impose fees or dues, hostile rejection of any leadership, and the moral priority given to direct confrontation with law enforcement (p. 376).

Further, the clandestine cell structure of ELF not only disallows awareness of activities by law enforcement, but also distributes power more evenly. Because the ability to make decisions is not concentrated in a minority of members, a dynamic of consensus decision making and collaborative organizing occurs (Rosebraugh, 2004; Starr, 2006). By embracing these anarchist ideologues, ELF members are able to self-manage and develop autonomy. These skills are essential, also, for students.

Understanding ELF's organizational structure can provide an alternate framework for education to counter the dominant industrial model. While the current system grooms children for their eventual positions in the boss/worker cycle, non-hierarchical leadership can also occur with teaching and learning. Drawing upon ELF operations, I am proposing that teachers relinquish their positions as authoritative figures to enforce arbitrary rules and disseminate knowledge. Rather, teachers can become guides and fellow learners. The structure of ELF also suggests that students can be encouraged to assume leadership positions with the goal of resolving actual rather than hypothetical problems (Weil, 2004). Students are also capable of activism and can collaboratively work toward organizing protests or other forms of resistance against injustices in their lives. In this way, educational spaces can work toward meeting both the immediate needs of students (such as hunger) and also validating marginalized youth cultures—two basic ways that ELF has developed community among members (Starr, 2006, p. 375).

Rejection of Punitive Justice Systems

Standard Western models of public schooling depend upon punitive systems of control in order to uphold larger capitalist structures. While the ELF is similarly affected by these systems, the goal of ELF is to resist and dismantle rather than perpetuate them. Schools do not exist in a vacuum; they are informed by particular cultural narratives (Bourdieu, 1973, 1974, 1979, 1991; Dewey, 1902; Pedersen, 2010) and therefore act as microcosms of society. The school's historical dependence upon the submission of students through violent and authoritative means is illustrated hauntingly in Sally Gardner's (2012) novel *Maggot Moon*:

> Little Eric was still laughing. Mr. Gunnell pulled the boy towards him by the ear then he started to beat him, first with the cane until it broke, then with his fists. He didn't stop, his punches coming harder and faster [...] The more Little Eric wept, the harder Mr. Gunnell went at him. We all watched paralyzed as gobbets of blood splashed on the pavement. Eric Owen wasn't moving, and I knew exactly what Mr. Gunnell was about to do as he lifted his army boot high above Little Eric's head. (pp. 77–78)

While the absolute brutality by which a teacher punishes and ultimately kills a student is no longer representative of Western schools, more subtle mechanisms of control function to push students into either compliant behaviour or the streets. This happens most evidently through the school-to-prison pipeline, a process that tracks students out of educational settings and tracks them directly into either juvenile or adult criminal facilities (Heitzeg, 2009,

p. 1). Although structural matters such as decreased funding, crowded classrooms, and high-stakes tests are inarguably contributors to the school-to-prison pipeline, the increasingly alarmist attitudes of the public and school administrators are largely instrumental in maintaining this pipeline. Heitzeg (2009) attributes false media representations combined with zero tolerance policies to the growing number of increasingly younger students at risk for punitive consequences. For instance, black and brown bodies are overrepresented as criminals or gang members in news programming, television, and films, which makes racialized students more vulnerable to being disproportionately targeted as troublemakers who must be punished (Heitzeg, 2009, p. 4). Schools have also adopted "zero tolerance policies" that have resulted in exaggerated responses to minor student infractions. For instance, a five-year-old child was handcuffed and arrested for knocking over papers in New York, another five-year-old was also handcuffed and arrested for disrupting a class in Florida, and a thirteen-year-old spent six days in jail for writing a scary story as part of a school assignment (Heitzeg, 2009, pp. 9–10). In essence, schools are modelled after the punitive justice systems of a wider social realm, and the expectation of students is to accept these measures.

Like students in the dominant system of public education, ELF members are also misrepresented in the media and severely punished for relatively minor crimes. Again, these false media assumptions operate covertly to create a discourse of public mistrust toward environmental activists and further the notion that they must be punished. The popularization of the term "ecoterrorism" in the media was a direct outgrowth of American federal law enforcement deciding that radical environmental and animal activists were the country's biggest domestic threat (Best & Nocella, 2004; Del Gandio & Nocella, 2014). Though the term "ecoterrorism" was not widely used in popular media prior to 2001, animal and environmental activists have been portrayed as dangerous and violent radicals since the 1980s (Cushnie, 2016; Varnell, 2016). In a 1998 Department of Justice report, for instance, terrorists were described as radical environmentalists (DOJ, 1998), and environmental activists have been "labeled as 'terrorists' by the federal government in press releases, congressional testimony, and other public discourse" (Cushnie, 2016, p. 17). This government characterization of activists as threatening terrorists has more recently been buttressed by news programs such as a 60-minute broadcast entitled "Burning Rage" (Bradley, 2005). The programme featured clips of burning buildings and other damaged property juxtaposed with references to the ALF and ELF as "environmental extremists," "so-called ecoterrorists," and the now cliché "biggest domestic threat" with very little clarification about the motives of these actions. As well, the

60 minutes report consistently referenced the potential for human harm (i.e. "luckily no one was injured"), implying that ELF actions are a threat to human safety. Similarly, season two of the popular Scandinavian television series *The Bridge* features a group of mysterious individuals who commit crimes in the interest of stopping businesses that are unjust or environmentally destructive. The show seems to be referencing the emergence of ELF cells in Europe, but inaccurately portrays the group as violent and unsympathetic to the well-being of individual people. In one episode the group breaks into the home of a woman involved in vivisection and holds her hostage in a cage, exaggerating and misrepresenting the goals of ALF. Characters also refer to the actions as "ecoterrorism" and "ecotage." Such language and imagery in media accounts and federal reports is misdirected, as ELF actions are meticulously orchestrated to ensure that damages are economically harmful to organizations, rather than physically harmful to people. Throughout the inception of ELF, there have been no reports of ELF actions being injurious to humans (Cushnie, 2016; Rosebraugh, 2004).

ELF activists also seem to experience inflated consequences for relatively minor actions. While verbally questioning a police officer about the unjust arrest of a fellow activist, Rosebraugh (2004) states that officers quickly resorted to physical violence and ultimately broke his arm (pp. 88–89). Rosebraugh's home and workplace (Liberation Collective) were also intermittently raided by FBI agents (p. 132). On a larger scale, Cushnie (2016) observes that over the last decade and a half, environmental activists have received higher penalties and convictions even though activities such as property destruction have remained fairly stable over the years (p. 10). Despite the possibility of these consequences, ELF and ALF activists have not resigned tactics that the movement was founded upon. In May 2014, 35 pheasants were released from a farm in Gervais, Oregon; in June and July 2014, activists poured bleach into the fuel tanks of three slaughter trucks in Battle Ground, Washington; in March 2015, 50 pheasants destined for canned hunting were released from a farm in Beavercreek, Oregon; and in June 2015, two trucks in Mississauga, Ontario, were set on fire to prevent the transport of animals used in vivisection (Perkowski, 2014; Rendleman, 2015; Rosella, 2015; Woodburn Independent, 2014). ELF's active presence despite the threat of federal authorities is an interesting contrast to the success that schools have had in pushing students away from education and toward incarceration (Heitzeg, 2009).

Examining the resilience of ELF in the face of punitive systems can help to challenge capitalist systems of education as the status quo for learning. While schools offer competition and hierarchy, ELF has fundamentally maintained that there would be no centralized leadership in the group. Because anyone can call themselves an ELF member or leader, Elves remain autonomous. This spirit of freedom and trust can be applied to schooling as well. Rather than providing required classes that students must attend, schools could be more inviting spaces with the opportunity for students to choose when and what they learn (Neill, 1960). While schools attempt to maintain homogeneity by tracking out students who are poor, minoritized, or disabled, ELF has provided an outlet for marginalized youths. By legitimizing difference, ELF has strengthened its capacity to understand and resist environmental harms that intersect with human oppressions. It is at this intersection that many students who find themselves disenchanted with traditional schooling will find opportunities for creating change and justice.

Spirituality and Intergenerational Knowledge

The current Western industrial paradigm for schooling is built upon upholding a system of hierarchy. While schools insist that collaboration and inclusion take place by virtue of state, district, and administrative mandates, these top-down procedures offer only a superficial remedy to a larger, structural issue. Schools attempt to divert attention away from the teacher as the sole authoritative bearer of knowledge by inviting parents and guest speakers into classrooms, as well as by taking students on field trips to experience new contexts. However, for most publicly funded schools, these opportunities are intermittent if they occur at all. The current model of K–12 education prioritizes the gaze of the teacher as a tool to enforce power and authority while overlooking non-Western ways of knowing and learning. As a central figure in a typical classroom, the panoptical surveillance of the teacher implicitly extinguishes opportunities for collaboration between adults and students, and reinforces competition rather than support among students. For instance, the gaze of the teacher is expected by both administrators and students to "catch" students in behaviours that deviate from the norm, creating an "us and them" binary between students and teachers (McCourt, 2005; Sadr-Kiani, 2014). Additionally, students are expected to report the "misbehaviour" of their classmates, which diminishes the community of support that students need. The centrality of the teacher in most classrooms also misses a connection to the Earth and nonhuman animals as important and valuable teachers. It would be of relevance for students to observe the behaviour of bees, for instance, as authorities in collaboration and disregard for arbitrary boundaries (Sadr, 2013).

The monolithic narrative of how teaching and learning occurs has been obstructed by ELF though its careful consideration of indigenous epistemologies. Becker (2006) observes that the goals of ELF have been consistent with indigenous resistance to colonialism and the capitalist destruction of sacred land and resources (pp. 84–85). ELF's support of and solidarity with indigenous activists is evident in communiqués that take up the language of indigeneity: "the horse nation," "mink and fox nations," and "wildlife nations" (Becker, 2006, p. 84) all reference a connection to nonhuman life and the Earth that indigenous groups have traditionally recognized. As well, Becker (2006) notes that Native American practice inherently values all forms of life; "even the rocks are acknowledged as the old ones who know everything because they have been here from the beginning" (p. 84). This movement to a world where humans, nonhumans, and the Earth are all harmoniously interconnected rather than exploited for profit and power is also a salient goal for ELF activists. ELF's strategies for sharing knowledge, information, and ideas among group members also draw upon indigenous ways of knowing. ELF rejects hierarchical titles that denote authority and power; decisions within individual cells and larger groups occur through sharing and consensus. Unlike a typical school, where students must show learning through an activity or test decided upon by the authoritative teacher, both ELF and indigenous groups invite members to express ideas in whatever form is meaningful to the individual (Becker, 2006, p. 87).

At the time of this writing, the Dakota Access Pipeline (DAPL) is a project that proposes to transport oil from North Dakota to Illinois. The pipeline extends through the Missouri river, and based on similar pipeline projects, the DAPL has great potential to leak and threaten the health of the river. In addition to wiping out plant and nonhuman animal life, the pipeline is a significant risk to the water supply of the Standing Rock Sioux Tribe and an imposition on sacred burial grounds (CBC news, 2016). The unification of Canadian and American indigenous activists in solidarity with the Sioux has disrupted Western understandings of what teaching can look like. The Sioux have brought international attention to this issue by employing traditional understandings through active resistance. Pellkey (2006) remarks that in the Native Youth Movement (NYM),

> The Young Warriors serve as the physical protectors, and the OGs (Original Guerrillas) act as the Advisor Warriors, giving direction through lessons, age old teachings, previous battles, and from the Spirits and our Ancestors who have passed on this responsibility of defending our Indian way. (p. 251)

This holistically driven determination to challenge colonial powers is evident in the intergenerational involvement of tribe members in protesting the DAPL. In an online petition started by a 13-year-old Sioux member, the narrative of water extends past the literal needs of people in a community to encompass the traditional teaching that water is the first medicine of every living being (Lee & Jean, 2016). The video accompanying the petition depicts Native American youths discussing the dangers of the pipeline through traditional teachings, information from contemporary sources, and the guidance of their community (Lee & Jean, 2016). This collaboration between multiple generations of a community demonstrates the effect of destabilizing the hierarchical teacher/student model in typical North American educational frameworks. Pellkey (2006) observes: "there are many Native youth that do not receive any type of direction, teaching, or values from their communities and families, leaving them a stranger to their own culture, land, and peoples" (p. 253). Activism can function as a pedagogical and spiritual tool to resist colonial systems that rely on conformity and assimilation. Furthermore, children and youths actively involved in protesting the DAPL witness activists employing the tactics of radical environmental groups such as ELF. Water protectors locked themselves to machinery, used their bodies to obstruct equipment, and spray-painted messages of decolonization and indigenous spirituality onto construction vehicles (Democracy now, 2016; Unicorn riot, 2016). While mainstream discourse may consider these actions inappropriate or illegal, it is evident that the interests of the state are considerably distanced from the protection of indigenous bodies and culture; North Dakota's position is a clear example of environmental racism, a state-sanctioned injustice. Environmental racism refers to the tendency for "human communities that experience [...] marginalization [to also] confront disproportionately intense exposure to pollution and other risks associated with industrialization" (Fitzgerald & Pellow, 2014). The DAPL places the Sioux's main water supply at risk in order to facilitate the transfer of oil. While indigenous activists literally risked their lives as well as risked fines and incarceration to protect their communities, Honor the Earth representative Tara Houska reported that state police officers watched as DAPL security took medicine, supplies, and water from protesters (Democracy now, 2016). Disturbingly, security also used pepper spray and dogs to physically injure activists, including at least one child (Democracy now, 2016).

This is the type of pedagogical moment that is so important and yet so absent from typical state-sponsored K–12 programs. The most confident insight that children and youths can glean from the DAPL activists is that justice is not always equivalent to or consistent with legality. What young children and youths learn as they witness their communities coming together to resist colonization and capitalism should form the basis for a radical paradigm of education: peacemaking is not always legal, the knowledge of our elders is immensely valuable, everyone has an important role regardless of arbitrary dividers like age, gender, or size, collaboration gives us a stronger voice, and we ought to protect the natural world that sustains us. The DAPL water protectors have also demonstrated a teachable moment in the realm of public pedagogy, as many informal online spaces (i.e. Twitter and Instagram) have pointed out the eerie similarities between tactics used by law enforcement in civil rights protests and the methods used against indigenous activists. Photographs of black activists being attacked by dogs almost seamlessly mirror images of indigenous activists in a haunting visual warning about history's tendency to replicate itself. Lupinacci (2015) writes "we perceive that to be in school, by situation of its location in society, means learning to function within, accept, and submit to the authority of a tremendously exploitative culture" (p. 181). The work of activists is fundamental in challenging this assumption; when those of us in the human community are able to actively resist destructive systems, we are teaching our children and youths that they are deserving of a world that is just and at peace with the Earth.

Conclusion

The radical environmental tactics of ELF function not only to draw attention to and combat the relentless destruction of the Earth by capitalist powers, but also to pedagogically guide humans. By closely considering the ways in which ELF has maintained its goals of Earth protection without violence or physical harm, educators can recognize that the current model of public schooling is in need of radical reform. Influenced by the Prussian industrial template for militaristic conformity, schools both implicitly and explicitly reinforce a destructive capitalist system. Students are fragmented and categorized, leading to an atmosphere of competitiveness and exclusion. They are expected to submit to the rules and information provided by an authority, such as a teacher or administrator; when students question these structures or take up positions outside the norm, they are ushered into spaces where they will not disrupt or expose the shortcomings of the system. An example of this is the school-to-prison pipeline, where schools work in tandem with law enforcement to track minoritized students into incarceration rather than institutions to further their education. The structure of ELF represents an alternative model of teaching and learning. By working with other groups and abandoning hierarchy, ELF members are empowered to use their own skills and knowledge to realize larger goals. ELF has also been resilient against the exaggerated responses of law enforcement, demonstrating a politics of solidarity and environmental stewardship in the face of physical and legal hardships. Finally, ELF has recognized the significance of spirituality and intergenerational sharing of knowledge through indigenous teachings. The model of ELF is inherently pedagogical; students would benefit enormously from quashing hierarchy, decentring the teacher, and inviting ways of knowing form their larger communities.

References

Becker, M. (2006). Ontological anarchism: The philosophical roots of revolutionary environmentalism. In S. Best & A. J. Nocella II (Eds.), *Igniting a revolution: Voices in defense of the earth* (pp. 71–91). Oakland, CA: AK Press.

Best, S., & Nocella, A. J., II. (2004). Introduction. In S. Best & A. J. Nocella II (Eds.), *Terrorists or freedom fighters? Reflections on the liberation of animals* (pp. 195–201). New York, NY: Lantern books.

Bourdieu, P. (1973). Cultural reproduction and social reproduction. In R. Brown (Ed.), *Knowledge, education and social change (pp. 71-112).* London: Tavistock.

Bourdieu, P. (1974). The school as a conservative force: Scholastic and cultural inequalities. In J. Eggleston (Ed.), *Contemporary research in the sociology of education.* London: Methuen.

Bourdieu, P. (1979). Symbolic power. *Critique of Anthropology*, *4*, 77–85.

Bourdieu, P. (1991). Epilogue: On the possibility of a field of world sociology. In P. Bourdieu & J. Coleman (Eds.), *Social theory for a changing society.* Boulder, CO: Westview Press.

Bradley, E. (2005). Burning rage: Reports on extremists now deemed biggest domestic terror threat [video]. Retrieved from https://www.cbsnews.com/news/burning-rage/

Crozier, M., Huntington, S. P., & Watanuki, J. (1975). *The crisis of democracy: Report on the governability of democracies to the trilateral commission.* New York, NY: New York University Press.

Cushnie, L. (2016, August). Mapping discursive and punitive shifts: Punishment as proxy for distinguishing state priorities against militant environmental activists. *Green Theory and Praxis Journal*, *9*(2), 8–28.

Del Gandio, J., & Nocella, A. J., II. (2014). Introduction: Situating the repression and corruption of the AETA. In J.D. Gandio & A. J. Nocella (Eds.), *The terrorization of dissent: Corporate repression, legal corruption, and the animal enterprise terrorism act (pp. xix-xxix).* Brooklyn, NY: Lantern Books.

Department of Justice. (1998). Terrorism in the United States. *Counterterrorism Threat Assessment and Warning Unit, National Security Division.* (1-29).

Dewey, J. (1902). *The child and the curriculum: Including, the school and society.* New York, NY: Cosimo, Inc.

Drew, L., & Socha, K. (2015). Anarchy for educational praxis in the animal liberation movement in an era of capitalist triumphalism. In A. J. Nocella II, R. J. White, & E. Cudworth (Eds.), *Anarchism and animal liberation: Essays on complementary elements of total liberation* (pp. 163–177). Jefferson, NC: McFarland & Company, Inc. Publishers.

Fitzgerald, A., & Pellow, D. (2014). Ecological defense for animal liberation: A holistic understanding of the world. In A. J. Nocella II, J. Sorenson, K. Socha, & A. Matsuoka (Eds.), *Defining critical animal studies: An intersectional social justice approach for liberation* (pp. 28–48). New York, NY: Peter Lang Publishing, Inc.

Freire, P. (1997). *Pedagogy of the oppressed.* New York, NY: Continuum.

Gardner, S. (2012). *Maggot moon.* Somerville, MA: Candlewick Press.

Heitzeg, N. A. (2009). Education or incarceration: Zero tolerance policies and the school to prison pipeline. In *Forum on public policy.*

Kahn, R. (2010). *Critical pedagogy, ecoliteracy, and planetary crisis: The ecopedagogy movement.* New York, NY: Peter Lang Publishing Inc.

Laws, C. (2006) Jains, the ALF, and the ELF: Antagonists or allies? In S. Best & A. J. Nocella II (Eds.), *Igniting a revolution: Voices in defense of the earth* (pp. 143–155). Oakland, CA: AK Press.

Lee, A, and B Jean. "Stop the Dakota access pipeline." *Change.org*, Oceti Sakowin Youth, 2015, www.change.org/p/jo-ellen-darcy-stop-the-dakota-access-pipeline. Accessed 5 Nov. 2018.

Lupinacci, J. (2015). Recognizing human-supremacy: Interrupt, inspire and expose. In A. J. Nocella II, R. J. White, & E. Cudworth (Eds.), *Anarchism and animal liberation: Essays on complementary elements of total liberation* (pp. 179–193). Jefferson, NC: McFarland & Company, Inc. Publishers.

McCourt, F. (2005). *Teacher man: A memoir.* New York, NY: Scribner.

Meschaninov, Y. (2012). *The Prussian-industrial history of public schooling* (pp. 1–10). N.p.: The New American Academy.

Molland, N. (2006) A spark that ignited a flame: The evolution of the Earth liberation front. In S. Best & A. J. Nocella II (Eds.), *Igniting a revolution: Voices in defense of the Earth* (pp. 47–58). Oakland, CA: AK Press.

ND national guard take over Dakota Access Pipeline checkpoints. (2016, September 8). In *Unicorn riot.* Retrieved from http://www.unicornriot.ninja/?p=8584

Neill, A. S. (1960). *Summerhill school: A new view of childhood.* New York, NY: Penguin Books.

Nocella, A. J., II, Sorenson, J., Socha, K., & Matsuoka, A. (2014). Introduction: The emergence of critical animal studies. In A. J. Nocella II, J. Sorenson, K. Socha, & A. Matsuoka (Eds.), *Defining critical animal studies: An intersectional social justice approach for liberation* (pp. 110–134). New York, NY: Peter Lang Publishing Inc.

Pedersen, H. (2010). *Processes and strategies in human-animal education.* West Lafayette, IN: Purdue university press.

Pellkey, K. (2006). Taking back our land: Who's going to stop us? In S. Best & A. J. Nocella II (Eds.), *Igniting a revolution: Voices in defense of the Earth* (pp. 251–256). Oakland, CA: AK Press.

Perkowski, M. (2014, July 16). Two more mobile slaughter units vandalized. In *Capital press: The west's Ag website.* Retrieved from http://www.capitalpress.com/Livestock/20140716/two-more-mobile-slaughter-units-vandalized

Rendleman, R. (2015, April 2). FBI, CCSO investigate "eco-terrorists" at Beavercreek farm. In *Portland Tribune.* Retrieved from http://portlandtribune.com/pt/9-news/255775-126185-fbi-ccso-investigate-eco-terrorists-at-beavercreek-farm

Rosebraugh, C. (2004). *Burning rage of a dying planet: Speaking for the Earth liberation front.* New York, NY: Lantern Books.

Rosella, L. (2015, June 8). Animal rights group claims responsibility for blaze that destroyed two trucks in west Mississauga. In *Mississauga news.* Retrieved from http://www.mississauga.com/news-story/5666768-updated-animal-rights-group-claims-responsibility-for-blaze-that-destroyed-two-trucks-in-west-missi/

Sadr-Kiani, A. (2014, September 29). The teacher's gaze: What we lose when we pathologize play. In *Edutopia.*

Sadr, J. (2013). Natural anarchism and ecowomanism: Crafting coalition-based ecological praxis. *Green Theory and Praxis Journal, 7*(1), 17–25.

Somma, M. (2006). Revolutionary environmentalism: An introduction. In S. Best & A. J. Nocella II (Eds.), *Igniting a revolution: Voices in defense of the Earth* (pp. 37–46). Oakland, CA: AK Press.

Starr, A. (2006). Grumpywarriorcool: What makes our movements white? In S. Best & A. J. Nocella II (Eds.), *Igniting a revolution: Voices in defense of the Earth* (pp. 375–386). Oakland, CA: AK Press.

Suspects sought in bird farm break-in. (2014, May 28). In *Woodburn independent.* Retrieved from http://www.pamplinmedia.com/wbi/152-news/222254-83613-suspects-sought-in-bird-farm-break-in

Tribe "disappointed" with judge's partial halt to North Dakota pipeline work. (2016, September 6). In *CBC news.* Retrieved from http://www.cbc.ca/news/business/dakota-access-pipeline-hearing-1.3749872

Varnell, J. (2016, August). Speaking about "eco-terrorists": Terrorism discourse and the prosecution of Eric McDavid. *Green Theory and Praxis Journal, 9*(2), 29–54.

Water protectors lock their bodies to machines to stop Dakota access pipeline construction. (2016, September 7). In *Democracy now.* Retrieved from http://www.democracynow.org/2016/9/7/water_protectors_lock_their_bodies_to

Weil, Z. (2004). *The power and promise of humane education.* Gabriola, BC: New Society.

12. *Those Mischievous Elves of Lore: The Legend and Legacy of Earth Liberation*

Alexander Reid Ross

With Spritely Grins

The modern animal rights and environmental movements have roots extending as far as the modern world, itself. The Earth Liberation Front's (ELF) Beltane communiqué obviated as much: "We take inspiration from the Luddites, Levellers, Diggers, the Autonome squatter movement, ALF, the Zapatistas, and the little people—those mischievous elves of lore" (ELF). In this chapter, we will look at each of these configurations, situating them within a *longue durée* that helps refine our understanding of the ELF and its historical place in a complex and often problematic tradition of trans-Atlantic social movements. Such a genealogy of the ELF will reveal a historical lineage of struggles against the state, religious repression, industrial exploitation, and capitalism that contains both right- and left-wing tendencies. To understand the complexity involved, we must gain greater insight into those forms of struggle that the ELF sought to emulate—from peasant insurgencies to autonomous networks—and their socioeconomic composition, as well as their geographic importance. Returning to the modern environmental movement through the lens of such archival analysis, we will discover the ideological paradigms involved in ecological direct action and the ways that the far-right compromises, co-opts, or deploys them on their own terms.

As the site of the origins of the Industrial Revolution, England perhaps boasts the earliest opposition to industrial civilization. Yet to investigate such origins, we must return to the prehistory of industrialization, even prior to the advent of capitalism and modern world systems. One can get a sense of

British life before the Romans through Strabo's commentary: "The forests are their cities; for they fence in a spacious circular enclosure with trees which they have felled, and in that enclosure make huts for themselves and also pen up their cattle—not, however, with the purpose of staying a long time" (1988, p. 496). The lives of Ancient Britons tended to be nomadic and the Irish even wilder in the eyes of Strabo, who described them and Iberians as "man-eaters" in an ironic foreshadowing of the fear of "Indian cannibals" prominent among those who colonized the Americas.

For the pagans of the pre-Roman era, the festive spring holiday of Beltane held the sacred properties of renewal and rebirth. Located at the midpoint between the equinox and the solstice, usually around the first of May, Beltane celebrations usually involved a "Maypole" around which merry revelers would dance and sing. Amid the food, wine, and celebrations, pagans believed, the boundary between the natural and supernatural would disappear. Out of the recesses of the world—the springs and caves—the mischievous, mystical Fae or faeries (also fairies) might emerge, for better or worse.

Often synonymous with fairies, the elves of Ancient Britain held powers unknown to man and typically used them against those who would settle into dormant and domesticated livelihood. Beautiful and seductive woodland spirits and nymphs, elves had magical powers and caused illness to livestock and person alike—if displeased. Humans would loathe to fall under an elf attack, elf disease, or elf-heartburn, according to Anglo-Saxon texts like the *Leechbook* (c. 950) and *Lacnunga* (c. 1050). If afflicted with elf-juices, "his eyes are yellow where they should be red. If you want to cure this person, consider his bearing, and know of which sex he is" (Jolly, 1996, p. 163). To ward off elfshot and heal potent magic, peasants would keep "rotund little shapes with spritely grins" as charms (p. 137).

In Icelandic tales, elves descend from matriarchal divinities—specifically, Eve's efforts to conceal her unwashed children from God (Ashliman, 2004, p. 118). According to an Icelandic tale, a farmer's sheep went missing, and his hands went out to find them. Fed up with waiting for their return, the farmer happened upon a mysterious lake, the dwelling of Valbjörg, an elf-woman. Beautiful and rich, she offers him the same bargain as his hands: stay and marry her or be murdered like them. The farmer accepts her proposal, and lives with Valbjörg learning elvish magic for three years. Before Christmas, the farmer appears in his father's dreams, instructing him to come to their elf-home on Christmas Eve with a well-trained priest. His father abides, and seeing Valbjörg holding their baby offers a Christian blessing. Valbjörg recoils in horror, throwing the baby down on the bed and running for the door, only to be caught in the arms of the priest and subdued. With Valbjörg's elven spirits exorcised, she forgets magic and embraces the Christian community (Bryan, 2011, pp. 182–183).

Mysterious and powerful, the elves exist in hidden spaces, different dimensions that open with the unlocking of clues, fetishes, charms. Their invitations can beckon from the woods, but hold dangerous portent. With the coming of Norman Conquest in 1066 and the transformation of England into a colonized state of new castles and roads, elves became all the more subversive. The mischievous German house-spirit, Hödekin ("little hat"), may have made his way up to England, joining Robin Goodfellow (aka "unsettled" Puck) to become Robin Hood of Nottingham Forest (Lee, 1908, p. 1152). Such elves, spirits, and fairies of the medieval times indicate complex subcultures of vagabonds, forest dwellers, and adventure seekers engaged in unrest against both lord and domestic peasant alike. Abiding by a kind of playful and dangerous individualism, they played tricks on wayward travelers and kept townspeople in their place.

Ther Ben No Fairies

Perhaps the elves helped townspeople watch over their communities while also keeping the forested wilderness the domain of outlaws and refugees. Yet with the seriousness of the Crusades and the Black Death sweeping Europe, a new class began to emerge. Those who fought in the Crusades returned with booty from their plunder of the Eastern Mediterranean and sought to continue the war against heretics like the Cathars and Waldensians within Europe. Grounded in the enclosures of the commons, the burgesses put down the old legends of the past in pursuit of a mechanical worldview. As Chaucer wrote,

> In th'olde dayes of the Kyng Arthur,
> Of which that Britons speken greet honour,
> All was this land fulfild of fayerye.
> The elf-queene, with hir joly companye,
> Daunced ful ofte in many a grene mede,
> This was the olde opinion, as I rede;
> I speke of manye hundred yeres ago.
> But now kan no man se none elves mo,
> For now the grete charitee and prayeres
> Of lymytours and othere hooly freres,
> That serchen every lond and every streem,
> As thikke as motes in the sonne-beem,
> Blessynge halles, chambres, kichenes, boures,
> Citees, burghes, castels, hye toures,
> Thropes, bemes, shipnes, dayeryes—
> This maketh that ther ben no fayeryes. (1903, p. 576)

Chaucer's Wife of Bath laments that the prevalence of holy friers, mendicant priests wandering throughout the countryside, as well as the bourses of the early stock exchanges and the burgesses, themselves, rid the world of its spiritual character. The Wife of Bath's despondence indicates further a connection between women and the patriarchal repression of peasant livelihoods that seemed healthier and happier in hindsight.

The depopulation of a third of Europe caused by the Black Death (1346–1353) would lead to increased hysteria surrounding the persecution of heretics, while the greater power of an organized peasantry against the landlords "stiffened people's determination to break the shackles of feudal rule" (Federici, 2004, p. 44). Vagabonds and outlaws joined with former soldiers, farmers, urban artisans, and even sectors of the burgher class to train their pikes on the nobility. This condition only worsened the nobility's hatred of peasants, articulated in stories like *Despit au Vilain*: "For they are a sorry lot, these villeins who eat fat goose! Should they eat fish? Rather let them eat thistles and briars, thorns and straw and hay on Sunday and peapods on weekdays" (Tuchman, 1978, p. 175). The equivalent of the poor person on food stamps buying smoked salmon at the grocery store, the medieval commoner was seen as spiritually untamed, overstuffed on luxury foods, lazy, and drunken.

From the French Jacquerie of 1358 to the Revolt of the Florentine Ciompi in 1378 to the English Peasant's Revolt of Watt Tyler and John Ball in 1381, peasants and poor laborers rose against the hierarchical landscape from urban to rural. The Jacqueries issued from a working class "who had begun with a zeal for justice, as it had seemed to them, since their lords were not defending them but rather oppressing them, turned themselves to base and execrable deeds," according to *The Chronicle of Jean de Venette* (Venette, 1953, pp. 76–77). Into the mouths of the Ciompi, Machiavelli places these words: "Strip us naked and we shall all be found alike; clothe us in their garments and them in ours, and be assured we shall seem noble and they the reverse; poverty and riches being the only causes of our disparity" (1906, p. 189).

As the commons rose up in revolt against feudalism, "the figure of the heretic increasingly became that of a woman, so that, by the beginning of the 15th Century, the main target of the persecution against heretics became the witch" (Federici, 2004, p. 40). The elves and their elfshot (arrows) would become the instrument of the witch, and she a symbol of tremendous class struggle shaking Europe of the feudal yoke through clandestine insurgencies jam-packed with all the techniques and tools of everyday resistance and breaching through the surface in dramatic and bloody wars (Hall, 2005, pp. 32–33; Van Meter, 2017). Cries for equality leveled at both crown and altar erupted into the 16th century but went unfulfilled, often with grave consequences. During the Friulian cruel Thursday of abundance, the poor actually stripped members of the wealthiest Italian families naked and dressed in their clothes, dragging their mutilated corpses through the streets as the town raged in riot and revolt. Such macabrely carnivalesque episodes signaled the early transvaluation of the sacred and profane that would become identified with the Protestant Reformation of Zwingli, the Hussites, Calvin, and Luther.

Witches and heretics would burn alike, while the burgesses hedged their bets between religions, political affiliations, and heresies, using witch trials as a means of subduing the ever-present threat of revolution. Hence, the peasant revolts that proceeded into the 17th century retained a crucial tension with the Reformation's highest leaders. If Thomas Müntzer's declaration, "The people will be free. And God alone will be lord over them," signaled the confluence of peasant and Protestant revolt, his calls for equality drove Martin Luther to condemn him as a heretic (2010, p. 2). Similarly, although Calvin's *Sermons on the Last Eight Chapters of the Book of Daniel* brought the French Huguenots to assert the Sovereignty of the People against Richelieu the Catholic monarchy, his continuation of the persecution of witches reassured the bourgeois that his doctrines would keep their hegemony over the peasants (Holt, 2005, pp. 78–79).

You Dissentious Rogues

The increasing division between burgesses and peasants only worsened with the complexity of religious schism. Predicated on the dual operation of colonialism and enclosures throughout the 16th and 17th centuries, the capitalist system launched the historical trajectory of the bourgeoisie while ruining the traditional livelihood of the peasants. Those who could enclose and profit from the commons had greater capital with which to invest and benefit from colonial expeditions, global trade, and the bourses of Northern Europe. Concomitant with the conversion to Protestantism, increased colonization of Ireland, and the expansion of the putting-out system through the Tudor period, commoners raged throughout the British Islands (Linebaugh & Rediker, 2000, pp. 18–19).

In an Ireland embroiled in unrest that would culminate in thousands of soldiers joining Tyrone's Rebellion to oust the British colonists by the end of the 16th century, "almost every large wooded glen bordering on the Englishry held a nest of human wasps, the Irish 'wood-kerne,' who lived by robbing the neighbouring colonists," according to one historian of Irish forests (Hore, 1858, p. 149). Speaking on their tactics, historian of these times,

Fynes Moryson, described how "Ulster, and the western parts of Munster, yield *vast* woods, in which the rebels, cutting up trees, and casting them on heaps, used to stop the passages" (1735, p. 370). In the English Midlands, increased enclosures frustrated the populous to the point of rising up en masse and pulling down the hedges and fences dividing up the commons. In response, Shakespeare took up his quill, issuing a warning to peasantry and aristocracy with the words of failed Roman military demagogue, Coriolanus: "What's the matter, you dissentious rogues, / That, rubbing the poor itch of your opinion, / Make yourselves scabs?" (Shakespeare, 1999, p. 967).

As with Shakespeare's Puck or the description of Queen Mab by Mercutio in *Romeo and Juliet*, the English of the Tudor and Stuart dynasties saw fairies and elves as representing, unleashing, and manipulating dangerous erotic urges. A student of magic, Elizabeth's successor James I penned the *Daemonologie* in 1597, a three-volume tome reflecting his obsession with repressing the phairies, sprites, and witches. Those who live as the royalty of the forest draw the sacred away from the sovereign and produce an upside-down law that James compares to "counterfeits God among the Ethnicks" (King James I, 2008, p. 44). As European colonialism expanded to Latin America and the Caribbean, the fairies and elves emerged in these new lands, often tied to indigenous spirituality and sexuality. Explorers like Amerigo Vespucci brought back tales of "Indians" who "live amongst themselves without a king or ruler, each man being his own master" (Federici, 2015, p. 351). To wit, those Puritans wishing to escape the Catholic Stuarts fled as far as the so-called "New World" to establish the first English colonies, taking with them a host of prejudices against nature, witches, and the savage.

Despite the Stuarts' interest in spiritual combat, a strong secular current beginning with the Tudors continued through the reign of James I. Solicitor General Francis Bacon labeled dispossessed men and women the "seed of peril and tumult in a state," insisting on the production of an orderly system of hospitals to destroy the plight of beggardom (1868, p. 252). Political theorist Thomas Hobbes would agree that such an industrious social system would encourage discipline: "men would be much more fitted than they are for civil obedience" if not for witches and superstition (1651, p. 11).

Though King James would prove adept at repressing the peasants, his successor Charles I fell to the New Model Army of Oliver Cromwell, which fought alongside the Protestant peasants organized as Diggers and Levelers. Those armies kept true to the precedent of the Peasants War of the 14th century, returning to that old demand of equality and the slogan, "When Adam delved and Eve span / Who was then the gentleman" (Morris, 1828,

p. 228). Popular tales of outlaws and rebels like Mol Flanders featured gender ambiguity and sexual dissidence and took on the fairy tale quality of that spritely Robin Hood (Defoe, 2011). In 1651, the people defeated and decapitated the monarch, yet Cromwell's grip on power only tightened. With Cromwell's son failing at sovereignty, Parliament restored the Stuart dynasty to the throne by inviting Hobbes's former pupil, Charles II, to return and rule, which he did in 1660.

Although resistance against Charles II remained significant, it often followed the leadership of the rising bourgeoisie, which came to rival the small nobility in power and prestige. Implicated in the Rye House Plot to overthrow Charles II, John Locke fled England to exile with other English and Scottish radicals, allegedly engaged in plotting Monmouth's Rebellion, and committed himself to outlining a schema for a Republican form of governance associated with propertarian rights (Ashcraft, 1986, pp. 416, 463–464). While Locke's work became incredibly influential among radicals throughout the North Atlantic—particularly the Carolina colonies for whom he would write official Constitutions—his language of equality referred to a burgeoning ruling class oriented toward property and levied against that tradition of popular revolt that found its bearings in the commons.

The transition of class struggle from peasants against lords to bourgeoisie against nobility came, in part, as a result of religious conflict. Rather than simple suppression of the spiritual community, Catholics or Protestants sought to crush peasant lore or co-opt it to raise greater armies and draw peasant affinities toward their side. For their part, peasant supporters on either Catholic or Protestant side often hoped for little more than the cynical dream of gaining a greater franchise upon victory. Yet the deeper logic of colonialism is what kept them poor, regardless. It is no coincidence that, during this period, whole forests of timber would be felled for the British fleet and Empire—particularly in Ireland—with the double effect and explicit intention of denying shelter to insurgents and paying for the debts incurred by the bourgeoisie's frivolities and excesses.

In a letter to Reverend William Mason dated September 3, 1773, English politician and man of letters, Horace Walpole, mused on the situation: "When the forests of our old barons were nothing but dens of thieves, the law in its wisdom made them unalienable. Its wisdom now thinks it very fitting that they should be cut down to pay debts at Almack's [casino] and Newmarket [racetracks]. I was saying this to the lawyer I carried down with me. He answered, 'The law hates a perpetuity.' 'Not all perpetuities,' said I; 'not those of lawsuits'" (1906, p. 500).

A Shred of Black Crape

Largely through the Lockean understanding of commons as wasted lands waiting for exploitation and capitalization, the colonization of the US interior following the Revolution of 1776 deepened and the institution of slavery enabled the development of European capitalism into a new "industrial revolution." While the bourgeoisie scrambled for the reigns of the French Revolution of 1789, peasants and the working poor took heart in the Rights of Man, influenced by Thomas Paine and the egalitarian current within the Revolution. Comprising Luddites, weavers, and conspiratorial revolutionists, a broad-based movement swept England at the turn of the 19th century, drawing from both revolutionary egalitarianism and the people's movements against the Stuart monarchy during the 17th century, which in turn relied on the tradition of peasant wars dating back to John Ball and Watt Tyler.

As peasants and workers thronged the streets, they flew the black banner not as the banner of a specific, honed ideology but an expression of the immiseration and struggle of the poor against displacement, starvation, disease, and stigma. During the Gordon Riots of 1780, anti-Catholic chauvinism mixed with workers' resentment, guided by the red and black flag flown by popular leader James Jackson (Thompson, 1966, pp. 71–72). A loaf of bread trailing a black ribbon came to signify peasant rebellion—for instance, amid protests against looming war with Spain in 1798 when someone shattered the King's carriage window either with pebble or bullet. During one 1812 women's march, historian J. F. Sutton describes "sticking a half penny loaf on the top of a fishing rod, after having streaked it with red ochre, and tied around it a shred of black crape, emblematic...of 'bleeding famine decked in Sackecloth'" (Sutton, 1880, p. 286; Thompson, 1966, p. 65). Becoming the symbol of poor people's movements in their horizontal, grassroots, and communal form, the black flag, imbued with the sacred tones of sackcloth, represented the fearsome specter of "anarchy," plague, and riot.

Anarchy became the watchword for the riotous people of Europe. Yet with new generations of Romantic poets and artists from Shelly to Wilde to Morris came renewed interest in ancient fairy tales and popular insurrection, and the term "anarchy" became more versatile. In 1813, Romantic poet Percy Bysshe Shelley penned a fairy tale about "Queen Mab," the fairy queen, rejecting the nations and authoritarian principles of industrial civilization. "Nature rejects the monarch, not the man," Shelley wrote, establishing his visions of a natural society peopled by those who scorned obedience in the name of the genius of truth (1822, p. 29). After the massacre of workers during a demonstration in 1819, Shelley responded with the famous poem, "The Mask of Anarchy," describing anarchy as the violent force of the ruling class: "On a white horse, splashed with blood; / He was pale even to the lips, / Like Death in the Apocalypse." In the final stanza, Shelley calls on the people: "Rise like Lions after slumber / In unvanquishable number— / Shake your chains to earth like dew / Which in sleep had fallen on you— / Ye are many—they are few" (Shelley, 1841, p. 231). Shelley shared his vision of a workers' revolution with his father-in-law, William Godwin, a philosopher inspired by conservative Edmund Burke who "provided the most coherent and comprehensive articulation of anarchist ideas around the time of the French Revolution" (Graham, 2016, p. 15). Soon, workers and intellectuals on the continent, like P. J. Proudhon, began to embrace the notion of "positive anarchy" as an alternative system that utilized the critique of representative government in support of ordinary people (Graham, 2016, p. 35).

Against the cynicism of the "dismal science" and the "ordered society" anarchism embraced popular upheaval and free association but faced an irreconcilable crisis of industrialism. Yet as the double-edged meaning of the word implies, anarchism and its poetic image of a prerational world offered a glimpse of a future that could take reactionary or egalitarian directions. Proudhon, for instance, retained sexist and anti-Semitic tendencies despite the clear leadership role that women took in forwarding the cause of the working class. In some ways, the world of anarchy and myth became a link between the reaction and egalitarian movements that would be constantly interrogated and engaged with over time. In other ways, the eradication of myth held the same quality.

Did anarchism mean the destruction of the industrial system or its expropriation? Could anarchists overcome representative parliamentarianism and transform it into a new, popular system, or should anarchists return to a preindustrial state? Most social movements fell somewhere between the two extremes of industrial utopianism and rural anti-industrialism. When in 1848 the Chartists took to the streets of English cities and towns, their songs raised the memory of the 14th-century Peasants Revolt: "For Tyler of old, / A heart-chorus bold, / Let Labour's children sing" (Buhle , p. 43). While Bakunin helped develop the collectivism that would inspire many early syndicalists in the 1860s and 1870s, his sensitivity to nature might be seen through his friendship with Élisée Reclus, a noted geographer and anarchist who rejected the binary categorization of "civilized" and "savage" (2013, pp. 215–216). When in 1886 the Federation of Organized Trades and Labor Unions chose Beltane, the first of May, as the beginning of the eight-hour workday, they enfranchised the popular celebration within the heart of the US working class, or, *mutatis mutandis*, the US working class in the world of myth and wonderment.

It was a motion toward the movement's development and fertility. While Pyotr Kropotkin and Ricardo Flores Mágon embraced peasants as leading figures in global revolution at the turn of the 20th century, like Godwin, they also demanded the modernization of food production necessary to further develop civilization (Mágon, 2005, p. 85). May Day also signified, to some, the refusal of work and the spirit of sabotage and vagabondage. Although Emma Goldman certainly expressed support for the syndicalism of the Industrial Workers of the World in the early 1900s, she remained a bohemian outlier more focused on issues of feminism and *Mother Earth*, which was the name of her periodical. Similarly, the Industrial Workers of the World championed free time over labor, romanticizing the lifestyles of the tramp and hobo admiring nature while hopping trains across North America (Rosemont, 2003).

Facing the rise of jingoist nationalism throughout the United States and Europe, anarchism's romantic streak extended to an international rejection of war, racism, and imperialism. Decadent poet Oscar Wilde wrote fairy tales that he had learned from his father, a medical doctor with a penchant for the spiritual world, with an ardor gleaned from his mother, a powerful player in the Irish nationalist movement. Waves of mendicant European mystics, like anarchist Gustav Landauer, fled cities and abandoned "civilization" as prophets searching for a simpler connection to nature and the universe. Paris became a hub for Chinese anarchists, while Indian anarchists integrated in radical left organizations from California to France, and on the subcontinent peasant hools (insurrections) collapsed the space between the vagabond and the rebel (Guha, 1999, p. 15, 154; Ramnath, 2011, pp. 65–67, 78–79). Some took the internationalism of decolonial struggle to mean drawing from ancient spiritual texts and stories from India to Central America. For some, the movement toward ecology against urban conditions manifested a nationalist return to blood and soil (Biehl & Staudenmaier, 1996). For those who hewed closely to the tested principle of equality, it held a close connection to the rejection of imperialism and capitalism, as exhibited in the aftermath of World War I, when Landauer participated in the overthrow of the government of Bavaria and the establishment of the short-lived Bavarian Soviet Republic of 1920.

The Branch of a Fir

After the Russian Revolution of 1917, Marxist-Leninism wielded significant power in workers' movements around the world, overwhelming with its serious materialism that sense of play and wildness associated with "superstition." Yet one can trace an influential fusion of the critique of capital and embrace of nature from the Bavarian Soviet Republic to the academic work of the Frankfurt School, beginning two years later with the creation of the Institute for Social Research (Jacobs, p. 2). Joining Freudian psychoanalysis with Marxism and Idealism, the Frankfurt School addressed sexual repression and industrial efficiency as crucial to the mechanisms of capitalism. Influenced by Max Weber's *Protestant Ethic and the Spirit of Capitalism* and a school of phenomenology called Existentialism growing around a professor named Martin Heidegger, members of the Institute for Social Research in Frankfurt identified alienation as central to the crises of the modern world through which the individual feels out of joint with time and purpose. Though they identified Protestantism with the movement of the bourgeoisie vis-à-vis the Industrial Revolution, the Frankfurt School offered new philosophical approaches to archaic myths and legends.

In the words of Frankfurt School thinker Theodor Adorno, "occultism is the metaphysics of dunces." For Adorno's metaphysics, the spirit could be distilled to something essential to thought—perhaps what fellow Frankfurt Schooler Walter Benjamin would describe as "profane illumination, a materialist, anthropological inspiration" (Adorno, 1978, pp. 238–244; Benjamin, 2005, p. 209). Benjamin linked such illumination to the codex of law and reason, on the one hand, and a response of "mythical violence" to it, on the other. In an ancient system whereby the fates bind humanity to law, the human spirit wills its independence. Through the "divine violence" that follows, universal forces unbind the laws of history and the world returns to the liminal space between reality and the supernatural (Benjamin, 1986, p. 294).

Benjamin's "profane illusions" might be seen throughout the writings of his Frankfurt School comrade, Ernst Bloch (2006), on eschatological, visionary, and prophetic strains of Christian revelation from Joachim of Fiore to Thomas Müntzer, and maintained direct connections to contemporary French avant-garde art movements against Catholicism and industrial rationalism—particularly Surrealism. Just as Russian Futurism emerged from the embers of the Symbolist "mystical anarchist" tendency, so had Surrealism joined the legacy of Romanticism, Dadaism, and other continental art movements seeking to challenge the narratives of Christianity, civilization, and modernity. Surrealists like Jacques Vaché and Andre Breton took profound interest in alchemy and magic associated in the poetry of Apollinaire with elves and witches, as in the text *L'hérésiarque et Cie*—"And everywhere, all round him, the elves of the *pouhons*, or fountains that bubble up in the forest, answered them..." (Green, 2005, p. 198; Palermo, 2015, p. 116; Rosemont, 2008a, p. 180). In his book *Nadja*, Breton depicts the fictional Madame Sacco as a *clairvoyante* and writes of "magnificent days of riot called 'Sacco-Vanzetti'" during which the Boulevard Bonne-Nouvelle "seemed to come up to my exception, after even revealing itself as one of the major strategic points I am looking for in matters of chaos, points which I persist in believing obscurely provided for me, as for anyone who chooses to yield to inexplicable entreaties, provided the most absolute sense of love or revolution are at stake and that this, naturally, involved the negation of everything else" (Breton, 1960, pp. 152–153). Breton disrupts the historical connections between places and times, insinuating the mythical violence of visionary poetry within everyday life. Despite the energetic spirit of revolution adopted by the Frankfurt School and Surrealists, the "heretical" interwar critique of industrial civilization and magical fascinations popular during the 1920s were never "owned" outright by the left.

Heidegger, himself, joined the Nazi Party, while the "conservative revolution" advocated by Ernst Jünger included a direct "critique of civilization" as stifling for the soul of the individual. The Frankfurt School ruthlessly criticized Carl Jung for deploying archaic tropes from mythology to justify the Nazi regime, while former Dadaist, Julius Evola, mutated the spiritual ideas of René Guénon into his own racist and anti-Semitic creed, calling for warrior elites to usher in a new spiritual imperium. The rise of the Third Reich in 1933 and subsequent conquest of France in 1940 forced "degenerate" left-wing movements like the Surrealists into exile, as the Nazis advocated a völkisch blood and soil movement and championed pseudo-spiritual ideas like those of Evola. However, the scattering of the seeds of revolution would only produce more vital, hybrid movements from Mexico to Algeria to New York City.

Perhaps the most popular theorist of the Frankfurt School, Herbert Marcuse, adopted a particularly libertarian mass political position linking the critique of industrial civilization to the overthrow of capitalism and decolonial struggles in the Global South led by various revolutionary Third World movements (Castro, 2016, p. 323). After the defeat of the Reich in 1945, the Frankfurt School developed a critique of the "authoritarian personality" tacit within the "pre-fascist individual" as an outcome of modernism, not a deviation from it. For the Frankfurt School, liberation from the sexual repression of everyday life under industrial civilization actualized the critique of capitalism. Because modern industry produces a decadent system of structural inequality, overproduction, and waste, Marcuse asserted, even the Soviet Union succumbed to its own form of commodity fetishism (1969, p. 254). The solution between the West and East became libertarian communism, the abandonment of bureaucratic systems, and the celebration of regenerative play.

We Were Like Elves

Alongside the Frankfurt School's critiques, and in similar relation to Heidegger's philosophy, the French Existentialist movement grew to tremendous popularity during the 1950s and 1960s. Also emerging with roots in the interwar avant-garde, Existentialism threw civilization into question by problematizing the content of the individual as a subject. "We were like elves," philosopher Simone de Beauvoir recalled. "Our life, like that of all petits bourgeois intellectuals, was in fact mainly characterized by its lack of reality" (Bair, 1990, pp. 155, 186). In fact, Existentialism provided another phenomenon for intellectuals and revolutionaries all over the world to engage in new ways of living and experiencing the world. According to Jean-Paul Sartre, Being-for-others manifests the predicate of true existence lived in accordance with freedom and against alienation (Catalano, 1985, pp. 124–125). The subject of alienation in the modern world taken up by the Frankfurt School and Existentialists recurs also in the work of French sociologist Henri Lefebvre, whose explorations of the urban environment and the critique of everyday life became tremendously important for a new generation of radicals emerging in the postwar period (Butler, 2012, p. 25).

In 1957, a group of Lefebvre's students created a small, avant-garde circle called the Situationists who produced films, art pieces, and tactics for resisting the patterns and procedures of quotidian repression. Contemporaneously, a host of intellectuals dissenting from the Structuralist ideology prevalent within France's institutions of higher learning buttressed the need to destroy the underlying logic of industrial civilization manifested in the alienating symbolic structures of everyday existence. These tendencies linked in the 1960s with widespread protest against the Vietnam War and in favor of Civil Rights and a strategy of Thoreauvian "civil disobedience"—an apparent direct reference to and inversion of the aforementioned Hobbesian resistance to nature.

The Port Huron Statement at the inception of Students for a Democratic Society (SDS) incorporated grassroots, nonhierarchical organizing, while groups as varied as the Yippies, Chicago Surrealists, Black Panthers, and Black Mask shared similarities with and/or drew inspiration from a variety of sources—from the Surrealists and Existentialists to the Situationists and the Frankfurt School (Hahne & Morea, 2011, p. 46, 152–153; James, 1973, p. 99; Rosemont, 2008b). Meanwhile, Civil Rights leader Martin Luther King and the Conference on Racial Equality were mutually informed by the anarchist–pacifist Bayard Rustin, as well as the non-violent tactics of Mohandas Gandhi, who in turn took influence not just from Thoreau or traditional anticolonial resistance but also international anarchism in the figure of Russian critic of industrial civilization, Leo Tolstoy (Cornell, 2016, pp. 165–167, 220). Toward the later part of the 1960s, gay men founded the Gay Liberation Front to advocate for sexual diversity in society, later taking on the identity of "Radical Fairies" to both expropriate the slur "fairy" and to articulate a form of subjectivity alien to modern, rational heteronormativity (Thompson, Roscoe, & Young, 2011).

By 1968, an intellectual synthesis of the economic, ecological, and philosophical rejection of industrial civilization in its imperialist form contributed to a broad-based revolt against oppressive structures around the world. The massive unrest taking place in the United States, France, Czech Republic, and England, to name a few places, during the wave of strikes, riots, and insurrections of 1968 carried over to Italy the next year, when a wildcat strike at a Fiat plant ignited a "hot summer" of strikes and factory occupations. Amid this mass movement against not just the company bosses but the control of the Communist Party and trade unions, a new movement of Autonomism emerged. Rather than mobilizing around key organizations like the Party, people organized in sites of everyday resistance—their homes, neighborhoods, factories, public spaces. They openly opposed capitalism as well as fascism and the forms of repression tacit within capitalism, including the development of new properties and increase of rents amid poor living conditions and the plenitude of available buildings and land lying empty and unused.

The Autonomist movement soon became contentious. Fascists attempted to clandestinely enter and distort the movement through a so-called "Strategy of Tension" inspired by fascist occultist Julius Evola. According to this strategy, terrorist attacks on civilian infrastructure would directly challenge the machinations of everyday life, thus drawing people closer to the state and further from the left (Bull, 2008, p. 19). The other side to the Strategy of Tension, however, was the attempt to draw people toward an ecological subculture beyond left and right, faithful to one another as Italians rather than political actors. Using the language of Tolkien and old tales of fairies and elves, Italian neo-fascists sponsored a two-day music festival called "Hobbit Camp." The festive atmosphere provided a break from ordinary views on fascism and the "Years of Lead" brought on by the Strategy of Tension, thus providing an important foray into attempts to exploit the autonomist milieu from a green and archetypal "third position" (Forlenza & Thomassen, 2016, p. 232).

Autonomism spread across the Alps to a new generation of leftists in Germany. Mobilizing through networks of squatters and decentralized groups against the reemergence of fascism and weapons-grade nuclear power, the

German Autonomists generated new tactics (for instance, wearing all black and donning masks to maintain anonymity) as well as an ecological concentration (Katsiaficas, 2006). This new movement, which the ELF would later call the "Autonome squatters movement," deployed their tactics in the struggle against fascists. However, the growing green movement also contained right-wing elements associated with the blood and soil ideology, perpetuating the ongoing conflict between right and left over issue-based movements and especially ecology (Lee, 2000, pp. 214–219).

Earth Night Outs

To defend the environment against the ongoing encroachment of highways throughout England, activists embracing this growing, horizontalist counterculture began to create long-term encampments in the countryside, blockading construction companies in what became known as the "Anti-Roads Movement." The decentralized, autonomous, and leaderless Animal Liberation Front (ALF) formed in the 1970s from a group of hunt saboteurs, vandals, and arsonists dedicated to taking militant direct action in defense of animals (Newkirk, 2000, p. 61). British anarchist punk bands influenced by Situationism and Autonomism, among other movements, celebrated and propagandized such movements—particularly Crass, who inveighed against war, and Conflict, who penned lyrics in favor of animal rights.

Yet this period also saw the deindustrialization of much of England and rising working-class resentment, accompanied by the growth of fascist organizations like the National Front. Punk and reggae bands came together to oppose racism through an outdoors music festival called Rock Against Racism, joined at first by a group called Crisis. However, the members of Crisis became disillusioned with punk and the left, drifting toward affiliations with fascist ideologues. The new musical genre of neo-folk drew from the same vein of Hobbit Camp, seeking a palingenetic desire for the rebirth of an organic, ultranationalist British spirit. By the mid-1980s, an officer of the Official National Front named Troy Southgate developed a new "revolutionary nationalist" group promoting the strategy of "entryism." Naming the ALF specifically, Southgate called on ultranationalists to join ecological and otherwise autonomous movements, steering them toward fascism or dismantling them from within (Macklin, 2005, p. 318).

Meanwhile, across the pond, anarchist groups like the Movement for a New Society helped organize a large antinuclear network called the Clamshell Alliance in 1976 through non-violent praxis taken up by decentralized networks of cooperative collectives that abided by vegetarian, free-love lifestyles and egalitarian, consensus-based decision-making processes (Cornell, 2011, pp. 41–42). As Movement for a New Society grew in the Northeast United States, environmentalists frustrated with large conservationist nonprofits and federal regulatory agencies produced a new group called Earth First! in the Southwest.

Explicitly organized along "anarchic" terms, Earth First! developed chapters throughout the United States united in taking direct action to stave off the deforestation of roadless areas and the destruction of habitat by mining and development, as well as dams, power plants, and hazardous agriculture. Though the catchphrase of "No compromise in defense of the Earth!" helped EF! grow manifestly, the group's non-hierarchical organizing strategy was compromised by cultural divisions between those who designated themselves "Rednecks for Wilderness" hostile to "urban issues" and those coming from the antiwar and antinuclear countercultures whom they deemed too "politically correct." Power struggles ensued over the direction of EF!, and in the late 1980s, the conservative faction began to abandon the group as the groups identified with the West Coast and supportive of prolabor feminist utopianism gained hegemony (Tokar, pp. 141–145).

Inspired by reports of a new group called the ELF, which had emerged from the anti-roads movement to form the basis for EF! in England by the early 1990s, the *Earth First! Journal* called for weekly "Earth Night Outs" where elves and fairies would sabotage logging equipment and developments. Though the "Earth Night Outs" resulted in relatively small-scale financial impact, EF! abandoned them in favor of alternative non-violent tactics like forest occupations consisting of "treesits." The language of "elves" and "fairies" identified radical imagination as transgressive against the mechanization of the mind brought about through the works of Descartes and Locke, Hobbes and Bacon. Rather than understand "elves or unicorns" as real, the figures of fantasy came to symbolize deeper imaginative readings of reality, time, space, and order (Graeber, 2009, p. 521).

When a timber company committed an arson in the Warner Creek area of Western Oregon in order to begin logging old growth, EF!ers hastily created a temporary forest occupation to halt logging. They gradually constructed a permanent camp fortified by walls from which activists could launch incursions against equipment and block roads using slash piles and rocks not unlike the Irish brigands of the 17th century. As treesits grew more permanent, and occupations like the Minnehaha Free State developed across the United States, a new conceptualization of earth liberation emerged. Deriving in no small part from the inspired occupation of a liberated, indigenous territory in Chiapas, Mexico, by a small guerrilla force known as the Zapatistas, EF!ers began to locate their positions within the ambit of the "free state." Against the logics of industrial man, clock time, and work, free states could become sites of practical decolonization and rewilding.

With the Zapatistas' emphasis on solidarity with indigenous peoples, those engaged in "free states" attempted to return to natural relations between human and environment held within ancient pagan spirituality and premodern social organization. Much of these efforts held the stain of colonial prejudices, but issued from genuine intentions to serve the land and be good stewards of it, as well as healing the centuries-deep wounds of genocide and slavery (King, 1996). After police brutally closed Minnehaha, the Warner Creek occupation, and others with pepper spray and pain compliance holds, some angry participants decided to take the tactics of anonymous sabotage to more significant levels.

Rage and Action

The ELF of the United States was thus born through a combination of movements, influences, and ideals. In their first communiqué, the ELF acknowledged this openly, doffing their proverbial cap to everyone from the Diggers and Luddites to the Autonomen. While EF! debated the goal of dismantling industrial civilization, the ELF openly called for the total destruction of industrial civilization through a growing wave of massive arsons and sabotage. Another central influence for the ELF, green anarchist John Zerzan found inspiration in the writings of Heidegger and the Frankfurt School, viewing not simply industrial civilization but agriculture as the manifestation of human alienation from nature (2008, p. 17).

Though Zerzan's influence actuated environmental direct action, particularly among the green anarchists of the Pacific Northwest during the 1990s, he also inspired reactionary traditionalists like Russian fascist Alexander Dugin and Southgate. For Dugin and Southgate (and Evola before them), civilization describes the modern world against traditional cultures that carry the true spiritual and linguistic content of a place and those whose ancestors first cultivated it. Like Hobbit Camp, Southgate joined with English green anarchist Richard Hunt to launch a Heretics Fair in attempts to draw members of the green community toward fascism. Although it was not as prominent a feature as with the early green movement of Germany and EF! itself, the ELF had a small portion dedicated to a twisted right-wing ideology identifiably approximate to Southgate's "national-anarchism." Two participants, Nathan "Exile" Block and Joyanna "Sadie" Zacher, lived on the fringes of overlapping alternative and fascist subcultures in which an interest in the occult mixed with Scandinavian black metal and the adoration of Charles Manson. Block would later create a Tumblr account dedicated to esoteric fascist imagery and quotations from figures like Jünger, Heidegger, and Evola (Ross, 2017).

Yet for the Radical Fairies and the autonomous movement peopled by squatters, punks, and other misfits, the fairy world remains a liminal world of escape against everyday repression. The world of elves and fairies provided a "space of exit" for radicals hoping to unsettle the conditions of history and industrial development, but it would prove elusive as ever (Grubačić, 2014). To unlock this concept of the "space of exit" it might help to draw an analogy to popular film. Suggesting the power of fairy tales as vehicles for escape, Guillermo del Toro's *Pan's Labyrinth* reveals the inner world of a girl caught up in the merciless forces of Generalissimo Franco during the Spanish Civil War. Despite or perhaps because of her unfortunate connection to the encroaching fascism, the girl gradually becomes enraptured by the fairy world of magic beyond the cold, authoritarian mercilessness of the Franco regime—a world populated not just by fairies but by antifascist guerrillas as well. Similarly, the anarchist collective Crimethinc.'s children's book, *The Secret World of Terijian*, tells of children who live on the outskirts of a forest falling under the ax. They find a bulldozer sabotaged with the word "ELF" scrawled on it and develop a fascination for the magical world of elves as it exists within the unknown depths of the forest. The fairy world could also reemerge publicly as an act of protest, as when the Radical Faeries joined the 2005 protests against the G8 in Gleneagles, Scotland.

Between the years 2004 and 2006, after more than five years of trying, the FBI ensnared the ELF in a massive operation named Operation Backfire, which activists called the "Green Scare" (Potter, 2011). However, the ELF's rhizomal structure spread throughout Latin America where actions have continued. In Chile, where a strong left and powerful student movement exist, the ELF have claimed a number of actions against developers, construction firms, mining companies, and banks. In 2009, activists in Uruguay called for a conference of ELF and ALF supporters but called it off after a visit from Interpol. Across the Panama Isthmus, in Mexico, the journal *Rabia y Accion* (Rage and Action) emerged to support political prisoners, disseminate anarchist ideology, and provide accounts of direct action (Rabia y Action, 2010, p. 78).

Although *Rabia y Accion* provided a format for merging political prisoner struggles with the earth liberation movement, other groups have challenged that solidarity. The group Individualidades tendiendo a lo Salvaje (ITS—Individualists Tending Toward the Wild, 2013) presents open opposition to all forms of "collectivism," insisting on "indiscriminate violence" against industrial civilization amplified by the prospect of collateral damage (Jacobi & Tepetli, 2016). ITS does not differentiate in their hatred of the political right and left, launching attacks against both (Individualists Tending toward the Wild, p. 72). "Nature is the good, Civilization is the bad," they proclaim, yet they appear puzzled at the same time: "we cannot conclude that Nature-Civilization are concepts that have credibility in time and space" (p. 54, 56). Their answer: to denude the world of spirit and restrict it to its absolute material base, while strangely reasserting the Manichean binary of nature and civilization, because "the best duality would center itself in morality" (p. 56). It is only because ITS's moralism happens to be sophistry that their notion of "Civilization" simply mobilizes the "reality" of "nature" fully realized and reified.

ITS's justification for sending letter bombs to groups like Greenpeace or murdering women for being civilized betrays the material consequences of such a paradoxically mechanistic attitude corresponding to "natural laws," since "everything in Wild nature has an order and because we say that we obey this order and these natural laws" (pp. 67, 96). Of course, they do not seem to notice or mind that the naturalization of "non-harmful authority"—especially vis-à-vis the return to a traditional family under "natural law"—characterizes colonial absolutism backed by bullets and bombs (pp. 95–96).

That their feint toward traditional, indigenous communities dissolves into an intransigent indifference toward "strangers" should not surprise anyone, regardless of their pretensions to quasi-rational moral instincts. What is frankly astonishing is that ITS actually seems to believe that such callousness stems from an individualist rejection of "unnatural" altruism and not a particular kind of estranged incoherence (p. 179). ITS fulfills its own Oedipal egomania, attacking what they claim to defend using ratiocination without reason while standing firmly on the anti-left, reactionary side of the ecological struggle. Their inane efforts at and critical relationship to Unabomber-style revolutionism (something like a "revolutionary traditionalism") would draw most of society toward a strong state rather than a long lineage of libertarian and egalitarian ecological and economic movements.

Another similarly senseless eco-group emerging in the United States following the worst of the Green Scare sought not only to provoke mass terror, but also to destroy the entire edifice of industrial civilization through a prolonged, militant campaign compared to the Allied "extermination bombing" of Germany during World War II. Calling itself Deep Green Resistance (DGR), this group grafts the history of anarchist organizing during the Spanish Revolution onto promises of a future egalitarian society following a race war concomitant to the collapse of industrial civilization. Foreseeing the effects of civilizational collapse they hope to bring about, DGR's leading writer Derrick Jensen predicts, "We will see an increase in violence against people of color.... My answer for people of color is, learn to defend yourself and form self-defense organizations" (Jensen, Keith, & McBay, 2011, p. 452). While criticizing the völkisch movement and various militant left-wing groups of the Vietnam War era, DGR promises to produce a feminist community aligned with natural hierarchies based on age and sex. For this reason, DGR rejects transgender people as the misogynistic reconstruction of patriarchal gender roles. Like ITS, DGR envisions a materialist lifeworld stripped of superstition, yet they support the Women's Liberation Front's efforts to join Christian dominionist group, Focus on the Family, in resisting pro-LGBT legislation and flirt with blood and soil ecology, suggesting that their proximity to white nationalist groups is not coincidental; they are actually representative of a deeply reactionary aspect of ecological thought that has existed at least since the 19th century (Matisons & Ross, 2015).

Other groups like Rising Tide, Radical Action for Mountain People's Survival, and remaining, persisting Earth First! chapters continue the struggle against oppression in all forms and have contributed to social mobilizations from Occupy Wall Street to Black Lives Matter. As long as industrialism represses people's basic needs and desires, social movements from every sector will continue to find the magic in expressing their power and establishing autonomy. As autonomous resistance proceeds in powerful formations against the frightful horizon of climate change, the revolutionary transformation of everyday life seems not only increasingly possible but also necessary.

Conclusion: We Must Listen to Poets

In this chapter, I have attempted to draw out the historical tradition of the Green movement that inspired the rise of the ELF, the critical role that the ELF held in advancing the struggle against industrial civilization in practical terms, and the ideological complexity tacit within that tradition, itself. From the Peasant Wars to the Protestant Reformation to the inchoate proletarian class struggle to the emergence of autonomism, social movements issue from cries against repression and destruction, yet also carry contradictory currents that reproduce the power dynamics they hope to destroy. The enemy, once attaining a specific identity, is too often located within, thus causing a self-destructive cycle of animosity and terror.

My hope is that the international movement for liberation from exploitation and oppression might continue through the current, tragic epoch of reaction into which much of the world has plunged, but cautiously, and accompanied by knowledge and understanding of the potential pitfalls of radical organizing. Direct action in the short term bodes long-term implications that must be understood strategically and from the spirit of egalitarianism rather than individualized and alienated animus toward humanity, in general. That means that vulgar materialism provides as much of an obstacle to intellectual growth as does any class or clear-cut.

Finally, civilization is a reflexive and relative concept that must be contextualized in terms of colonialism and domination, as well as liberation, mutual aid, and community solidarity. To totalize particular subjects is to lose sight of their dynamic relation to other subjects, producing unbeatable enemies and risking mutually assured totalitarianisms. Hence, the struggle against the oppressive structures of industrial civilization must present viable alternatives, offer imaginative, though reasonable, solutions, and inspire not only agency among participants but general cogency among those seeking to foster similar sustainable and adaptive systems.

We have yet to find a proper remedy for the ailments of modern society that has not run aground on state repression or self-destruction (or both). Perhaps there is an intellectual technology we continue to develop that will bring about our collective empowerment against oligarchy and the greed that feeds it. Until then, we do still have fairy tales. Perhaps "those mischievous elves of lore" still play in a surreal landscape or through some hidden, trans-dimensional flight. Perhaps they will return with a little conjuring. To quote the luminary philosopher Gaston Bachelard, "The hidden in men and the hidden in things belong in the same topo-analysis, as soon as we enter into this strange region of the *superlative*, which is a region that has hardly been touched by psychology. To enter into the domain of the superlative, we must leave the positive for the imaginary. We must listen to poets" (1994, p. 89).

References

Adorno, W. (1978). *Minima moralia* (E. F. N. Jephcott, Trans.). London: Verso.

Ashcraft, R. (1986). *Revolutionary politics and Locke's Two Treatises of Government*. Princeton, NJ: Princeton University Press.

Ashliman, D. L. (2004). *Folk and fairy tales: A handbook*. Westport, CT: Greenwood Press.

Bachelard, G. (1994). *Poetics of space* (M. Jolas, Trans.). Boston, MA: Beacon Press.

Bacon, F. (1868). *The letters and the life of Francis bacon including all his occasional works, vol. IV* (J. Spedding, Ed.). London: Longmans, Green, Reader, and Dyer.

Bair, D. (1990). *Simone de Beauvoir: A biography.* New York, NY: Simon and Schuster.

Benjamin, W. (1986). *Reflections* (E. Jephcott, Trans., P. Demetz, Ed.). New York, NY: Schocken Books.

Benjamin, W. (2005). *Selected writings, Volume 2: 1927-1934* (R. Livingstone, Trans., M. W. Jennings, H. Eiland, & G. Smith, Eds.). Cambridge, MA: Belknap Press.

Biehl, J., & Staudenmaier, P. (1996). *Ecofascism: Lessons from the German experience.* Retrieved from theanarchistlibrary.org

Bloch, E. (2006). *Traces* (A. Nassar, Trans.). Stanford, CA: Stanford University Press.

Breton, A. (1960). *Nadja* (R. Howard, Trans.). New York, NY: Grove Press.

Bryan, E. S. (2011). Conversion and convergence: The role and function of women in post-medieval Icelandic folktales. *Scandinavian Studies, 83*(2), 165–187.

Buhle, P. (2011). *Robin Hood: People's Outlaw and Forest Hero; A Graphic Novel.* Oakland, CA: PM Press.

Bull, A. C. (2008). *Italian Neofascism: The strategy of tension and the politics of neoreconciliation.* New York, NY: Berghahn Books.

Butler, C. (2012). *Henri Lefebvre: Spatial politics, everyday life and the right to the city.* New York, NY: Routledge.

Castro, J. S. (2016) *Eros and revolution: The critical philosophy of Herbert Marcuse.* Boston, MA: Brill.

Catalano, J. (1985). *A commentary on Jean-Paul Sartre's being and nothingness.* Chicago, IL: University of Chicago Press.

Chaucer, G. (1903). *The complete works of Geoffrey Chaucer.* New York, NY: Henry Frowde.

Cornell, A. (2011). *Oppose and propose!: Lessons from movement for a new society.* Oakland, CA: AK Press.

Cornell, A. (2016). *Unruly equality: U.S. anarchism in the twentieth century.* Berkeley, CA: University of California Press.

Defoe, D. (2011). *Moll Flanders.* Oxford: Oxford University Press.

Federici, S. (2004). *Caliban and the witch.* New York, NY: Autonomedia.

Federici, S. (2015). Global anarchism: Provocation. In B. Maxwell & R. Craib (Eds.), *No gods, no masters, no peripheries.* Oakland, CA: PM Press, pp. 349-358.

Forlenza, R., & Thomassen, B. (2016). *Italian modernities: Competing narratives of nationhood.* New York, NY: Palgrave MacMillan.

Graeber, D. (2009). *Direct action: An ethnography.* Oakland, CA: AK Press.

Graham, R. (2016). *We do not fear anarchy, we invoke it.* Oakland, CA: AK Press.

Green, C. (2005). *Picasso: Architecture and vertigo.* New Haven, CT: Yale University Press.

Grubačić, A. (2014). Exit and territory: A world-systems analysis of non-state spaces. In Alexander Reid Ross (Ed.), *Grabbing back: Against the global land grab.* Oakland, CA: AK Press, pp. 159-175.

Guha, R. (1999). *Elementary aspects of peasant insurgency in colonial India.* Durham, NC: Duke University Press.

Hahne, R., & Morea B. (2011). *Black mask & up against the wall motherfucker: The incomplete works of Ron Hahne, Ben Morea and the black mask group.* Oakland, CA: PM Press.

Hall, A. (2005, April). Getting shot of elves: Healing, witchcraft and fairies in the Scottish witchcraft trials. *Folklore, 116*(1), 19–37.

Hobbes, T. (1651). *Leviathan: Or, the matter, forme, & power of a common-wealth ecclesiasticall and civil.* London: Andrew Crooke at the Green Dragon.

Holt, M. P. (2005). *The French wars of religion, 1562–1629.* Cambridge: Cambridge University Press.

Hore, H. F. (1858). Woods and fastnesses in ancient Ireland. *Ulster Journal of Archaeology, 1*(6), 145–161.

Individualists Tending toward the Wild. (2013). *The collected communiqués of individualists tending toward the wild* (War on Society, Trans.). Retrieved from https://interarma.info/wp-content/uploads/2014/01/its.pdf

Jacobi, J., & Tepetli, M. (2016, July 13). *Dialogue on wildism and eco-extremism.* Retrieved from http://wildism.org

Jacobs, Jack. (2015). The Frankfurt School: Jewish Lives and Antisemitism. Cambridge: Cambridge University Press.

James, J. (1973). *New abolitionists: The (neo)slave narratives and contemporary prison writings.* Albany, NY: SUNY Press.

James I. Daemonologie. (2008). *Project Gutenberg Ebook.* Retrieved from http://www.gutenberg.org

Jensen, D., Keith, L., & McBay, A. (2011). *Deep Green Resistance: Strategy to save the planet.* New York, NY: Seven Stories Press.

Jolly, K. L. (1996). *Popular religion in late Saxon England: Elf charms in context.* Chapel Hill, NC: University of North Carolina Press.

Katsiaficas, G. (2006). *The subversion of politics: European autonomous social movements and the decolonization of everyday life.* Oakland, CA: AK Press.

King, E. (1996). *LISTEN: The story of the people at Taku Wakan Tipi and the reroute of highway 55 or the Minnehaha free state.* Tucson, AZ: Feral Press.

Lee, M. A. (2000). *The beast reawakens: Fascism's resurgence from Hitler's Spymasters to today's Neo-Nazi groups and Right-Wing extremists.* New York, NY: Routledge.

Lee, S. (1908). Robin Hood. In Leslie Stephen (Ed.), *Dictionary of National Biography, 1885–1900, volume 9* (pp. 1152–1157). New York, NY: The MacMillan Company.

Linebaugh, P., & Rediker, M. (2000). *The many-headed hydra: Sailors, slaves, commoners, and the hidden history of the revolutionary Atlantic.* Boston, MA: Beacon Press.

Machiavelli, N. (1906). *The Florentine history, volume 1* (Nina Hill Thomson, Trans.). London: Archibald Constable and Co.

Macklin, G. (2005, September 1). Co-opting the counter culture: Troy Southgate and the national revolutionary faction. *Patterns of Prejudice, 39*(3), 301–326.

Mágon, R. F. (2005). *Dreams of freedom: A Ricardo Flores Mágon reader* (Chaz Bufe & Mitchell Cowen Verter, Eds.). Oakland, CA: AK Press.

Marcuse, H. (1969). *Soviet Marxism.* New York, NY: Columbia University Press.
Matisons, M. R., & Ross, A. R. (2015, August 9). *Against Deep Green Resistance. Perspectives on anarchist theory.* Retrieved from https://anarchiststudies.org
Morris, W. (1828). *The collected works of William Morris, volume 16.* New York, NY: Longmans Green and Company.
Moryson, F. (1735). *A history of Ireland: From the year 1599, to 1603, volume 2.* Dublin: George Ewing.
Müntzer, T. (2010). *Sermon to the princes* (M. G. Baylor, Trans.). New York, NY: Verso.
Newkirk, I. (2000). *Free the animals: The amazing story of the Animal Liberation Front.* New York, NY: Lantern Books.
Palermo, C. (2015). *Modernism and Authority: Picasso and his milieu around 1900.* Berkeley, CA: University of California Press.
Potter, W. (2011). *Green is the new red: An insider's account of a social movement under siege.* San Francisco, CA: City Lights.
Rabia y Action. (2010, November 1). Rage and action. *Earth First! Journal, 31*(1), 78.
Ramnath, Maia. (2011). *Decolonizing anarchism.* Oakland, CA: AK Press.
Reclus, É. (2013). *Anarchy, geography, modernity: Selected writings of Elisée Reclus* (J. P. Clark, C. Martin, Ed. & Trans.). Oakland, CA: PM Press.
Rosemont, F. (2003). *Joe Hill and the making of a revolutionary working class counter culture.* Chicago, IL: Charles H. Kerr.
Rosemont, F. (2008a). *Jacques Vaché and the roots of surrealism: Including Vaché's war letters & other writings.* Chicago, IL: Charles H. Kerr.
Rosemont, P. (2008b). *Dreams and everyday life: André Breton, surrealism, rebel worker, SDS & the seven cities of Cibola.* Chicago, IL: Charles H. Kerr.
Ross, A. R. (2017). *Against the fascist creep.* Oakland, CA: AK Press.
Shakespeare, W. (1999). *The complete works of William Shakespeare.* Hertfordshire: Wordsworth Editions.
Shelley, P. B. (1822). *Queen Mab.* London: R. Carlile.
Shelley, P. B. (1841). *Poetical works of Percy Bysshe Shelley.* Paris: A. & W. Galignani & Co.
Strabo. (1988). *Geography: Books 3-5.* Cambridge, MA: Harvard University Press.
Sutton, J. F. (1880). *The Date Book of Nottingham.* Nottingham: Nottinghamshire County Council.
Thompson, E. P. (1966). *The making of the English working class.* New York, NY: Vintage Books.
Thompson, M., Roscoe, W., & Young, B. (Eds.). (2011). *The fire in moonlight: Stories from the radical faeries: 1975–2010.* Brooklyn, NY: White Crane Books.
Tokar, B. (1997). Earth For Sale: Reclaiming Ecology in the Age of Corporate Greenwash. Boston: South End Press.
Tuchman, B. (1978). *A distant mirror: The calamitous 14th century.* New York, NY: Random House.

Van Meter, K. (2017). *Guerrillas of desire: Notes on everyday resistance and organizing to make a revolution possible*. Oakland, CA: AK Press.

Venette, J. (1953). *The chronicle of Jean de Venette*. New York, NY: Columbia University Press.

Walpole, H. (1906). *The letters of Horace Walpole: Fourth earl of Orford, volume 5* (Peter Cunningham, Ed.). Edinburgh: John Grant.

Zerzan, J. (2008). *Twilight of the machines*. Port Townsend, WA: Feral House.

13. *Magic Kills Industry: Reclaiming ELF and Witch Deviance as Ecoqueer and Anticapital*

MARA PFEFFER AND BETHANY RICHTER

Introduction

In a world filled with injustice, we find ourselves re-enchanted to this world through the actions of ecoqueer liberation movements, elven actions in defense of mother earth, abolitionist actions, and by the crafting of poetry and art in social movements. In this chapter, we define and embrace an ecoqueer approach to resistance through an analysis of the modes of expression embraced by the Earth Liberation Front (ELF). We also seek to use this approach to remember the deviance of those constructed as witches in the ELFs lineage and examine what is needed in order for magic to truly be ecoqueer and anticapital. Heavily inspired by the visionary craft and writing of adrienne maree brown, this chapter weaves together prose, creative writing, and analysis. We utilize this style in order to uproot the monoculture of colonial and capitalistic noise that has seeded our minds and lands and embrace an ecoqueer peculiarization of our writing and the world. "The peculiarization of the world…not only opens the mind and disconnects the human brain from the machine of ideology, but it also breaks the shop windows of all commercial chains, negates authority and shouts with a clear and pristine voice, ENOUGH!" (Sepúlveda, 2005).

Ecoqueer

To be ecoqueer is to embrace the magic of the natural world, the moon tides in balance with the sun's rays, to revel in the spectrum of how seasons fluctuate between the masculine sun and the feminine moon, and how each movement is non-definitive and ever-changing. Ecoqueer seeks to unseat patriarchy, but celebrates feminine masculinities, masculine femininities, and every queer expression of gender, gender bending, natural fluidity, and magical being that can be imagined or unimagined. Ecoqueer sees the incredible, undefinable queerness of the natural world and resists the process of boxing through colonial systems of identity and academic study. Ecoqueer is in the rich tradition of Ecofeminist theory of Vandana Shiva, Judith Plant, Irene Diamond, Carol Adams, and Catriona Sandilands. Ecofeminism is a theoretical and activist movement that understands feminism and ecological justice as interdependent struggles and centers a critique of capitalist patriarchy.

> We see the devastation of the earth and her beings by the corporate warriors, as feminist concerns. It is the same masculinist mentality which would deny us our right to our own bodies and our own sexuality, and which depends on multiple systems of dominance and state power to have its way. (Mies & Shiva, 1993, p. 14)

This foundational theory has also been thoroughly critiqued by queer theorists, critical race theorists, and Indigenous scholars for centering white feminist interpretations of the natural world. We recognize the need for intersectional analysis, queer presence, decolonial and anti-racist work in conjecture with ecofeminism to engage the ever-shifting and responsive systems of oppression that attack the natural and built world. From the strong roots of intersectional analysis and creative process, ecoqueer seeks to center the process over the outcome. It privileges the creative over the defined. It recognizes the need to resist systems of knowing and to instead embrace the unintelligible.

Our starting point of reference is a critique of colonial capitalism and the process of enclosing. This process removed what was communal and imposed concepts of property and ownership on these spaces and living beings. Ecoqueer rejects the capitalist, heteropatriarchal logic that living beings, plants, organisms, cells, DNA, and bodies could ever be owned, named, or claimed by any.

Emotion

Colonial heteropatriarchy holds emotions as a weapon to undermine and delegitimize any who offer alternative ways of being. As we actively resist systems of violence, we embrace the full spectrum of emotion and vulnerability. With the medicalization of emotions, ableism of mental health standards, embracing emotion and radical vulnerability becomes key to resistance.

Animal and social justice advocates are often critiqued "for caring about 'little things,' like individuals and beings with feelings," through the assumptions that the nonhuman world does not have feelings, and that human feelings are not valid ways of knowing (Davis, 1995, p. 202). In turn, many within the animal rights and other movements are frequently dismissive of the roles of imagination, emotion, and compassion—instead relying heavily on theory to justify its existence (Socha, 2012). In the tradition of many activists before us, we recognize that political resistance brings emotions out of the "private" (or the feminine, closed off, not appropriate) and into the "commons" (or the communal, the shared, the cooperative, the masculine). Claudia Rankine (2016) argues for the use of public mourning as a political tactic by Mamie Till Mobley, mother of Emmett Till, and the continued use of this tactic through Black Lives Matter is an important political act to state "the condition of black life is one of mourning" (p. 148).

We bear witness to Valerie Castille's powerful refusal to hide her grief and anger away and instead to alter the collective consciousness of the violent continuation of racism and a politics of death. In the 1980s and 1990s ACT UP demonstrated how queer communities were being politically murdered through the refusal of society to publically acknowledge the AIDS epidemic. Through carrying caskets through the streets, and even to the White House lawn, community members dared the public to feel the grief with them.

> Americans are terrified of death. Death takes place behind closed doors and is removed from reality, from the living. I want to show the reality of my death, to display my body in public; I want the public to bear witness. We are not just spiraling statistics; we are people who have lives, who have purpose, who have lovers, friends and families. And we are dying of a disease maintained by a degree of criminal neglect so enormous that it amounts to genocide. (Mark Lowe Fisher, 1992)

Ecoqueer politics is one that embraces the queer, communal nature of emotion, the lived reality of its magic, and the ways that emotions are a key tactic of resistance and re-creation.

Emergence

A critically engaged, creative resistance insists that we liberate our imaginations by embracing uncertainty as we imagine liberation, "creating the possible out of the impossible" (Esteva & Prakash, 1998, p. 205). Embracing a politics of uncertainty means acknowledging that this process of transformation is messy and imperfect, and there are no prescribed pathways. Expression of this kind shatters any illusion of itself as a finished product and negates how a finished product is a measure of success. This queer and uncertain politics is one that celebrates all steps of discovery, growth, failure, rest, not just the perceived successes. It holds precious the "contingent connection and the hiddenness of unfolding," an unfolding of critical activism which can express itself in shape-shifting and responsive ways, cultivating the "conditions of a less predictable and more productive politics" (Gibson-Graham, 2006, p. xxxi). While productivity is not our goal, we queer our understanding of what is productive, what is re-productive, and what is creative.

We want to embrace this and the kind of emergence that adrienne maree brown demonstrates through the weaving together of spells, conversations, and poetry, embracing a state of discovery and wonder over white colonial ways of "knowing."

> Emergence is beyond what the sum of its parts could even imagine… Cells may not know civilization is possible. They don't amass as many units they can sign up to be the same. No—they grow until they split, complexify. Then they interact and intersect and discover their purpose—I am a lung cell! I am a tongue cell! And they serve it. And they die. And what emerges from these cycles are complex organisms, systems, movements, societies.
>
> Nothing is wasted, or a failure. Emergence is a system that makes use of everything in the iterative process. (brown, 2017, pp. 13–14)

Our celebration of ecoqueer centers the emergent processes of uncertainty and nonlinearity, recognizing that "social change occurs in fits and starts, and movements experience as many or more 'reversals' and setbacks as they do 'progress'" (Pellow, 2014, pp. 255–256). This means rejecting a "linear" approach that says that liberation must come in fragmented moments of growth that require the sacrifice of some. It also means rejecting the positioning of any "ultimate freedom movement" that assumes "all other freedom struggles have been won" (pp. 255–256). Emergence suggests that we "sense ourselves in a stream [or nonlinear swirl!] of activism" that began before us, exists all around us, and will continue after us (Ledgerwood, 2017). Ecoqueer approaches embrace the mystical nature of nonlinear, emergent processes of becoming. As we become, so may we emerge.

Peculiarity

Ecoqueer embraces the peculiar, the queer, the weird, and the unknown. Vital to this process of resistance and remembrance is that we resist the normalizing nature of colonial heteropatriarchy. We reject the notion that what is found outside of the normal is inherently without value. While we employ the tactic of the moral imperative and value the balance of light and dark, good and evil, we also recognize that moralistic arguments of what is "natural" and "normal" behavior are often tools of violence and the policing of marginalized beings. There is an important and perhaps subtle difference in these uses of the "moral" and we emphasize the critical nature of their distinction. Queer theory offers an important interjection into this dialogue and emphasizes the political nature of resisting normalcy and embracing the peculiar, or the queer.

Expressing peculiarities (as one being, collective, or movement) disrupts the omnipresence of monoculture, standardization, and assimilation. We embrace peculiarity with the understanding that "Social change is messy, and that notion should be both humbling and emboldening: there is a great deal of work to be done, so there must be many forms of activism and many types of activists" (Pellow, 2014, p. 256). Peculiarity also deploys a queer and playful approach that rejects dominant norms of propriety and proper ways of "behaving." To be queer is to be rejected by the state. Colonial heteropatriarchy was formed at the exclusion of queer bodies and requires queer, brown, black, and Indigenous bodies to sacrifice for the capitalist machine. We are looking to the birth of new systems. "Being taken seriously means missing out on the chance to be frivolous, promiscuous, and irrelevant" (Halberstam, 2011, p. 6). Peculiarity embraces the unknown, the unintelligible, and the value of relationships over productivity.

Ecoqueer as a Futility, Illegible, and Anticapital

The act of destruction through a lens of queer theory offers an important alterity to what dominant capitalistic society labels as violent or destructive. The prosecution of activists and folks in resistance through the passage of bills such as the American Enterprise Terrorism Act (AETA) of 2006 demonstrates the continued centering of property as being of supreme value over the dynamic and magically living natural world. AETA states that any threat to capital is an act of terrorism, specifically whenever it disrupts profit. The SHAC 7 (Stop Huntingdon Animal Cruelty USA, and the activists associated with SHAC) were convicted for running a website that advocated the use of legal methods to end animal research and exploitation at Huntingdon Life Sciences. These activists were convicted and sentenced for using what were arguably traditional activist tactics and by many standards would not be considered radical.

Capitalism is the same system that actively profits from bodies, the exploitation and violent attacks against nature, and that terrorizes human and nonhuman animals, labels anything that halts capitalism as terror, and property as being the utmost value system. What is labeled violence and nonviolent are structured in a way to uphold those systems. Ecoqueer politics rejects these labels of violence, in solidarity with the many activists and communities who are in shared resistance.

We celebrate and embrace the queer, political act of being unintelligible. When colonial heteropatriarchy seeks to label, to study, and to name, this often comes at the violent dismemberment of the subject of study. One example among thousands is the violent history of gynecology as a field of study that required the sacrifice of enslaved, black, feminized bodies, most notably Archana, Lucy, and Betsey (Vedantum, 2017). The colonial drive to "define" and to "know" has a very specific and violent history for those who are outside of the prescribed norm.

> Illegibility may in fact be one way of escaping political manipulations… Illegibility has implications for all kinds of subjects who are manipulated precisely when they become legible and visible to the state (undocumented workers, visible queers, racialized minorities)… We may in fact want to think about how to see *unlike* the state; we may want new rationales for knowledge production, different aesthetic standards for ordering or disordering space, other modes of political engagement than those conjured by the liberal imagination. (Halberstam, 2011, p. 10)

It is through failure that we learn, grow, and also experience a natural part of life. Failure is not just a stepping stone to the road to success, but also has value in and of itself. Linear approaches to progress and activism often fall prey to fragmenting activist movements and pitting activists against each other. Using dominant value systems of progress, activist tactics such as art, riot, property destruction, spirituality, and ritual are often dismissed as not being strategic or effective.

The ELF actively resists the enclosure of expression through embracing emotion (through its expressing the planet's rage and its calls for building loving community and kinship with all creatures), emergence (through its embodiment in small activist cells and sensing themselves in a "stream" of uprisings going as far back as the peasant revolts), nonlinearity (in its rejection of dominant, movement building measurements of progress and of the capitalist concept of Progress itself), and peculiarity (in its use of playful elven imagery, defending the world's polycultures, and calling for a variety of decentralized, diverse tactics to do so). Through the use of radical approaches, the ELF actively works to resist colonial heteropatriarchy through local, community-oriented tactics of resistance.

> A tight community of love is a powerful force.
> Recon—check out targets that fit your plan and go over what you will do
> Attack—
> powerlines: cut supporting cables, unbolt towers, and base supports, saw wooden poles.
> transformers: shoot out, bonfires, throw metal chains on top, or blow them up.
> computers: smash, burn or flood buildings.
> Please copy and improve for local use.
> ELF Communiqué (2007)

To refuse enclosure is to refuse the premise of property, ownership, and violence. To halt, interrupt, and destroy tools of violence is a tactic of futility, of illegibility, and a rejection of capitalism. This instills terror in a capitalist system and has the potential to shake its foundation.

Property Damage for Total Liberation

We feel it's important to explore the ELF's strategy of property damage in the context of "the diversity of tactics" that they call for. Often, the fixation on whether or not the ELF's use of property damage is an effective strategy takes it out of this much larger, emergent context. What is a tongue cell on its own? Or even a tongue without the context of the body (brown, 2017)? Rather than call for one strategy that is best and most effective, we echo the calls for a polyculture of activism, and so does the ELF and those who call for total liberation, "which grasps the need for, and the inseparability of, human, nonhuman animal, and Earth liberation and freedom for all in one comprehensive, though diverse, struggle" (Best, Nocella, Kahn, Gigliotti, & Kemmerer, 2007, p. 2).

The guidelines of the ELF link environmental and social justice, calling for total liberation for earth and its creatures:

> 1. To cause as much economic damage as possible to a given entity that is profiting off the destruction of the natural environment and life for selfish greed and profit.
> 2. To educate the public on the atrocities committed against the environment and life.
> 3. To take all necessary precautions against harming life.
>
> (Pellow, 2014, p. 55)

Pellow illustrates the guidelines of the ELF and the messages left in their communiqués reveal the four pillars of the total liberation framework:

1. an ethic of justice and anti-oppression linking all beings,
2. anarchism,
3. anti-capitalism, and
4. direct action. (pp. 53–54)

Pellow argues that by linking liberation movements, "The ELF ha(s) deliberately distanced itself from many activists in the first generation of Earth First!ers who were unwilling or unable to articulate links between environmentalism and social justice" (2014, p. 55). For many Elves, the artificial human separation and elevation from nature is part of an interlocking web of white supremacy, heteropatriarchy, capitalism, industrialism, ableism, nationalism, and colonialism. These connections can be understood as being interdependent with black, queer, trans, indigenous, and women's liberation movements. But often the impetus for white radical environmentalists beginning this dialogue is birthed in a mixture of longing for and listening to the nonhuman world and in the transformation of non-activists into activists by the voices of this more than human resistance:

> The threats to ecosystems and nonhuman animals produce an interpellation (a call) that beckons earth and animal liberation activists to take action individually and collectively…inanimate and nonhuman actors spur activists. Threatened wildernesses and genetically engineered chickens exert agency and impact the imaginations, motivations, and actions of activists…. (Pellow, 2014, p. 30)

While the membership of the ELF is anonymous, it has been noted that the movement is likely largely white. Pellow notes that this claiming of kinship by "white, middle class, heterosexuals" of all human beings with the nonhuman is in many ways problematic when it ignores the history of dehumanization faced by oppressed groups (2014, p. 252). In this knocking down of barriers between human and nonhuman, there is often a "glossing over" of the facts that "some of these barriers were actually already flattened and broken down via centuries of European and Euro-American racism, a class system, and heteropatriarchy…" (Pellow, 2014, p. 252).

We understand this to be true and see this pointing not only to the various black, indigenous, queer liberation movements that have inspired white militant activist for total liberation rather than an embrace of white respectability politics, but also to the importance of "reckoning with the specific character of white middle class dissent" (Thompson, 2010, p. 15), asking the questions:

> How is it that a militant movement seemed to emerge spontaneously from white middle class spaces like the campus and the suburb-spaces where "oppression" can often seem like an abstract category? How did' the "dirty kids" get angry and why did they feel so ill at ease in their world of plenty despite the undeniable privilege their circumstance afforded? (Thompson, 2010, p. 14)

In Starhawk's book, *City of Refuge,* the character Bird says in a meeting, "We're planning for a war, when what we need to plan for is a mutiny" (2016, p. 161). We imagine the Birds of the nonhuman world sending out their interpolations to the human world, and we wonder if the actions of the ELF could be one form of mutiny, in this case from white supremacy's war on the earth. As Thomson suggests, the Elves work destroying property has been discussed as a means of white people to cast out their whiteness, becoming a racial deviant or race traitor, "who defies the rules of whiteness so flagrantly as to jeopardize his or her ability to draw upon the privileges of the white skin" (Ignatiev, 1994, p. 177):

> Threats of jail time means privileged activists risk facing some of the "subhuman" treatment that the majority working-class and people of color prison population faces everyday. These radicals then are racial deviants in two ways: as white activists who are labeled "terrorists" and as human activists who are antihumanist and antidominionist. (Pellow, 2014, p. 13)

Anthony Nocella (2014) clarifies that white, middle-class, able-bodied activists carrying out this deviance are also not really able to shed their whiteness so completely: "Not a single radical animal liberation activist has been assassinated, put on Death Row, shot by police, or given a life sentence.... I suspect that if a group of Black youths bombed a McDonalds for political reasons in the name of the ALF, they would likely receive much harsher penalties than their white peers" (p. 29).

We hold the white desire for perfection of anti-racist politics and its impossibility in tension with the need for whites to do battle with and deconstruct all the forms of our constructed supremacy. An interdependent approach calls into question heteropatriarchy and to stand in solidarity with Black Lives Matter, Missing and Murdered Indigenous Women, Standing Rock, Black Mesa Solidarity Network, and the countless others who are currently in an environmentally racist and colonial resistance. In the midst of these struggles, in the words of Alicia Garza: "We need you defecting from White supremacy and changing the narrative of White supremacy by breaking White silence" (Showing Up for Racial Justice, Nashville, 2017).

To truly "defect from" and challenge white supremacy that pervades our movements, those of us with white privilege must examine how we unintentionally engage in and perpetuate white supremacy and continuously work to develop counterpractices. This is a continual act of defecting from dominant systems of oppression. Through the use of queer, eco-spirituality, folklore, and magic, Elves and witches share the struggle against violent systems of oppression.

Caliban and the Elf? Elves and Witches in Solidarity

Expressing Francis Bacon's fears, the members of the ELF have made many direct comparisons of their resistance to the Peasant Revolts of the Middle Ages, destroying industry property and leaving anti-capitalist communiqués bursting with pagan imagery and "dark green religion" in their wake (Taylor, 2008). The acronym ELF itself "provided a rubric for the most radical of actions that was good public relations: elves are viewed positively in western literature as playfully mischievous, not malicious...the idea of elves in the woods cohered with the pagan spiritualities commonly found in radical environmental movements" (Taylor, 2005, p. 6). In an ELF communiqué released in 1997, members present themselves as one of many manifestations of centuries of revolt against enclosure, capital, the desecration of the sacred, identifying with the elves of European folklore.

> Welcome to the struggle of all species to be free.
>
> We are the burning rage of this dying planet. The war of greed ravages the earth and species die out every day. ELF works to speed up the collapse of industry, to scare the rich, and to undermine the foundations of the state...
>
> Since 1992, a series of earth nights and halloween smashes has mushroomed around the world.... We take inspiration from Luddites, Levellers, Diggers, the Autonome squatter movement, the ALF, the Zapatistas, and the little people—those mischievous elves of lore.
>
> Authorities can't see us because they don't believe in elves. We are practically invisible. We have no command structure, no spokespersons, no office, just many small groups working separately, seeking vulnerable targets and practicing our craft.
>
> Many elves are moving to the Pacific Northwest and other sacred areas. Some elves will leave surprises as they go. Find your family! And let's dance as we make ruins of the corporate money system.
>
> Form "stormy night" action groups, encourage friends you trust. A tight community of love is a powerful force. (III publishing, 1997)

And in another example from 1998, from a communiqué released claiming an arson committed in the Medford US Forest Industries office,

ELF members present themselves as Santa's elves sabotaging a corporatized Christmas:

> To celebrate the holidays we decided on a bonfire. Unfortunately for US Forest Industries it was at their corporate headquarters office. On the foggy night after Christmas when everyone was digesting their turkey and pie, Santa's ELFs dropped two five-gallon buckets of diesel/unleaded mix and a gallon jug with cigarette delays; which proved to be more than enough to get this party started. This was in retribution for all the wild forests and animals lost to feed the wallets of greedy fucks like Jerry Bramwell, USFI president. This action is payback and it is a warning to all others responsible, we do not sleep and we won't quit. (quoted in Rosebraugh, 2004, p. 60)

The stories of elves or the little, hidden people of European lore, fairy tales, and Tolkien novels are varied and distinct, reimagined and pieced together by ELF members and Earth First!ers. Much has already been written about this. Some Earth First!ers openly express that they are inventing religions, while others express it as "resurrecting old ways" (Taylor, 2002, p. 47):

> Gnomes and elves, fauns and faeries, goblins and ogres, trolls and bogies...[must infiltrate our world to] effect change from the inside... [These nature-spirits are] running around in human bodies...working in co-ops...talking to themselves in the streets...spiking trees and blowing up tractors...starting revolutions...[and] making up religions. (Young Buck cited in Taylor, 2002, p. 47)

There are many tensions in embracing and "inventing" earth-based spiritualities as white settlers. And the elves of the ELF are presumably largely, if not entirely made up of white settlers. While many call for total liberation, linking social and environmental justice, and take action and make statements in solidarity with indigenous, anti-imperialist, and anti-racist struggles, the pagans of the radical environmental movement have often been found guilty of cultural appropriation of indigenous spiritualities and practices in our longing to re-member and re-enchant ourselves to the earth, its creatures, and ultimately ourselves. The resurrection of European folklore by white environmentalists could be a response to this, as an expression of white desire for a post-/pre-capitalist, decolonized, earth-based spirituality that is non-appropriative and recognizes much of European earth-based spiritualities were systematically destroyed through the witch hunts of early colonial capitalism. "The extension of the witch-hunt to the American colonies...was...a deliberate strategy used by the authorities to instill terror...It was also a strategy of enclosure" (Federici, 2014, p. 220).

The seeking of these traditions also reflects the white perfectionist desire for *purity* that permeates white organizing. As evidence of this desire and the company it keeps, the elves of the ELF and other radical environmentalist pagans and wiccans are not the only ones reclaiming these traditions. Soldiers of Odin and Asatru Folk Assembly are two current groups deeply embedded with the white nationalist movement seeking to reclaim European earth-based spiritualities. As Taylor writes:

> Since nature mysticism does permeate radical environmental subcultures, and sometimes the racist right, it does make sense to inquire about possible linkages and to wonder whether the cultural "tent" of the cultic milieu is pitched so broadly that radical environmentalists, and those from the racist right, might cross paths underneath it. (2002, p. 26)

Some radical environmentalists seeking, reimagining, and practicing European nature mysticism confront ourselves in this, calling attention to the similarities and differences between our desires and that of the white nationalist movement:

> When extreme white supremacists needed a religion exclusive of other races they found Paganism. Specifically the Paganism related to their "heritage" of Germanic and Scandinavian descent—Odinism or Asatru. They felt this could be the ultimate "white" religion because it's not from a foreign land such as Islam from the Middle East or Buddhism from India, etc. (Canfield, 2017)

Many Wiccans and Pagans are speaking out against these groups, in groups such as the Heathens United Against Racism, Pagans Against Racism, Witches Against Fascist Totalitarianism, and "*Declaration 127*, which was a document signed by 170 organizations across almost 20 countries that came together to denounce white nationalist paganism" (Canfield, 2017). However, we believe it's not only important for white radical environmentalists to take actions like these against white nationalist groups, but to also acknowledge and examine the less overt ways we too engage in and perpetuate white supremacy. We must be with the tension that purity is unattainable while also developing liberatory practices, and we must also "develop the ability to identify, name, and appreciate what's right" as we identify and learn from our mistakes (Okun & Jones, 2001).

We turn to these tasks by first looking at what we find exciting about the ELF's embrace of European folklore, beginning with the ELF's allusions to the Witch's Sabbat and then identifying the kinds of politics that make their use of folklore liberatory.

"Wilderness Rendezvous" and the Witch's Sabbat

Within radical environmental scholarship, there is much written exploring the ELF's allusions to elvish folklore and the peasant revolts of the Middle Ages, as well as paganism. And yet there is not as much written connecting this gathering of modern Elves to the story of the Witch's Sabbat.

But the allusions to the Sabbat-like practices are there—in the ELF's references to the flight of elves to the northwest and "other sacred areas," along with "earth nights and Halloween smashes," "practicing our craft," having bonfires, dancing, community building, forming "stormy night action groups," "in retribution for all the wild forests and animals lost to feed the wallets of greedy fucks." All these things whisper of the "wilderness rendezvous" of Earth First!, "where activists gather in remote places to conduct workshops, bond, and engage in revelry and ritual" (Taylor, 2002, p. 32) as well as the concept of the Witch's Sabbat of early capitalist nightmares and propaganda.

Sabbat means "to cease working." The Witches Sabbat of the Middle Age propaganda referred to a time when witches would cease working to plot the liberation of the commons and themselves—amidst orgies, feasting (we imagine sometimes on the rich), crime time storytelling, and ritual promises to the devil to rebel against all masters (Federici, 2014, pp. 176–177). In this we hear loud and clear "the echo of the secret meetings the peasants held at night, on lonesome hills and in the forests, to plot their revolts" (Federici, 2014, p. 176).

This makes sense, as the witch hunts took place in the context and aftermaths of peasant revolts (often initiated by women):

> Uprisings against the "enclosures" in England (in 1549, 1607, 1628, 1631), when hundreds of men, women and children, armed with pitch-forks and spades, set about destroying the fences erected around the commons, proclaiming that *"from now on we needn't work anymore."* (Federici, 2014, p. 174)

The connections of elves to the anticapital, anti-industrial peasant revolts of the Middle Ages are made loud and clear by members of the radical environmentalist movement, and Earth First!ers are even said to sing a song called "Turning the World Upside Down" in honor of the Diggers at wilderness rendezvous (Taylor, 2002, p. 51).

But we find the use of The Witch's Sabbat to be a fruitful connection in need of being made, as it is an extension of these movements against capitalism, authority, and property as well as being distinctly queer, magic, and nature oriented. That the Sabbat happened at night has been interpreted as "a violation of the contemporary capitalist regularization of work-time, and a challenge to private property and sexual orthodoxy as the night shadows blurred the distinctions between the sexes and between 'mine and thine'" (Parinetto, 1998, in Federici, 2014, p. 177).

The Sabbat also blurs the lines between human and nonhuman. Witches of the Middle Ages are often depicted surrounded by animals, dancing and holding hands with the devil in ceremony we can presume is the Sabbat. Witches were constructed as being too close to animals, often accused of shapeshifting and the crime of bestiality; and animals in turn were vilified (Federici, 2014). "Such was the presence of animals in the witches' world that one must presume that they too were being put on trial.... In an era that was beginning to worship reason and to dissociate the human from the corporeal, animals, too, were... reduced to mere brutes, the ultimate 'Other'" (Federici, 2014, p. 194).

We understand the Witch's Sabbat as being ecoqueer in its celebration of corporeality, animality, and desire and "turning of the world upside down" to restore intimacy and kinship with earth.

In many ways, the ELF is one living embodiment of this tradition of "turning the world upside down" in its embrace of elvish and Sabbat imagery, its expression of kinship with the earth and its creatures, and its plots and actions carrying out the property smashing that rulers in this system have had nightmares about since its beginning.

In addition, the elves of the ELF are recognized by the state as terrorists for their actions of property damage, as their crimes are intended "to inflict economic damage," a category which served a similar purpose to the charge of witchcraft in its time, which was a punishment for attacks on property and theft (Federici, 2014, p. 200). And like the crime of terrorism, "the very vagueness of the charge—the fact that it was impossible to prove it, while at the same time it evoked the maximum of horror—meant that it could be used to punish any form of protest and to generate suspicion towards the most ordinary aspects of daily life" (Federici, 2014, p. 170).

These elves have more in common with the Witches Sabbat than is acknowledged. We imagine the earth liberation elves of today and the peasant witches of the Middle Ages dancing in timeless solidarity with each other as they plot their next revolt, recounting tales of prior clandestine activities, and pledging their allegiance to the earth and its creatures, and to liberation.

Magic as Anticapital

> The world had to be "disenchanted" in order to be dominated.
>
> — Silvia Federici

> When I tell people I am a prison abolitionist and that I believe in ending all prisons, they often look at me like I rode in on a unicorn sliding down a rainbow.
>
> — Walida Imarisha

The early masters of capitalism imagined not just the Sabbat but magic itself as anticapital. Federici illustrates how the very use of magic is incompatible with a capitalist work discipline (2014, p. 143). Not only were the crimes of property damage and theft considered witchcraft and a threat to capital, so too was the existence of an enchanted world and those who believed in it. Federici demonstrates: "…it was not just the 'bad witch' who cursed and allegedly lamed cattle, ruined crops, or caused her employer's children to die, that was condemned. The 'good witch' who made sorcery her career, was also punished, often more severely" (2014, p. 20).

One could argue that to believe in magic requires imagination, and imagination in combination with action for liberation is a threat to those who abuse power. Walida Imarisha explains the power of imagination and the visionary sci-fi that allows, beginning with a quote by Ursula Leguin: "We live in capitalism. Its power seems inescapable. But then, so did the divine right of kings. Any human power can be resisted and changed by human beings." Imarisha continues: "The only way we know we can challenge the divine right of kings is by being able to imagine a world where kings no longer rule us—or do not even exist…" (2015).

But magic today and the imagination it requires is not inherently anticapital or anti-racist. Both can be co-opted by capitalism and white supremacy. For instance, the "heathens" of the alt right believe in "chaos magic" which they use in efforts to bring about a white state (Spencer, 2016). Federici argues that the witch hunts and now well-grown capitalism have destroyed the "subversive potential" of witchcraft, and that the system is no longer threatened by the domesticated bits of European magic we have left (2014, p. 205). What makes magic liberatory rather than reactionary is its politics and its purpose, and what actions it is linked to.

By disrupting resource extraction, the elves of the ELF are rekindling magic in a way that does in fact threaten and "kill industry" and positions itself against white supremacy. We perceive the practice of property damage in this context as not simply "destructive" of oppressive systems, but imaginative and "reconstructive" of possibility outside these systems: seeing a wall, slaughterhouse, dam, seeing the possibility of tearing it down, and taking it. In the ELF's conjuring images of the Elves of lore, the world becomes alive and enchanted, and their participation in resistance is both an attack on capitalist white supremacy and a possibility to re-enchant others to their imaginations and the Natural world.

As we grow our tactics and practice, here are just a few subversive practices shared by witches and elves (besides having magical powers) illuminated by placing them in conversation together:

- Forming "stormy night action groups" (ELF, 1997)
- Refusing to "work" for capitalists
- Refusing to obey laws of enclosure
- Disrupting the concept of private property
- Destroying of enclosures of land, animals, water
- Disenchanting others to labor-power and the capitalist heteropatriarchy
- Experiencing the nonhuman world as enchanted and possessing desire
- Re-enchanting others to inherent value in the corporeal world

Elves, Witches, and Reproductive Justice

Early capitalism crafted the witch as the criminal, an enemy of the state and the people, of creation, an abomination, to be destroyed. Witches were not only women who worked with earth magic, but they were also folks who acted outside of heterosexual marriage—midwifes, queers, prostitutes, adulterers—women who rebelled (Federici, 2014, p. 184). The witch was constructed as violent and non-normative, her actions and body *revolting*, while the violence committed against her was constructed as natural and necessary.

It was through the witch hunts, in the name of production and capital, the state claimed control of reproduction. The sexual activity of women was enclosed as "work" in order to subordinate women to capitalism's demands for birth laborers and attack a source of female and community power. "Central to the process was the banning, as anti-social and demonic, of all non-productive, non-procreative forms of female sexuality" (Federici, 2014, pp. 192–194).

At the same moment that the interests of state, ruling class, and church coalesced to enclose control over reproduction and construct this control as violence in the hands of women, they also enclose all acts of destruction themselves, constructing arson and property damage as violence in the hands of the people. And as this is unfolding, "witch-hunting and charges of devil worshipping were brought to the Americas to break the resistance of the local populations, justifying colonization and the slave trade...providing for capital the seemingly limitless supply of labor necessary for accumulation" (Federici, 2014, p. 198). The witch hunts in Europe provided a model for the colonization of the Americas, a means of subordinating women to reproduction of labor-power and disrupting European peasants' relationships with earth and their labor, which was then brought to the African and American continents in the form of conquest and the slave trade.

Destroying property and reclaiming control of reproduction are thus constructed as criminal acts of violence by the state because both ultimately threaten enclosure, production, and white supremacy, while acts of state-sanctioned mass starvation, imprisonment, slavery, torture, and genocide are normalized because they enable enclosure and are how the system maintains itself.

Witches of the Middle Ages were women who resisted professional medicine's enclosure of their bodies, reproductive systems, healing practices, midwifery, herbal medicine, knowledge, and land. Earth Liberation Elves explicitly resist enclosure of the earth and its creatures and have taken action against the rape of the earth by sabotaging genetically engineered crops, slaughterhouses, vivisection labs, and in the face of racist rantings about population control, redirecting the attack back to capitalism. We see these acts of resistance by Elves and witches as acts of reproductive justice "that make critiques of capitalism and criminalization central to the analysis rather than simply expand either pro-choice or pro-life frameworks" (Smith, 2005). The "heathens" of white nationalism instead align themselves with a politics of rape, "might makes right," misogyny, eugenics, criminalization, and genocide (Alderman, 2017). They construct their violence and hatred as natural and good, and the birth of black and brown babies as white genocide. They are not proponents of reproductive justice.

In order for magic to be liberatory, it must be in alignment with reproductive justice.

Movements that truly challenge capitalism must always be movements for reproductive justice.

We demand an anti-capitalist, anti-racist, pro-planet, pro-earthling redefinition of labor and our lives. We refuse to "work" for capitalism and white supremacy. We co-labor another world.

References

Alderman, J. (2017, January 19). The rape-promoting, misogynist harassers who are a key part of Trump's "alt-right" alliance. *Media Matters for America*. Retrieved from https://www.mediamatters.org/research/2017/01/19/rape-denying-misogynist-harassers-who-are-key-part-trumps-alt-right-alliance/215055

Best, S., Nocella, A. J., II, Kahn, R., Gigliotti, C., & Kemmerer, L. (2007). Introducing critical animal studies. *Journal for Critical Animal Studies, 5*(1), 1–2.

brown, a. m. (2017). *Emergent strategy: Shaping change, changing worlds*. Chico, CA: AK Press.

Canfield, N. (2017, January 4). Racism and Paganism: Exposing the racist sects, prejudice, and discrimination in Paganism. *Exemplore*. Retrieved from https://exemplore.com/paganism/Racism-and-Paganism-Exposing-the-Racist-Sects-Prejudice-and-Discrimination-in-Paganism

Davis, K. (1995). Thinking like a chicken: Farm animals and the feminine connection. In C. J. Adams (Ed.), *Animals and women: Feminist theoretical explorations* (pp. 192-212). Durham, NC: Duke University Press.

Earth Liberation Front. (1997). Communiqué. *III Publishing*. Retrieved from http://www.iiipublishing.com/elf.htm

Esteva, G., & Prakash, M. S. (1998). *Grassroots postmodernism: Remaking the soil of cultures*. London: Zed.

Federici, S. (2014). *Caliban and the witch: Women, the body and primitive accumulation*. Brooklyn, NY: Autonomedia.

Fisher, M. (1992). Bury me furiously. *ACT UP NY*. Retrieved from http://www.actupny.org/diva/polfunsyn.html

Gibson-Graham, J. K. (2006). *A postcapitalist politics*. Minneapolis, MN: University of Minnesota Press.

Halberstam, J. (2011). *The queer art of failure*. Durham, NC: Duke University Press.

Ignatiev, N. (1994, June/July). *Treason to whiteness is loyalty to humanity: An interview with Noel Ignatiev of Race Traitor magazine*. Excerpted from the anarchist tabloid THE BLAS.

Imarisha, W. (2015, February 11). Rewriting the future using science fiction to re-envision justice. *Bitch Media*. Retrieved from https://www.bitchmedia.org/article/rewriting-the-future-prison-abolition-science-fiction

Ledgerwood, M. (2017). Sustainable communities public presentation.

Mies, M., & Shiva, V. (1993). *Ecofeminism*. Halifax: Fernwood Publications.

Nocella, A. J., II. (2014). Anarchist criminology against racism and ableism and for animal liberation. In A. J. Nocella II, R. J. White, & E. Cudworth (Eds.), *Anarchism and animal liberation* (pp 40-58). Jefferson, NC: McFarland & Company, Inc.

Okun, T., & Jones, K. (2001). White supremacy culture. In *Dismantling racism: A workbook for social change groups*. ChangeWork. Retrieved from http://www.cwsworkshop.org/PARC_site_B/dr-culture.html

Pellow, D. N. (2014). *Total liberation: The power and promise of animal rights and the radical earth movement.* Minneapolis, MN: University of Minnesota Press.

Rankine, C. (2016). The condition of black life is one of mourning. In J. Ward (Ed.), *The fire this time* (pp. 145-155). New York, NY: Scribner.

Rosebraugh, C. (2004). *Burning rage of a dying planet: Speaking for the Earth Liberation Front.* New York, NY: Lantern Books.

Sepúlveda, J. (2005). *The Garden of peculiarities.* Los Angeles, CA: Feral House.

Showing Up for Racial Justice, Nashville. (2017). Retrieved June 20, 2017 from https://surjnashville.org/

Smith, A. (2005) Beyond pro-choice versus pro-life: Women of color and reproductive justice. *Feminist Formations, 17*(1), 119–140.

Socha, K. (2012). *Women, destruction, and the avant-garde: A paradigm for animal liberation.* Amsterdam: Rodopi.

Spencer, P. (2016, November 18). Trump's occult online supporters believe "meme magic" got him elected. *Motherboard.* Retrieved from https://motherboard.vice.com/en_us/article/pgkx7g/trumps-occult-online-supporters-believe-pepe-meme-magic-got-him-elected

Starhawk. (2016). *City of refuge.* San Francisco, CA: Califia Press.

Taylor, B. (2002). Diggers, wolfs, ents, elves and expanding universes: Bricolage, religion, and violence from earth first! and the earth liberation front to the antiglobalization resistance. In J. Kaplan & H. Lööw (Eds.), *The cultic milieu: Oppositional subcultures in an age of globalization* (pp. 26-74). Lanham, MD: Altamira/Rowman and Little field.

Taylor, B. (2005). Earth First! and the Earth Liberation Front. *The encyclopedia of religion and nature.* Retrieved from http://www.religionandnature.com/ern/sample/Taylor—EF!andELF.pdf

Taylor, B. (2008). From the ground up: Dark green religion and the environmental future. In D. K. Swearer (Ed.), *Ecology and the environment: Perspectives from the humanities* (pp. 89-107). Cambridge, MA: Harvard University Press.

Thompson, A. K. (2010). *Black mask, white riot: Anti-globalization and the genealogy of dissent.* Oakland, CA: AK Press.

Vedantum, S. (2017). Remembering Anarcha, Lucy, and Betsy: The mothers of modern gynecology. *Hidden Brain.* NPR. Retrieved from http://www.npr.org/2017/02/07/513764158/remembering-anarcha-lucy-and-betsey-the-mothers-of-modern-gynecology

14. *Problematising Non-violent "Terrorism" in an Age of True Terror: A Focus on the Anarchic Dimensions of the Earth Liberation Front*

RICHARD J. WHITE

Introduction

As the 21st century progresses it has, regrettably, become something of a truism to say that we are living through a time of crisis. Undoubtedly, the intersectional struggles that animate social and spatial justice approaches for liberation are responding to a catastrophic set of unprecedented economic, social, environmental, and political turmoil that pose a real threat to end the world as we know it (Shannon, 2014). The latest geological epoch, the Anthropocene, emerges as the terrifying realisation of generation upon generation upon generation of reckless anthropocentric violence and—in its true meaning—terrorism that has, and continues to, plundered and ravaged the living world, and devastated those vital support systems which sustain all life on Earth. Here it is also important to emphasise that the vast majority of this desecration that haunts humans, nonhuman animals, and the natural world has been legal. To think therefore that the injustices and crises in the world can be solved by appealing to the State, a political elite that is in the thralls of capitalism, is dangerously utopian, naive, and futile (Wade, 2003). As Best and Nocella (2006, p. 8) observed:

> Barely out of the starting gates, on the hells of the bloody and genocidal century that preceded it, the 21st century already is a time of war, violence, environmental disasters, and terrorism against human populations, animals and the Earth as a whole. This omnicidal assault is waged by powerful and greedy forces, above all,

> by transnational corporations, national and international banks, and G8 alliances. Stretching their tentacles across the Earth, they hire nation states as their cops, juntas, hit men, dictators, and loan sharks to extract natural resources, enforce regimes of total exploitation, and snuff out all resistance. These menacing foes are part of a coherent system rooted in the global capitalist market currently in the final stages of privitization and commodification of the natural and social worlds.

This suppressed reality has long been understood by critical minority of the population, particularly by anarchists who have long concluded that pre-figurative praxis and other forms of non-violent direct action (legal or illegal) are the only option left to challenge and confront the commodification of life and emancipate space (see Springer, 2016). Yet, in rejecting the spectacle of democracy, the subversive path of direct action is a dangerous and precarious one to walk down. It is no surprise that amidst this Anthropogenic and dystopian nightmare, those who are motivated in ways that promise to end these cycles of violence, by bringing new, transformative, and healing forms of justice, compassion, care, love, and liberation into human and more than human worlds, are demonised and vilified by a hostile agenda set by those political and economic elites whose vested interests are threatened by this type of direct activism. The chapter draws particular attention towards one extremely powerful word, one that is appallingly abused and misappropriated from its true meaning: terrorist. For decades now—but more powerfully over the last ten years (see Hirsch-Hoefler & Mudde, 2014; Joosse, 2012; Loadenthal, 2013a)—those who engage in non-violent direct action are repeatedly accused as adopting an "extreme" stance. This works powerfully in impressing upon the (Anglo-American) public imaginary that these social justice activists are extremists; people to be treated with suspicion, a potentially dangerous, feared, and a plausible threat (to safety and security) and thus deserving of political and corporate repression and punitive justice. Indeed, in our increasingly Orwellian world any distinction between identification with extremism and terrorism has been severely eroded, just as social justice activists are successfully rebranded and repackaged as extremists; then more illegal or unlawful forms of direct activism constitute a terroristic threat (Joosse, 2012; Leader & Probst, 2003). A deeply troubling element here though is that vast majority of these illegal forms of social justice activism are explicitly and unconditionally non-violent. A critical reading of legalistic narratives of terrorism, and definitions of terrorism, exposes a—quite frankly—appalling and morally abhorrent easy-going equivalence in place between "violence against persons" and "violence against property". The understanding of domestic terrorism in this testimony of James F. Jarboe to the FBI (2012, n.p.) illustrates the point:

> Domestic terrorism is the unlawful use, or threatened use, of violence by a group or individual based and operating entirely within the United States (or its territories) without foreign direction, committed against persons or property to intimidate or coerce a government, the civilian population, or any segment thereof, in furtherance of political or social objectives.

The testimony continues:

> During the past several years, special interest extremism, as characterized by the Animal Liberation Front (ALF) and the Earth Liberation Front (ELF), has emerged as a serious terrorist threat. Generally, extremist groups engage in much activity that is protected by constitutional guarantees of free speech and assembly. Law enforcement becomes involved when the volatile talk of these groups transgresses into unlawful action.
>
> Special interest terrorism differs from traditional right-wing and left-wing terrorism in that extremist special interest groups seek to resolve specific issues, rather than effect widespread political change. Special interest extremists continue to conduct acts of politically motivated violence to force segments of society, including the general public, to change attitudes about issues considered important to their causes. These groups occupy the extreme fringes of animal rights, pro-life, environmental, anti-nuclear, and other movements. Some special interest extremists—most notably within the animal rights and environmental movements—have turned increasingly toward vandalism and terrorist activity in attempts to further their causes. (Jarboe, 2012, n.p.)

This chapter focuses explicitly on the "special interest group", the Earth Liberation Front (ELF), which has "become the most active and the most destructive environmental terrorist [sic] group in the United States". It begins by emphasising the key ideologies, approaches, and forms of organisation that are congruent with anarchist praxis, another woefully abused and misunderstood tradition. In emphasising its anarchic dimensions, the aim here is to encourage greater critical understanding and awareness of the ELF in ways that problematises and counters the visceral attacks made on it by state-capitalist organisations. The second and greater part of the chapter makes the case for deconstructing dominant narratives of terrorism (Loadenthal, 2013b) that uncouples its attribution to the ELF and related radical non-violent social justice movements. This interrogation of the misuse of terrorism is embedded on moral grounds: calling for "property damage" to be excluded from future definitions of terrorism. For Ackerman (2003, p. 162) observes: "While the ELF has caused millions of dollars' worth of property damage, it has not yet intentionally (or even unintentionally) brought harm to anyone."

Should this truth no longer be the case in future, then all bets are off. Should any person be deliberately injured or worse through the tactics of social justice activism, and that true terrorist tactics "against people" would be applicable, this would be a devastating turn of events. While the moral line between people and property is held as an absolute then, I believe, there is at this time a real opportunity to (i) expose the ugly and unjustifiable connection of property and people currently in place; (ii) move social justice activism out of the considerable shadow that "terrorism" cast, and (iii) strengthen and extend their support among a much wider and more mainstream societal base. To this end eco-activists—to be consistent with anarchist praxis of non-violence—need to redouble all efforts to ensure that in the face of utmost provocation they maintain their tremendous commitment to non-violence and non-coercive forms of direct action. Given the need to avoid any element of doubt or risk here, the chapter argues that the use of arson and incendiary devices as part of repertoire of non-violent direct action must be re-considered.

To seek effective ways that carry with the promise of liberation of "social justice activists" from the accusation of "terrorism" is not a trivial concern. Indeed the actions of the ELF, in common with all social justice activists, are done so in the hope of changing hearts and minds of those across Western and North American society, whose political and economic elites continue to be at the epicentre of such destruction unleashed on life. Given this, then unshackling them from false accusation of terrorism is of the greatest priority, insofar as public reappraisal here may inspire the cumulative wider social changes necessary to move towards a post-capitalist/post-crisis world hoped for. The need to urgently revisit dominant narratives of terrorism is also a moral imperative in a society that is being traumatised by deliberate human on human acts of violence, in other words "true" terrorism. Any definition, or application, of terrorism that equates "people and property" needs to be called out for the sham it is.

The Earth liberation Front and Anarchism

> Focus on one problem and put your heart and soul into that one thing. Don't rat out your comrades and do no harm to all living beings; that includes Mother Earth. If you do choose to practice civil disobedience, be prepared to go to jail if you're busted. But keep in mind, you won't be an effective "ecommando" or activist behind bars. Think for yourself! Don't follow leaders. Good luck.... (ELF webpage, 2017, n.p.)

The Environmental Life Force, or "original ELF", was founded by John Hanna and Carla Susan Olander in March 1977 and disbanded in 1978 (Anon, 2011). The contemporary ELF emerged in the UK in the early 1990s; its initial communiqué stands both as a powerful critique of the violence unleashed against the Earth and as a clear-cut raison d'être for the ELF:

> Beltane, 1997
> Welcome to the struggle of all species to be free. We are the burning rage of this dying planet. The war of greed ravages the Earth and species die our every day. ELF works to speed up the collapse of industry, to scare the rich, and to undermine the foundations of the state.... We embrace social and deep-ecology as a practical resistance movement.
>
> We have to show the enemy that we are serious about defending what is sacred. Together we have teeth and claws to math our dreams. Our greatest weapons are imagination and the ability to strike when least expected.
>
> Since 1992 a series of Earth-nights and Halloween smashes has mushroomed around the world. 1,000s of bulldozers, power lines, computer systems, building and valuable equipment have been composted. Many ELF actions have been censored to prevent our bravery from inciting others to take action.
>
> We take inspiration from Luddites, Levellers, Diggers, the Autonome squatter movement, the ALF, the Zapatistas, and the little people—those mischievous elves of lore. Authorities can't see us because they don't believe in elves. We are practically invisible. We have no office, just small groups working separately seeking vulnerable targets and practicing our craft.
>
> Many elves are moving to the Pacific Northwest and other sacred areas. Some elves will leave surprises as they go. Find your family! And let's dance as we make ruins of the corporate money system. (quoted in Pickering, 2009, p. 163)

The ELF is one of a constellation of so-called radical or dissent groups and "organisations" (though, as Loadenthal (2013b, n.p.) observes: "The ELF is not an organization in the traditional sense and is more akin to a movement of informal networks") that change the world through engaging in non-violent direct action. Their raison d'être is simple: to protect life on earth from being exploited and terrorised and violated (by humans). Other related radical environmentalist movements (REM) here would include the Animal Liberation Front (ALF; who have often issued joint communiqués and expressions of solidarity with the ELF), and others ranging from Earth First!, the Sea Shepherd Conservation Society, and hunt saboteurs (Somma, 2006). Interestingly in the context of the chapter, what defines key differences between these dissident groups are often their contrasting response to the question as to "what constitutes appropriate and justifiable forms of direct action". This was certainly the case in the recent history of the ELF, as Best and Nocella (2006, p. 19) note:

> Breaking from the constraints of U.K. Earth First! in order to employ ALF-style sabotage tactics, The Earth Liberation Front formed in the early 1990s, and spread like bushfire throughout Ireland, Germany, France, Eastern Europe, Australia, the U.S. and elsewhere.

In important ways, some more explicit and pronounced than others, the approach and success of the ELF (in common with other radical environmental and animal liberation activists) can be understood as an expression of, and a testament to, anarchy in action (Ward, 1973). Certainly a close identification of anarchism as key influence within the ELF has been noted many times (Nocella, White, & Cudworth, 2015). But what is meant by anarchism in this context? The observations given by Pellow (2014) following his empirical research on radical activists are instructive:

> The type of anarchism most interviewees expressed to me was not stereotypical—the public protest often dismissed as youthful rebellion, outfitted with black clothing, red bandanas, and passionate shouts. These anarchists oppose the state, but primarily because they reject authoritarian rule, repression, and the primacy of property rights over the needs of all living being. Instead, they prioritize democratic decision making and cooperation, mutual aid and assistance, and community building among ordinary people. (pp. 94–95)

Of course, anarchism, by virtue of its indomitable spirit of revolt and freedom, has long been vilified and abused by the propaganda spewed by those in position of authority and hierarchy, and those who seek to profit from the exploitation and oppression of those weakened by the inequalities of power that flow from anarchist structures (see Goodway, 1989; Mac Laughlin, 2016). The abuse of language is—unsurprisingly—prevalent here; one needs to think only of the common reading of anarchy and anarchism as a synonym for chaos, violence, nihilism, and of course terrorism. Yet for many others, the term anarchist has brought much needed solidarity, strength, and support to those who desire to advance social and spatial justice in the here and now, and offer new visions of hope and possibilities for post-crisis, post-capitalist worlds (see Souza, White, & Springer, 2016; Springer, White, Souza, 2016; White, Springer, & Souza, 2016). Indeed, the influence of anarchist praxis here—in comparison to all the other so-called radical and dissident traditions—is not unexpected: anarchism has consistently emphasised and recognised the intersectional tapestry of violence and oppression that weaves its common threads through life—whether human, nonhuman animal, or more than human worlds (see Nocella et al., 2015). The framing of violence in this intersectional way and the call to end these forms of oppression neither by appealing to the state nor by adopting a politics of waiting but through direct action and pre-figurative praxis; non-violence; small-scale autonomous and horizontal forms of organisation are all identifiable within the ELF. "The lack of organization also seems to fit the anti-authoritarian orientation of many ELF activists" (Leader & Probst, 2003, p. 39).

Far from the popular stereotype of being disorganised, anarchist praxis emphasises horizontal forms of voluntary organisation and commitment that are voluntary, in contrast to hierarchical modes of organising maintained by appealing to authority (and its power to threaten, intimidate, coerce, tyrannise, and indeed terrorise). This praxis—the appeal to autonomous, self-organisation, and cooperation—mirrors the organisation of the ELF. Expanding on how this works in practice, the North American Earth Liberation Front Press Office (NAELFPO, n.d., pp. 2–3) state:

> The ELF is organized into autonomous cells which operate independently and anonymously from one another and the general public. The group does not contain a hierarchy or a sort of leadership. Instead the group operates under an ideology. If an individual believes in the ideology and follows a certain set of guidelines she or he can perform actions and become a part of the ELF.

This radical, de-centralised mode of organising through leaderless communiqués and the rejection of a single figure-head has been a considerable advantage within the ELF, as it sidesteps and avoids being repressed by the conventional approach adopted by mainstream government and intelligence agencies. As part of the same testimony quoted earlier, Jarboe (2012, n.p.) notes the acknowledgement of the success of these (anarchist) modes of organising (alongside the easy-going language that brings "eco-terrorism" and "criminal enterprises" together) and how these descriptions stand or fall by the discussion on rationales that follow shortly:

> Currently, more than 26 FBI field offices have pending investigations associated with ALF/ELF activities. Despite all of our efforts (increased resources allocated, JTTFs, successful arrests and prosecutions), law enforcement has a long way to go to adequately address the problem of eco-terrorism. Groups such as the ALF and the ELF present unique challenges. There is little if any hierarchal structure to such entities. Eco-terrorists are unlike traditional criminal enterprises which are often structured and organized.

The small temporal nature of such social and voluntary organisation, based on strong bonds of trust and familiarity, insulates not only against easy infiltration from police and state agencies, but also from activists who turn informants.

This argument was strongly emphasised in *Resistance Magazine: Journal of the Earth Liberation Movement* (cited in Deshpande & Ernst, 2012, p. 3):

> Most every indictment of earth and animal liberationists has come about through snitches and government informants. This makes it all the more important for one to carefully select who s/he decides to work with. As a simple matter of statistics, you're most likely to be betrayed by someone you've worked with...and the fewer cooks in the kitchen the fewer people there are to stab you in the back.

Another darker and ethically disturbing reality is worth drawing attention to in this context: that of undercover police operations, which has frequently involved police infiltrating activist groups by posing as activists. Focused on covet police practices involving animal and environmental activist groups over the last forty years, Lubbers (2015, p. 338) notes that these have practices including "withholding of exculpatory evidence; the tricking of women (and men) into intimate or even sexual relationships with undercover agents; the siring of subsequently unsupported children by undercover officers under false identities (Wistrich, 2013, pp. 1-2); identity theft from dead children (Home Affairs Committee, 2013); and active planning of and participation in serious crimes, including arson" (Lucas, 2012). To appreciate the sense of anger, torment, violation, and devastation felt by those whose trust has been abused, I'd encourage readers to look up Bob Lambert, sent by the UK Metropolitan Police to infiltrate the animal rights and radical environmental movement in the 1980s (see Campaign Opposing Police Surveillance, 2016; Casciani, 2014; Evans & Lewis, 2011; Loadenthal, 2014). Lambert entered into a long-term sexual relationship with "Jacqui", a young animal rights activist, fathered a child with her, and then disappeared. It was 24 years later that she discovered, via the newspapers, that he was a police infiltrator working a few miles away from her. Jacqui (speaking on behalf of herself and other women who have been similarly conned by undercover police) has said: "We are psychologically damaged; it is like being raped by the state" (Lewis, Evans, & Pollak, 2013, n.p., italics added).

It would be dreadfully remiss of me not to draw attention—not least as a precursor to the focus of the next section—to the fact that an ongoing Scotland Yard investigation is due to conclude later this year as to whether Lambert was responsible for planting an incendiary device in a high street store in Harrow London. This was one of three devices simultaneously planted in three Debenhams stores in July 1987, in protest at their selling of fur. The incendiary devices were designed to be set off at night when the stores were closed, with the intention of triggering the sprinkler system. The intention was to cause economic damage by ruining the stock of (fur) clothes and garments that the store sold at that time. An estimated £8,000,000–£9,000,000 pounds in store damage and lost revenue resulted from this action. Debenhams stopped selling furs. Based on Lambert's information, and subsequent raids by anti-terrorist police, two other activists involved in planting the devices were arrested and prosecuted (Evans, 2017, n.p.). The third activist has never been caught.

Earth Liberation Front and Non-violent Direct Action

The ELF has always held a clear and unconditional respect and reverence for life and have strived to ensure that their strategies and tactics reflect this. As Ackerman (2003, p. 145) observes:

> [The ELF] has a long-held belief in not causing harm to any life—an ideology to which many radical environmentalists subscribe, teaches that all life (including that belonging to human beings) is sacred and cannot be harmed. The ELF's guidelines explicitly state that members must take "all necessary precautions against harming life". The ELF has long held that it is not a violent organization, a belief that is probably still regarded as central by many of its members.

Popular types of activism engaged by the ELF have euphemistically been referred to as monkey wrenching (Weignant, 2017). Monkey wrenching would include:

> acts of sabotage and property destruction against industries and other entities perceived to be damaging to the natural environment. "Monkeywrenching" includes tree spiking, arson, sabotage of logging or construction equipment, and other types of property destruction. (Long, 2004, p. 259)

As far as sentient life is concerned then ELF forms of direct action have—by any reasonable and proper definition of violence—been of a non-violent nature. Given this, the ongoing accusations of engaging in extremist or terrorist tactics, at a time when deliberate acts of violence are being used to terrorise (human) lives, are—as previously mentioned—morally reprehensible. Indeed when the economic and moral rationale(s) that inform ELF actions are understood, then the accusation of terrorism held against them becomes even less justifiable.

Regarding economic rationales, given the defendable argument that the mass exploitation and destruction of nature/the natural world are driven by a capitalist imperative, that is, the need to make profit, then a rational economic response would be to underpin these profit margins. Thus, as Leader and Probst (2003, p. 37) argue: "Their tactics emphasize attacks on property not people and include arson, sabotage, and vandalism designed to cause significant economic damage." The moral argument in comparison draws on the need to aid, support, and protect life that is under threat and unable to defend itself/themselves. Given the argument that some (humans) are quite literally bringing war—and terrorism—to nature then both economic and moral arguments fuse closely together, as Pickering (2007, pp. 89–90) observes:

> The Earth Liberation Front does not commit merely symbolic acts to simply gain attention to any particular issues. It is not concerned merely with logging, genetic engineering, or even the environment for that matter; its purpose is to liberate the earth.
>
> The Earth, and therefore all of us born to it, are under attack. We are under attack by a system which values profit over life, which has, and will, kill anything to satisfy it's never ending greed. We have seen a recent history rich in the destruction of peoples, cultures, and environments. We have seen the results of millions of years of evolution destroyed in the relative blink of an eye.

Thus, "in defence of the Earth, the ELF burned down housing complexes under construction, torched SUVs and ski lodges, and ripped up biotech crops" (Best & Nocella, 2006, p. 19). Here the focus on destroying property can be justified on both economic and moral grounds. Unpacking the (less obvious perhaps) moral appeal, if an object is designed to facilitate and damage and bring suffering to life in future, then is there not a moral obligation to destroy or disable these infrastructure and objects and the wider infrastructures that support them in the present?

When read against the key rationales for direct action, to protect (future) loss of life and environmental devastation, and maximise economic damage to those companies that prosper and profit from environmental destruction and devastation, then the tactics of the ELF have had a significant impact where they have taken place:

> The ALF and the ELF have jointly claimed credit for several raids including a November 1997 attack of the Bureau of Land Management wild horse corrals near Burns, Oregon, where arson destroyed the entire complex resulting in damages in excess of four hundred and fifty thousand dollars and the June 1998 arson attack of a U.S. Department of Agriculture Animal Damage Control Building near Olympia, Washington, in which damages exceeded two million dollars. The ELF claimed sole credit for the October 1998, arson of a Vail, Colorado, ski facility in which four ski lifts, a restaurant, a picnic facility and a utility building were destroyed. Damage exceeded $12 million. On 12/27/1998, the ELF claimed responsibility for the arson at the U.S. Forest Industries Office in Medford, Oregon, where damages exceeded five hundred thousand dollars. Other arsons in Oregon, New York, Washington, Michigan, and Indiana have been claimed by the ELF. Recently, the ELF has also claimed attacks on genetically engineered

> crops and trees. The ELF claims these attacks have totaled close to $40 million in damages. (Jarboe, 2012, n.p.)

Arguably the most destructive and controversial practice that the ELF has used is that of arson. It should also be recognised though that when arson has been employed as an ELF tactic, particularly as this truth is (deliberately) excluded when reporting on this in the public domain, meticulous pre-activity surveillance and planning have been undertaken to ensure that the fire does not harm human (or nonhuman) life.

> [Here the] ELF took extraordinary measures to avoid loss of life or injury. The devices were designed so only the low-yield detonators would fire. The napalm mix had been allowed to solidify so it could not catch fire. The fuses were timed to ignite at 2:00 am. I waited nearby until all the detonators exploded. If someone would have happened by, I was prepared to warn him or her off, even at the risk of capture. Later in the day, a communiqué was dropped at the local newspaper. ELF listed viable alternatives to the excessive and inappropriate use of pesticides on our food. (Anonymous, 2001, n.p.)

It is not then through sheer luck or fortune that ELF tactics—even the use of arson—have not resulted in the physical harm of a single human being. As the NAELFPO note:

> The guidelines for the ELF specifically require members to take all necessary precautions to ensure no one is physically injured. In the history of the ELF internationally no one has been injured from the group's actions and that is not a coincidence. Yes, the use of fire as a tool is dangerous but when used properly it can tremendously aid in the destruction of property associated with the killing of life. (NAELFPO, n.d., p. 27)

However, the element of unpredictability intrinsic to arson/responding to arson is sufficient to argue ever more strongly against its use in the future. Before continuing it is important to note that I am all too aware (having seen violence against vulnerable populations and ecosystems first hand) that the discipline to maintain non-violent actions when bearing witness to the human web of violence, abuse, suffering, and desecration weaved by relentless (capitalist) exploitation of the life takes incredible strength and heroic restraint. However, the need at this time of true terrorism to uncouple ELF from this shameful stigma and cause a popular reappraisal of this group necessitates towards avoiding any elements of doubt and risk. The focus here on arson echoes that of one of the founders of the original ELF, John Hannah, that he gave in an interview.

> Question: What would you like to say to the ELF today?
>
> Answer (John Hannah): If I transport myself back to when I was Underground, I don't think I would have listened to an old fart like me. Most likely a lot of the people who make up today's ELF weren't even born when ELF was founded. So I'm not too optimistic that the current cadre will listen. But here's my request: Stop the violence. It's only a matter of time before someone gets injured or killed. Arson can get out of hand very quickly. Who would want an innocent firefighter to get killed doing his or her job? I'm so thankful no one was hurt during my activities. I couldn't live with myself had that happened.

As anarchists and others have repeatedly pointed out, violence often plays into the hands of the state: the state—a violent entity in itself—knows how to fight violence. As Hannah also observed:

> Regardless of the frustration we all feel about the enormous perils facing our Mother Earth, engaging the perceived wrong-doers with threats, intimidation and destructive tactics will always fail. Fighting fire with fire will get you burned. (Anonymous, 2017, n.p.)

To continue to hold the moral grounds of non-violence, just as decentralised (anarchist) forms of organisation in the ELF have confounded the ability of "the state" to effectively suppress and close down, holds a significant tactical advantage. As part of this discussion it is also important to state that: "the moment violence enters the equation of whatever social action is being called forth under the name of 'anarchism', it ceases to actually be anarchism" (Springer, 2014, p. 86). Avoiding violence and rejecting coercion, maintaining consistency between the means and the end is absolutely central: as the Italian anarchist Baldelli (1971, p. 20) argues, "Let the tree be judged…by what it feeds upon, the so-called means."

This argument notwithstanding, one should never forget that to be accused of engaging in non-violent action deemed "terrorism" in the eyes of the state comes with a huge cost vis-à-vis imprisonment, which once again suggests that economic sabotage against property is a more serious offence than violence against people. The extensive period of incarceration that is brought to those found guilty of eco-terrorism not only affects the individuals involved, but also the wider networks, such as family, friends, and so on (see Deshpande & Ernst, 2012; Harper, 2003). As Hannah noted—pay particular attention to the references to "the Feds" (Anonymous, 2017, n.p.),

> Those who are now serving prison sentences are effectively removed from the battle to save our planet. We are all losers in that regard, even the Feds. If anything can be learned from Operation Backfire, it is the necessity to channel our frustrations and concern for earth's welfare into positive direct action. Build—

> don't destroy. Build consensus and public support. Get an education and build a better world and future.

Securing wider public support is absolutely central if the suffering and violence that haunts the Earth is to end. The darker truths that the ELF (and other REM) has unearthed through their activism carry with them the real potential of encouraging greater support and solidarity in ways that can have this revolutionary impact. As NAELFPO (n.d., p. 4) states,

> If people are serious about stopping the destruction and exploitation of all life on the planet then they must also be serious about recognizing the need for a real direct action campaign and their own personal involvement.

In this regard, as well as firmly apportioning blame for the devastation of the human and natural worlds at the door of particular individuals and companies, the ELF has also stimulated a deep intersectional awareness among social justice groups: as Becker (2006, p. 77) argues:

> (T)he ELF is among the few groups the few groups to forcefully bring to the attention of millions of people such a basic and important insight regarding the techno-corporate matrix. Consistently, the ELF, ALF, and other revolutionary forces criticise single-issues environmental organisation for failing to understand the systemic and urgent nature of the homicidal assault on the Earth by corporate technics.

Successful forms of activism and protest in the 21st century will undoubtedly be those that recognise the interconnected natures of social and spatial struggles for justice and liberation. The time to embrace and engage with a politics of total liberation, which refers to "the theoretical process of holistically understanding movements in relation to one another, to capitalism, and to other modes of oppression, and to the political process of synthetically forming alliances against common oppressors, across class, racial, gender, and national boundaries, as we link democracy to ecology and social justice to animal rights" (Best, 2010, n.p.) is now. In this respect, arguably the most successful and inspirational forms of direct action that embraced an intersectional praxis have been those carried out by the ELF and ALF (see Nocella, Sorenson, Socha, & Matsuoka, 2010).

Undoubtedly, one of the real threats of re-thinking the definition of terrorism, in a way that fully differentiates between "property" and "people", is that this will encourage a more positive reappraisal of the ELF, what it stands for, its advocacy of non-violence (i.e. non-terrorist activity) by the mainstream media (see Joosse, 2012) and everyday citizens. This is a re-imagining that in turn promises deeper insights into not only the need for direct action to protect (innocent) life on this planet from harm and violence from both capitalism; but also in the complicity in creating, and profound limits in preventing, these violent geographies to wreck such havoc and destruction. It was always the case that, "the great emancipatory gains for human [and more than human] freedom have not been the result of orderly, institutional procedures but of disorderly, unpredictable, spontaneous action cracking open the social order from below" (Scott, 2012, p. 141).

Conclusion

Drawing reference to the (anarchist) praxis consistent with the ELF, from its organisational structure to its use of communiqués, to its championing of direct action as an important moral and political strategy, it is hoped that a better broader understanding of the group has been made. The principal thrust of the chapter though has been to maintain a critical focus on how the ELF, despite its explicit narrative and history of non-violent direct action, continues to be criminalised in the most extreme way by law enforcement agencies, and judged and condemned by a wider public by being labelled a "terrorist" organisation. In the present Orwellian environment of double speak, where anarchism is used as a synonym for violence and chaos and nihilism, and government stands for peace, freedom, prosperity and justice, it is of little surprise to see how those who transgress the "accepted" ways of changing the world (e.g. through state-sanctioned representative democracy) continue to be vilified, abused, and condemned. But this must end! How can any definition of "terrorism" that creates equivalence of human life with property be allowed to stand? The present moment in time provides an opportunity to draw attention to this, particularly with the intent of sparking a new consciousness among a mainstream audience. The ELF, by any fair and just definition of terrorism, cannot be considered a terrorist organisation.

As I write this conclusion, on 6 June 2017, BBC news coverage of the death of seven civilians in London plays out on the television. Their deaths were the result of a terrorist attack, involving a van used deliberately to run people down on the sidewalk, and then three men stabbing anybody in their vicinity, with the single intent to wound and kill. This latest act of terror follows (20 May 2017) the atrocity caused by a suicide bomber who targeted young children attending an Ariana Grande concert at Manchester Arena, UK. His actions resulted in the death of at least 22 people and injured 116 more. Across Europe more generally, since 2015 other terror attacks have taken place in Paris (20 April 2017), Stockholm (7 April 2017), London (22 March 2017), Paris (3 February 2017), Berlin (19 December 2016),

Normandy (26 July 2016), Nice (14 July 2016), Brussels (22, March 2016), and Paris (13 November 2015). The Islamic State militant group (ISIS) has claimed responsibility for all these attacks, which stand as appalling examples of true terrorism, which continues to cast a dark shadow across the (Western) world. At this time of crisis, when true terrorist tactics are increasingly present in the lived realities of (Western) citizens and urban society it is absolutely critical that the ELF, consistent with anarchist appeals, stays true to commitment to respect and not endanger (human) life. Almost fifteen years ago, Ackerman (2003, p. 62) pointed out that:

> While the ELF has caused millions of dollars worth of property damage, it has not yet intentionally (or even unintentionally) brought harm to anyone. With the plethora of current threats to national security in the US, it is essential to devote our limited investigative and law enforcement resources towards addressing the most pressing threats.

Faced with an unprecedented threat of true terrorism in 2017, and a public urgency to respond effectively to this, the attention and resources (time, money, intelligence) previously been invested by Western governments into pursuing the ELF are untenable. Things must change. Indeed, despite the toxic propaganda that has created equivalence between radical, non-violent social and environmental justice movements with terrorism in the public imaginary, there is much to be optimistic about here: the cracks are becoming increasingly apparent. Ultimately, as Becker (2006, p. 71) argued, it will become clear that:

> History is on the side of the Earth Liberation Front. ELF communiqués demonstrate a fundamental critique of contemporary technology and global capitalism, and a radical reassessment of human relations with one another and the natural world.

Looking confidently towards the future, such a radical reassessment holds two promises vis-à-vis re-thinking the definition of terrorism. Removing "property" will mean that the revised definition of terrorism falls well short of the ELF and other intersectional social justice activists. But this is a minimal revision. A radical reassessment will see the definition extended in a different way. This will, ultimately, recognise the intrinsic rights of all sentient beings to life, and the natural world more generally. Crucially we are not indulging in utopian thought. In 2010 Bolivia passed The Law of the Rights of the Mother Earth, the world's first laws granting equal rights of all nature to humans (Vidal, 2011). More recently, in 2017 the Ganges River in India and the Whanganui River in New Zealand were granted the same legal rights as human beings (Roy, 2017). Granting rivers rights might inspire us to revision what counts as terrorism and, perhaps, not only legitimatize the ELF, and other so-called "radical" social and environmental justice moments, but also provide more positive awareness and support for this type of direct action? Of course, it will also condemn and criminalise those whose actions profit from the intentional abuse, exploitation, and terrorising of nonhuman and more than human worlds. They will, quite rightly, as those will before them engaging in forms of terrorism. Acknowledging this truth would be a monumental step forward towards achieving post-capitalist, post-crisis worlds built on mutual relationships animated by compassion, love, and beauty and, of course, justice and non-violence.

References

Ackerman, G. A. (2003). Beyond arson? A threat assessment of the Earth Liberation Front. *Terrorism and Political Violence, 15*(4), 143–170.

Anonymous. (2011). *Interview with ELF founder, John Hanna*. Retrieved from http://www.originalelf.com/interview_with_elf_founder.html

Anonymous. (2017). *Earth liberation front*. Retrieved from http://earth-liberation-front.com/

Baldelli, G. (1971). *Social anarchism*. London: Penguin.

Becker, M. (2006). Ontological anarchism: The philosophical roots of revolutionary environmentalism. In S. Best & A. J. Nocella II (Eds.), *Igniting a revolution: Voices in defense of the earth* (pp. 71–91). Edinburgh: AK Press.

Best, S. (2010). *Total liberation: Revolution for the 21st century*. Retrieved from https://drstevebest.wordpress.com/2010/12/31/total-liberation-revolution-for-the-21st-century-4/

Best, S., & Nocella, A. J., II. (2006). A fire in the belly of the beast: The emergence of revolutionary environmentalism. In S. Best & A. J. Nocella II (Eds.), *Igniting a revolution: Voices in defense of the earth* (pp. 8–30). Oakland, CA: AK Press.

Campaign Opposing Police Surveillance. (2016). *Bob Lambert investigated by police over Firebombing*. Retrieved from http://campaignopposingpolicesurveillance.com/2016/04/20/bob-lambert-investigated-by-police-over-firebombing/

Casciani, D. (2014). The undercover cop, his lover, and their son. *BBC News*. Retrieved from http://www.bbc.co.uk/news/magazine-29743857

Deshpande, N., & Ernst, H. (2012). *Countering Eco-Terrorism in the United States: The Case of "Operation Backfire."* Final Report to the Science and Technology Directorate, U.S. Department of Homeland Security. College Park, MD: National Consortium for the Study of Terrorism and Responses to Terrorism. Retrieved from http://start.umd.edu/sites/default/files/files/publications/Countermeasures_OperationBackfire.pdf

Evans, R. (2017). Probe into claim that police spy set fire to Debenhams could end by July. *The Guardian*. Retrieved from www.theguardian.com/uk-news/undercover-wi

th-paul-lewis-and-rob-evans/2017/jan/16/probe-into-claim-that-police-spy-set-fire-to-debenhams-expected-to-end-by-july

Evans, R., & Lewis, P. (2011). Undercover police: how "romantic, attentive" impostor betrayed activist. *The Guardian*. Retrieved from https://www.theguardian.com/uk/2011/oct/23/undercover-police-animal-liberation-front

Goodway, D. (Ed.). (1989). *For anarchism: History, theory, and practice*. London: Routledge.

Harper, J. (2003). Facing the agents of omnicide. In S. Best & A. J. Nocella II (Eds.), *Igniting a revolution: Voices in defense of the earth* (pp. 232–240). Edinburgh: AK Press.

Hirsch-Hoefler, S., & Mudde, C. (2014). "Ecoterrorism": Terrorist threat or political ploy? *Studies in Conflict & Terrorism, 37*(7), 586–603.

Home Affairs Committee. (2013, February 26). Thirteenth Report. Undercover Policing: Interim Report. Retrieved from http://www.publications.parliament.uk/pa/cm201213/cmselect/cmhaff/837/83702.htm

Jarboe, J. F. (2012). Testimony. *Federal Bureau of Investigation, FBI*. Retrieved from https://archives.fbi.gov/archives/news/testimony/the-threat-of-eco-terrorism

Joosse, P. (2012). Elves, environmentalism, and "eco-terror": Leaderless resistance and media coverage of the Earth Liberation Front. *Crime, Media, Culture, 8*(1), 75–93.

Leader, S. H., & Probst, P. (2003). The Earth Liberation Front and environmental terrorism. *Terrorism and Political Violence, 15*(4), 37–58.

Lewis, P., Evans, R., & Pollak, S. (2013). Trauma of spy's girlfriend: "Like being raped by the state". *The Guardian*. Retrieved from https://www.theguardian.com/uk/2013/jun/24/undercover-police-spy-girlfriend-child

Loadenthal, M. (2013a). Deconstructing "eco-terrorism": Rhetoric, framing and statecraft as seen through the Insight approach. *Critical Studies on Terrorism, 6*(1), 92–117.

Loadenthal, M. (2013b). The Earth Liberation Front: A social movement analysis. *Radical Criminology, 2*, 15–45.

Loadenthal, M. (2014). When cops "go native": Policing revolution through sexual infiltration and panopticonism. *Critical Studies on Terrorism, 7*(1), 24–42.

Long, D. (2004). Ecoterrorism. New York, NY: Facts On File.

Lubbers, E. (2015). Undercover Research—Corporate and police spying on activists. An introduction to activist intelligence as a new field of study. *Surveillance & Society, 13*(3/4), 338–353.

Lucas, Caroline. (2012, June 13). Commons Debate, Daily Hansard–Westminster Hall.

Starting at Column 96WH. Retrieved from http://www.publications.parliament.uk/pa/cm201213/cmhansrd/cm120613/halltext/120613h0001.htm

Mac Laughlin, J. (2016). *Kropotkin and the anarchist intellectual tradition*. London: Pluto Press.

Nocella, A. J., II, Sorenson, J., Socha, K., & Matsuoka, A. (2010). Introduction. In A. J. Nocella II, J. Sorenson, K. Socha, & A. Matsuoka (Eds.), *The emergence of critical animal studies: The rise of intersectional animal liberation.* (pp. ix–xxxvi). New York, NY: Peter Lang.

Nocella, A. J., II, White, R. J., & Cudworth, E. (Eds.). (2015). *Animals and animal liberation: Essays on complementary elements.* Jefferson, NC: McFarland Press.

North American Earth Liberation Front Press Office. (n.d.). *Frequently asked questions about the Earth Liberation Front (ELF).* Retrieved from http://www.animalliberationfront.com/ALFront/ELF/elf_faq.pdf

Pellow, D. N. (2014). *Total liberation: The power and promise of animal rights and the radical earth movement.* Minneapolis, MN: University of Minneapolis Press.

Pickering, L. J. (2007). *The Earth Liberation Front 1997–2002.* Portland, OR: Arissa Media Group.

Pickering, L. J. (2009). The Earth Liberation Front and revolution. In C. Rosebraugh (Ed.), *This country must change: Essays on the necessity of revolution in the USA* (pp. 159–172). Portland, OR: Arissa Media Group.

Roy, E. A. (2017, March 16). New Zealand river granted same legal rights as human being. *The Guardian.* Retrieved from https://www.theguardian.com/world/2017/mar/16/new-zealand-river-granted-same-legal-rights-as-human-being

Scott, J. C. (2012). *Two cheers for anarchism.* Princeton, NJ: Princeton University Press.

Shannon, D. (2014). *The end of the world as we know it? Crisis, resistance, and the age of austerity.* Oakland, CA: AK Press.

Somma, M. (2006). Revolutionary environmentalism: An introduction. In S. Best & A. J. Nocella II (Eds.), *Igniting a revolution: Voices in defence of the earth* (pp. 37–70). Oakland, CA: AK Press.

Souza, M. L. de, White R. J., & Springer, S. (2016). *Theories of resistance: Anarchism, geography and the spirit of revolt.* Lanham, MD: Rowman & Littlefield.

Springer, S. (2014). War and pieces. *Space and Polity, 18*(1), 85–96.

Springer, S. (2016). *The Anarchist roots of geography: Towards spatial emancipation.* Minneapolis, MN: University of Minnesota Press.

Springer, S., White, R. J., & Souza, M. L. de. (Eds.). (2016). *The radicalization of pedagogy: Anarchism, geography and the spirit of revolt.* Lanham, MD: Rowman & Littlefield.

Vidal, J. (2011, April 10). Bolivia enshrines natural world's rights with equal status for Mother Earth. *The Guardian.* Retrieved from https://www.theguardian.com/environment/2011/apr/10/bolivia-enshrines-natural-worlds-rights

Wade, J. (2003). Radical environmentalism: Is there any other kind? In S. Best & A. J. Nocella II (Ed.), *Igniting a revolution: Voices in defence of the earth* (pp. 280–283). Oakland, CA: AK Press.

Ward, C. (1973). *Anarchy in action.* London: Aldgate Press.

Weignant, C. (2017). Occupation, arson, and terrorism. *Huffington Post.* Retrieved from http://www.huffingtonpost.com/chris-weigant/occupation-arson-and-terr_b_8926124.html.

White, R. J., Springer, S., & Souza, M. L. de. (2016). *The practice of freedom: Anarchism, geography and the spirit of revolt.* Lanham, MD: Rowman & Littlefield.

Wistrich, H. (2013, January 23). Explaining the judgment over secret tribunal. Police Spies Out of Lives, Support group for women's legal action against undercover policing. Retrieved from http://policespiesoutoflives.org.uk/explaining-the-judgment-byharriet-wistrich/.

Contributors

Michael Becker is a lecturer in the Political Science Department at California State University, Fresno. He has a doctorate in political science from Purdue University. His research interests include radical environmentalism, anarcho-primitivism, earth and animal liberation, and critical interpretation of technology. Becker has been active locally with the traditional Choinumni Tribe, Occupy Fresno, and with Earth First!

Lawrence J. Cushnie received his doctorate in political science from the University of Washington. His dissertation considers property destruction as political participation in American resistance movements—both historical and contemporary. Lawrence's research and teaching interests include political theory, social movements, American politics, race, radical resistance, labor studies, and American political development. He lives in Seattle, WA, with his partner and twin children. In his spare moments he enjoys live music, bowling, cooking, and getting lost in record stores.

Carolyn Drew engages in imaginary battles liberating animals or protecting the great forests which had dominated her childhood games have come full circle. She spends her nights tracking down shooters to try and stop the carnage unleashed on kangaroos. During the day, she teaches those who have been disadvantaged by the system in programs which enable them access to higher education. Carolyn has a master's in adult education and is a co-author of "The Harvest" published in *The Southerly*, the journal of the English Association, Australia. She lives with her family in Canberra, Australia.

Stephanie Eccles is an activist-scholar currently based in Montreal, Quebec. Stephanie is currently working toward her master's degree in critical animal geographies at Concordia University. Her research will focus on the lives of the undesired nonhuman animals in urban spaces, grasping the lively commodities they become in the pest industry, and the biopolitical management of urban ecologies. These lives have woven themselves not only into Stephanie's academic career, but in the ways by which she conducts her everyday life. She works with pit bull breeds by day and is a companion to her beautiful "elderbull" pit bulls, Clementine and Eleanor. Stephanie has found much to be learned, experienced, and understood when it comes to the lives of whom we deem pests.

Amber E. George, Ph.D., is an activist scholar who teaches philosophy at Galen College. She has taught undergraduate courses in ethics and social philosophy and presented her research at numerous colleges and universities. Dr. George is a member of the Eco-ability Collective and Executive Director of Finance of the Institute for Critical Animal Studies (ICAS). She is the editor of *Journal of Critical Animal Studies.* She is currently working on writing numerous books and book chapters concerning the issues of non/human animal liberation, disability studies, and critical theory. She has also served as an editor on several projects for the *Social Advocacy and Systems Change Journal* and is on the review board of the *Green Theory and Praxis Journal* and Transformative Justice book series.

Paul Joosse, Ph.D., is Assistant Professor of Criminology in the Department of Sociology at the University of Hong Kong. His empirical work has examined landowner movements that resist oil and gas extraction, green anarchist movements and "ecoterrorism," foreign fighter recruitment in the Somali-Canadian diaspora, and the formation of charismatic bonds in new religious movements (colloquially, "cults"). He has published in the *British Journal of Criminology*, *Terrorism and Political Violence*, *Sociology of Religion*, and *Crime, Media, Culture,* among others. He can be contacted at pjoosse@hku.hk.

Michael Loadenthal, Ph.D., is Visiting Professor of Sociology and Social Justice at Miami University of Oxford, Ohio, and Executive Director of the Peace and Justice Studies Association. He has taught courses on political violence, terrorism, and sociology at Georgetown University, George Mason University, University of Cincinnati, University of Malta, and Jessup Correctional Institution. Michael has served as the dean's fellow for George Mason's School for Conflict Analysis and Resolution, a practitioner-in-residence for Georgetown's Center for Social Justice, and a research fellow at Hebrew Union College's Center for the Study of Ethics and Contemporary Moral Problems.

Anthony J. Nocella II, Ph.D., is Assistant Professor of Criminal Justice and Criminology in the Institute of Public Safety and the Department of Criminal Justice at Salt Lake Community College. He is the editor of the

Peace Studies Journal, the *Transformative Justice Journal*, and the book series Poetry Behind the Walls, along with being a co-editor of five book series, including Critical Animal Studies and Theory with Lexington Books and Hip Hop Studies and Activism with Peter Lang Publishing. He is the National Coordinator of Save the Kids, Executive Director of the Institute for Critical Animal Studies, and Director of Academy for Peace Education. He has published over fifty peer-reviewed book chapters or articles and over forty books. He has been interviewed by *Houston Chronicle*, *Durango Herald*, *Fresno Bee*, *Los Angeles Times*, *Washington Post*, CNN, CBS, Fox, and *New York Times.*

Sean Parson is Assistant Professor of Politics and International Affairs and Sustainable Communities at Northern Arizona University. His current research is on radical environmental political theory, anarchist social movements, and resource politics. In addition to academia, Sean has been involved in Food Not Bombs, the Northwest Forest Defense movement, climate justice activism, and is currently working with the Repeal Coalition. He is finishing a book manuscript on Food Not Bombs and their struggle against the City of San Francisco. When not writing, grading, or teaching he spends most of his time taking his best friend, Diego, out on hikes.

Mara Pfeffer is a social and environmental justice organizer and educator living in Flagstaff, Arizonza. She teaches courses in human, animal, and earth liberation movements; community organizing; liberation psychology, radical mental health, historical trauma, and healing; and has coordinated student-led research of regional ranches, farms, animal sanctuaries, and shelters. She is a part of SPARCC (Speaking uP and Resisting Racism and Colonialism), Refund the Plateau, Flagstaff Animal Liberation Now, A Queer Affinity, and Flagstaff Community Coalition. She is excited about emergent strategies, trans-species traumatology & healing, and struggle and sanctuary for people, land, and animals in the borderlands.

Alexander Reid Ross worked with the Earth First! journal as an editorial collective member (2009–2011) and co-founding moderator of the EF! Newswire (2010–2015). He edited the anthology *Grabbing Back: Essays Against the Global Land Grab* (AK Press, 2014) and wrote *Against the Fascist Creep* (AK Press, 2017). He is currently Lecturer in Geography and Ph.D. student in the Earth, Environment, and Society program at Portland State University. His work has appeared in the *Association of American Geographers Review of Books*, *In These Times*, *Jacobin*, *ROAR! Magazine*, and *Upping the Anti*.

Meneka Repka is a teacher, scholar, activist, and artist living in Calgary, Alberta. She is currently a doctorate candidate at the University of Calgary; her dissertation employs focus group research to investigate the school experience of teenagers who are vegan. Meneka is primarily interested in transforming educational spaces to reflect values of intersectionality, inclusivity, and justice. She is also a co-director of Students for Critical Animal Studies (SCAS), a student-run component of the Institute for Critical Animal Studies that works to unify and provide spaces for student scholar-activists.

Bethany Richter is a community organizer and an instructor of women's and gender studies. Both through her studies and in her time teaching critical theory, Bette continues to explore the dynamics of power, art, and activism and the ways that community and emergent strategies offer alternatives to marginalized communities. Having grown up in farming communities of Central Texas, Bette gained a fascination for sustainability, anti-racist, and decolonial work, as well as interdependent approaches to issues facing our world today. She is a curious gardener, accidental chef, and passionate about the work of total liberation.

Colin Salter is an educator and social and political theorist with an interest in conflict and social change. His research focuses on relations of power and disputes within and across cultures, including the policing of dissent and its governmental implications. Colin's current research draws on political theory, critical animal studies, postcolonial studies, international relations, and subcultural theory to explore the dispute over whaling in the Southern Ocean. He is particularly interested in the relational construction of cultural *Others* and how these are mobilized as a performance of individual and national identities.

Mark Seis is Emeritus Professor of Sociology at Fort Lewis College in Durango, Colorado. He has published on a variety of topics ranging from juvenile death penalty, to environmental topics, including the Clean Air Act, global warming, ozone depletion, and acid rain, to various types of environmental crime, to globalization and the environment, and to issues concerning radical environmentalism. His primary research interests include sustainable communities, all things environment, anarchist studies, and radical pedagogy.

John Sorenson is Professor of Sociology at Brock University in Canada, where he teaches critical animal studies. His past research was on war, nationalism, and refugees. His books are *Constructing Ecoterrorism*; *Critical Animal Studies: Thinking the Unthinkable*; *Defining Critical Animal Studies: An Intersectional Social Justice Approach for Liberation*; *Animal Rights*; *Ape*; *Culture of Prejudice: Arguments in Critical Social Science*; *Ghosts and Shadows: Construction of Identity and Community in an African Diaspora*; *Imagining Ethiopia: Struggles for History and Identity in the Horn of Africa*; *Disaster and Development in the Horn of Africa*; and *African Refugees.*

Joshua M. Varnell is a doctoral candidate in political science with a concentration in comparative politics and political theory at Western Michigan University. His research interests include animal rights and environmental issues, political violence, democratic theory, and social movements. Joshua is currently finishing his dissertation, which is a critical discourse analysis of federal prosecutions of animal rights and environmental activists. Joshua teaches courses in political theory and comparative politics, served two years as president of the Political Science Graduate Student Association, and was honored in 2012 with the Z.D. Shilling Graduate Student Award.

Matthew J. Walton is the Aung San Suu Kyi Senior Research Fellow in Modern Burmese Studies at St Antony's College. His research focuses on religion and politics in Southeast Asia, with a special emphasis on Buddhism in Myanmar. Matt's first book is*Buddhism, Politics, and Political Thought in Myanmar* (Cambridge University Press, 2016). He has also published academic articles on Buddhism, ethnicity, and politics in Myanmar. Matt is one of the co-founders of the Myanmar Media and Society project and of the Oxford-based Burma/Myanmar blog *Tea Circle.*

Richard J. White is Reader in Human Geography at Sheffield Hallam University, United Kingdom. Greatly influenced by anarchist praxis and critical animal studies, White is particularly interested in (1) deconstructing the ways in which the exploitation of human and nonhuman animals intersects in society and (2) developing a new geographic imaginary based on peace and nonviolence. He can be contacted at Richard.White@shu.ac.uk.

Index

A

Abbey, Edward, 161, 182
academic repression, 4
American Indian Movement, 111, 186, 191
anarchism, xiv, 17, 18, 19, 36, 40, 44, 47, 51, 55, 199, 223, 234, 237, 238, 277, 278, 282, 285, 302, 318, 320, 326, 328
anarcho-primitivism, 65
Animal Liberation Front, 2, 3, 14, 42, 80, 85, 104, 130, 140, 145, 166, 181, 188, 201, 255, 283, 317, 319
Ayers, Bill, 111

B

Black Lives Matter, 288, 297, 303
Black Panther Party, 7, 111, 196

C

California, 43, 45, 95, 96, 98, 140, 149, 207, 210, 223, 231, 232, 278
capitalism, 2, 3, 6, 8, 22, 26, 28, 45, 48, 52, 53, 54, 55, 56, 63, 64, 65, 66, 67, 107, 110, 112, 117, 134, 179, 180, 182, 185, 186, 187, 188, 189, 209, 218, 236, 253, 254, 264, 269, 276, 278, 279, 280, 282, 296, 299, 300, 301, 302, 305, 307, 309, 310, 311, 315, 327, 328, 329
cattle, 187, 269, 309
Chomsky, Noam, 111, 208, 218
climate change, 1, 46, 142, 143, 221, 288
Coronado, Rod, 20, 32
corporate repression, 316
critical animal studies, 3, 118, 253

D

Department of Homeland Security, 87
Department of Justice, 192, 259

E

Earth First!, 7, 14, 39, 81, 160, 161, 162, 163, 164, 165, 168, 170, 172, 181, 182, 199, 226, 255, 284, 288, 302, 305, 307, 319, 320
ecofeminism, 296
eco-sabotage, 68
environmental justice, 103, 305, 329, 330

F

factory farming, 142
feminism, 278, 296
fishing, 137, 276
Fur Commission USA, 142, 188

G

global warming, 51, 59, 66, 128, 143
Goldman, Emma, 278
government repression, 96, 189
green anarchism, 40, 47, 51, 199
Greenpeace, 134, 166, 287

H

Hampton, Fred, 111
Harper, Josh, 183
hate, 4, 8, 31, 67, 150, 202, 240, 275
hooks, bell, 4
hunting, 43, 163, 260, 311

I

informant(s), 7, 196, 198, 204, 221, 223, 224, 225, 226, 227, 228, 231, 232, 241, 242, 321, 322

J

Jensen, Derrick, 8, 47, 48, 49, 56, 59, 61, 68, 69, 70, 71, 72, 73, 74, 75, 229, 232, 288

K

King Jr., Martin Luther, 111
Ku Klux Klan (KKK), 4, 141, 202, 240

L

LGBT+, 4, 8, 288
Lovitz, Dara, 80, 143, 227
Luers, Jeffrey, 168, 184, 185

M

Marshall, Critter, 168, 184
Marxism, 17, 39, 55, 279
McDavid, Eric, 7, 204, 205, 206, 217–243
McGowan, Daniel, 240
mining, 55, 62, 63, 284, 286

N

National Association for the Advancement of Colored People (NAACP), 111
Nazi, 4, 185, 280
Nocella II, Anthony J., 1, 2, 4, 6, 8, 179, 217, 227, 253, 259, 301, 303, 315, 320, 324, 328

P

pacifist, 14, 73, 281
Parson, Sean, 5, 39, 199, 202, 210
PATRIOT Act, 103, 108, 115, 116, 187, 191, 194, 207, 208, 209, 226
Pellow, David, 263, 298, 299, 301, 302, 303, 320
Pickering, Leslie James, 39, 49, 50, 51, 52, 53, 54, 160, 319, 324
political repression, 5
porn, 188

R

revolutionary environmentalism, 5, 14, 17, 19, 30, 33, 40, 44
Rosebraugh, Craig, 29, 49, 89, 90, 96, 158, 160, 161, 167, 168, 184, 188, 253, 257, 260, 305

S

sexual repression, 279, 280
Shiva, Vandana, 296
Showing Up for Racial Justice, 303
Sierra Club, 162, 187
snitch, 322

speciesism, 220
Socha, Kim, 253, 297, 327
Standing Rock, 262, 303

T

Texas, 187
The Family, 57, 171, 196, 288
total liberation, 4, 8, 104, 105, 119, 301, 302, 305, 327

V

vegan, 52, 115, 129, 163, 228

X

X, Malcolm, 111

Z

Zapatistas, 5, 7, 13–36, 183, 186, 269, 284, 285, 304, 319
Zerzan, John, 48, 49, 53, 285

RADICAL ANIMAL STUDIES AND TOTAL LIBERATION

Anthony J. Nocella II, SERIES EDITOR

The **Radical Animal Studies and Total Liberation** book series branches out of Critical Animal Studies (a field co-founded by Anthony J. Nocella II) with the argument that criticism is not enough. Action must follow theory. This series demands that scholars are engaged with their subjects both theoretically and actively via radical, revolutionary, intersectional action for total liberation. Founded in anarchism, the series provides space for scholar-activists who challenge authoritarianism and oppression in their many daily forms. **Radical Animal Studies and Total Liberation** promotes accessible and inclusive scholarship that is based on personal narrative as well as traditional research, and is especially interested in the advancement of interwoven voices and perspectives from multiple radical, revolutionary social justice groups and movements such as Black Lives Matter, Idle No More, Earth First!, the Zapatistas, ADAPT, prison abolition, LGBTTQQIA rights, disability liberation, Earth Liberation Front, Animal Liberation Front, political prisoners, radical transnational feminism, environmental justice, food justice, youth justice, and Hip Hop activism.

To order other books in this series please contact our Customer Service Department:

(800) 770-LANG (within the US)
(212) 647-7706 (outside the US)
(212) 647-7707 FAX

To find out more about the series or browse a full list of titles, please visit our website:

WWW.PETERLANG.COM

Lightning Source UK Ltd.
Milton Keynes UK
UKHW051306040722
405352UK00019B/287

9 781433 159930